Lichens to Biomonitor the Environment

Vertika Shukla • D.K. Upreti
Rajesh Bajpai

Lichens to Biomonitor the Environment

Vertika Shukla
Environmental Sciences
Babasaheb Bhimrao
  Ambedkar University
Lucknow, UP, India

D.K. Upreti
Lichenology Laboratory
CSIR - National Botanical
  Research Institute
Lucknow, UP, India

Rajesh Bajpai
Lichenology Laboratory
CSIR - National Botanical
  Research Institute
Lucknow, UP, India

ISBN 978-81-322-1502-8   ISBN 978-81-322-1503-5 (eBook)
DOI 10.1007/978-81-322-1503-5
Springer New Delhi Heidelberg New York Dordrecht London

Library of Congress Control Number: 2013944053

© Springer India 2014
This work is subject to copyright. All rights are reserved by the Publisher, whether the whole or part of the material is concerned, specifically the rights of translation, reprinting, reuse of illustrations, recitation, broadcasting, reproduction on microfilms or in any other physical way, and transmission or information storage and retrieval, electronic adaptation, computer software, or by similar or dissimilar methodology now known or hereafter developed. Exempted from this legal reservation are brief excerpts in connection with reviews or scholarly analysis or material supplied specifically for the purpose of being entered and executed on a computer system, for exclusive use by the purchaser of the work. Duplication of this publication or parts thereof is permitted only under the provisions of the Copyright Law of the Publisher's location, in its current version, and permission for use must always be obtained from Springer. Permissions for use may be obtained through RightsLink at the Copyright Clearance Center. Violations are liable to prosecution under the respective Copyright Law.
The use of general descriptive names, registered names, trademarks, service marks, etc. in this publication does not imply, even in the absence of a specific statement, that such names are exempt from the relevant protective laws and regulations and therefore free for general use.
While the advice and information in this book are believed to be true and accurate at the date of publication, neither the authors nor the editors nor the publisher can accept any legal responsibility for any errors or omissions that may be made. The publisher makes no warranty, express or implied, with respect to the material contained herein.

Printed on acid-free paper

Springer is part of Springer Science+Business Media (www.springer.com)

Preface

Among the different organisms, lichens are considered as one of the best bioindicators of environmental changes either due to natural or man-made disturbances. Lichens show differential sensitivity towards wide range of pollutants. Certain species are inherently more sensitive, while some species show tolerance to high levels of pollutants. These characteristics make certain lichen species suitable for being utilised as an indicator species (based on their sensitivity and tolerance) for various pollutants including metals, metalloids, radionuclides and organic pollutants. Lichens as indicators possess an undeniable appeal for conservationists and land managers as they provide a cost- and time-efficient means to assess the impact of environmental disturbances on an ecosystem.

Information about the sensitivity of lichens towards pollutants and their use in pollution monitoring is available since the sixteenth century; however, more systematic studies utilising lichens in biomonitoring studies were initiated in the 1960s in Europe. Lichen biomonitoring studies were introduced in India in the 1980s, and it was after 1995 that more systematic studies were carried out in different regions of the country. Owing to vast geographical area and rich lichen diversity of the country, few such studies are available in context with India that too scattered in different reports, journals or other published scientific articles. Thus, to popularise the subject in the country, the present attempt has been made.

This book is a valuable reference for both students and researchers interested in environmental monitoring studies involving lichens and from different fields of science including environment, botany and chemistry.

Lucknow, India	Vertika Shukla
Lucknow, India	D.K. Upreti
Lucknow, India	Rajesh Bajpai

About the Book

The book embodies the detailed account about unique symbionts, i.e. *lichens* in ecosystem monitoring. The first chapter deals with the unique characteristic features of lichens which facilitate their survival in extreme climates and make them an ideal organism for ecosystem monitoring. Biosynthesis of secondary metabolites is known to protect lichens against increasing environmental stresses; therefore, the second chapter provides insight into various chromatographic and modern spectroscopic techniques involved in separation and characterisation of lichen substances.

The third chapter elaborates the criteria for selection of biomonitoring species and characters of host plant that influence lichen diversity and detail about different lichen species utilised for biomonitoring.

One can retrieve preliminary information about the air quality based on the lichen community structure and distribution of bioindicator species as lichen communities/indicator species provide valuable information about the natural-/anthropogenic-induced changes in the microclimate and land-use changes due to human activity. Therefore, for identification of species, a key to genera and species provides concise information to identify the lichen species based on their morphological and anatomical characters and chemicals present. Keys provided in Chap. 4 will help the beginners to identify some common lichen species based on the distribution in different climatic zones of India. The section also provides comprehensive information about the bioindicator communities and bioindicator species from India.

Chapter 5 provides the details of factors affecting the ecosystem (natural as well as anthropogenic disturbances), and the role of lichens in ecosystem monitoring in India has been discussed in detail.

Chapter 6 discusses the need and utility of indicator species especially lichen biomonitoring data in sustainable forest management and conservation.

The content about lichens in biomonitoring will be a valuable resource for researchers from different fields and will provide an essential reference for people interested in lichens and its role in ecosystem monitoring. The book will also hopefully popularise lichenological studies in India and will generate more active participation of lichen biomonitoring studies in management and conservation of natural resources in India.

Contents

1 Introduction 1
    1.1 Introduction 1
    1.2 About Lichens 3
        1.2.1 Thallus Morphology and Anatomy 3
        1.2.2 Anatomical Organisation of Algal
              and Hyphal Layer 5
    1.3 Development and Establishment 8
        1.3.1 Colonisation 8
        1.3.2 Growth 9
        1.3.3 Succession 9
        1.3.4 Competition 10
    1.4 Role of Lichens in Biodeterioration
        and Soil Establishment 10
        1.4.1 Biodeterioration (Rock Weathering) 10
    1.5 Effect of Microclimatic Condition on Lichen Diversity 17
    References 17

**2 Secondary Metabolites and Its Isolation
and Characterisation** 21
    2.1 Environmental Role of Lichen Substances 22
    2.2 Biosynthetic Pathways of Lichen Secondary Metabolites 27
    2.3 Isolation and Modern Spectroscopic Techniques Involved
        in Characterisation of Lichen Substances 28
        2.3.1 Spot Test 29
        2.3.2 Microcrystallography 29
        2.3.3 Chromatography 29
        2.3.4 Spectroscopic Techniques 35
    References 42

3 Selection of Biomonitoring Species 47
    3.1 Introduction 47
    3.2 Criteria for Selection of Biomonitoring Species 49
    3.3 Biomonitoring Species (World and India) 50
        3.3.1 Are Lichens Suitable Phytoremediator? 54
        3.4 Host Plants for Lichen Colonisation 55
    References 57

4 Lichen Diversity in Different Lichenogeographical Regions of India ... 61
- 4.1 Introduction ... 61
- 4.2 Distribution and Diversity of Widespread and Rare Lichen Species in Different Lichenogeographical Regions Along with Its Pollution Sensitivity ... 62
  - 4.2.1 Western Himalayas ... 63
  - 4.2.2 Western Dry Region ... 67
  - 4.2.3 Gangetic Plain ... 67
  - 4.2.4 Eastern Himalayas ... 68
  - 4.2.5 Central India ... 69
  - 4.2.6 Western Ghats ... 70
  - 4.2.7 Eastern Ghats and Deccan Plateau ... 70
  - 4.2.8 Andaman and Nicobar Islands ... 71
- 4.3 Key for Identification of Toxitolerant and Common Lichens in Different Lichenogeographical Regions ... 71
- 4.4 Bioindicator Lichen Species ... 76
- 4.5 Bioindicator Lichen Communities ... 88
- References ... 93

5 Ecosystem Monitoring ... 97
- 5.1 Introdcution ... 97
- 5.2 Natural and Human Disturbances/Disasters ... 99
  - 5.2.1 Heavy Metal ... 100
  - 5.2.2 Arsenic (Metalloids) ... 103
  - 5.2.3 Polycyclic Aromatic Hydrocarbons (PAHs) ... 106
  - 5.2.4 Radionuclides ... 112
  - 5.2.5 Climate Change ... 118
  - 5.2.6 Assessment of Paleoclimatic Conditions (Lichenometry) ... 121
  - 5.2.7 Loss of Biodiversity ... 123
- 5.3 Solely Human Disturbances/Disasters ... 130
  - 5.3.1 Urbanisation, Expanding Cities and Industrialisation ... 132
  - 5.3.2 Power Plants ... 147
  - 5.3.3 Persistent Organic Pollutants (POPs) ... 149
  - 5.3.4 Peroxyacyl Nitrates (PAN) ... 152
  - 5.3.5 Ozone (O_3) ... 152
  - 5.3.6 Increasing Tourism ... 153
- 5.4 Conclusion ... 155
- References ... 155

6 Management and Conservational Approaches ... 171
- 6.1 Introduction ... 171
- References ... 176

Glossary ... 179

About the Authors ... 185

Introduction

Lichens are composite organisms comprised of a fungus and one or more algae living together in symbiotic association in which the algal partner produces essential nutrients for the fungal partner through photosynthesis, while the fungal partner provides mechanical support to the algal partner. Development and establishment of lichen on a substratum is achieved by fruiting bodies (apothecia) produced by the fungal partner, which must germinate and find an algal partner before they can form a new lichen thallus or may produce minute fragments (as finger-like outgrowths, isidia or sugar-like granules, soredia) containing both partners, which can disperse quickly and colonise available habitats.

Being pioneers on rock surface lichens are important component of the ecosystem that establishes life on rock and barren disturbed sites. As lichens colonise rocks, they trap dust, silt and water which leads to biogeophysical and biogeochemical weathering of the rock surface leading to soil formation.

Lichens occur in all available substrata and in all possible climatic conditions, but the lichen diversity of an area of interest or substratum is highly dependent on prevalent microclimatic conditions. Apart from morphology and anatomy of lichens, the high success of lichens in extreme climates has been attributed to the secondary metabolites produced by the fungal partner to protect the algal partner.

This chapter discusses the unique characteristics about lichens which facilitate their survival in extreme climates and makes them an ideal organism for ecosystem monitoring.

1.1 Introduction

The word lichen has a Greek origin, which denotes the superficial growth on the bark of tree, rock as well as soil. Theophrastus, the father of botany, introduced the term 'lichen' and this group of plants to the world. Lichen species collectively called as 'stone flower' in English, 'Patthar ka phool' in Hindi, 'Dagad phool' in Marathi, 'Kalachu' in Karnataka, 'Kalpasi' in Tamil, 'Richamkamari' in Urdu and 'Shilapushpa' in Sanskrit.

Lichen is a stable self-supporting association of a mycobiont (fungus) and a photobiont (alga) in which the mycobiont is the exhabitat (Hawksworth 1988). The plant body of lichen is called 'thallus', and it had been considered a single plant till 1867 when Schwendener (Swiss botanist) demonstrated the lichen thallus to be a composite body made up of fungus and an alga, and he propounded his well-known dual hypothesis for this body. This view was not accepted by the prominent lichenologist of the time, but later the composite nature of lichen thallus was universally accepted, and the two components were considered to be in a symbiotic association. The mycobiont predominates and the gross morphology of lichen thallus is generally determined by mass, nature and modifications of the fungal hyphae. The photobiont (Fig. 1.1), on the other

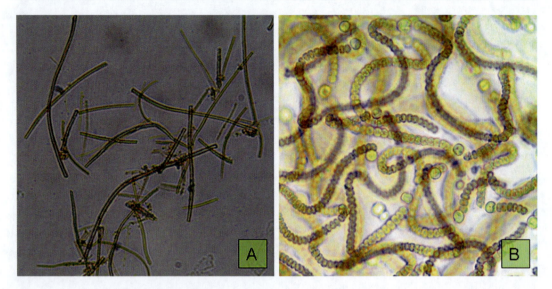

Fig. 1.1 Common photobionts which facilitate lichenisation with compatible fungal partner resulting in formation of lichen thallus. (**a**) *Trentepohlia* and (**b**) *Nostoc*

hand, also plays an important role in the development of the thallus. The resultant thallus is unlike the two symbionts in morphological appearance and physiologically behaving as a single autonomous biological unit (Awasthi 2000). About 85 % of lichens have green algae as symbionts, approximately 10 % have blue-green algae and/or cyanobacteria, and less than 5 % of lichens have both green algae as primary symbionts and blue-green algae as secondary symbionts. The lichen fungus is typically a member of the Ascomycota or rarely Basidiomycota and hence termed as ascolichens and basidiolichens, respectively. The hyphae of mycobiont are septate, branched and thin or thick walled and possess either a single or three septate through which the cytoplasmic contents of the adjoining cells are connected by plasmodesmata; rarely, the septum may be multiperforate (Honegger 2008).

Repeated septation, branching and different degree of compactness, coalescence or conglutination result in the formation of diverse type of tissue. The development of such tissues is also necessary for a durable and proper functioning of the symbiotic relationship in a lichen thallus. It is necessary that the photobiont, which is basically and essentially aquatic in nature, remains protected from desiccation in a lichenised terrestrial condition.

Most of the cyanobacterial photobiont possess a mucilaginous sheath, which protects them from desiccation to some extent, but no such protective sheath is present in the green-algal photobiont, which predominate in lichen taxa. The protection, however, is provided by the mycobiont through the development of specialised hyphal tissues in the form of a cortex over the stratum of the photobiont (Ahmadjian 1993).

Lichens have a poikilohydric nature to survive in various climatic conditions as they have no mechanism to prevent desiccation; they desiccate and remain dormant when their environment dries out but can rehydrate when water becomes available again. Lichens usually absorb water directly through their body surface by aerosol, mist and water vapours; due to this nature, lichens live long in dry areas even on stones and rocks used for construction of monuments and other building artefacts.

The role of alga in lichen partnership is that of producing food for themselves and their partner, the fungus. The fungus (mycobiont) in a lichen partnership cannot live independently. The fungus partner is able to obtain and hold water, which is essential for alga. The mycobiont attached firmly to the algal cell (photobiont) cannot be easily washed away by water or blown by wind. Due to this type of relationship, lichens are eminently successful and enjoy worldwide distribution and occur in every conceivable habitat, growing on a variety of substrata, including most natural substrata as well a host of human-manipulated or manufactured substrata. The common natural substrata on which lichens can colonise and grow successfully include all categories of rock (saxicolous), trees (corticolous), soil (terricolous) wood (lignicolous) and leaves (foliicolous), while manmade substrata include rubber, plastic, glass, stonework, concrete, plaster, ceramic, tiles and brick.

1.2 About Lichens

Lichens are perennial plants with very slow growth rate which is mainly attributed to the growth of the mycobiont. There is no supply of nutrients from the central part to the growing part, and the food produced by the photobiont at the growth site is used by the mycobiont. The pattern of growth in general is centrifugal, apical and marginal. In crustose and foliose lichen radical growth occurs, while in fruticose lichens there is an increase in length. The annual radial increase has been found to be 0.2–1.0 mm in crustose and 1.0–2.5 mm in foliose, while in fruticose lichens 2.0–6.0 mm growth can be seen in a year (Awasthi 2000).

1.2.1 Thallus Morphology and Anatomy

Growth forms (Fig. 1.2) are categorised mainly on the basis of morphology without direct reference to ecological adaptations, yet distribution of growth form primarily reflects competition and adaptation to abiotic environmental conditions, mainly water relations (frequency and intensity of periods of water shortage). Based on the external morphology (growth form), lichen thalli exhibit three major growth forms: crustose, foliose and fruticose. Crustose lichens are tightly attached to the substrate with their lower surface, which cannot be removed without destruction. Water loss is restricted primarily to the upper, exposed surface only. Crustose lichens may be subdivided as leprose, endolithic, endophloeodic, squamulose, peltate, pulvinate, lobate, effigurate and suffruticose crusts. Majority of crustose lichens grows directly on the surface of the substrate and is referred to as episubstratic, while a small minority grows inside the substratum called endosubstratic. The episubstratic thallus consists of a crust-like growth adherent or attached to the substratum throughout its underside by hyphae and cannot be detached without destruction (Büdel and Scheidegger 2008).

The leprose thallus is effuse and consists of a thin layer of uniformly and loosely disposed photobiont cells intermixed with hyphae forming a yellow, greyish-white or reddish powdery mass on the surface of the substratum. In terms of complexity, the powdery crust, as found in lichen genus *Lepraria*, is the simplest and lacks an organised thalline structure. Algal structure is embedded in the loose fungal hyphae, having no distinct fungal or algal layer.

Endolithic (growing inside the rock) and endophloeodic (growing underneath the cuticle of leaves or stems) are more organised as compared to leprose lichens. In most cases, an upper cortex is developed. The upper cortex can consist of densely conglutinated hyphae forming a dense layer named 'lithocortex' as seen in *Verrucaria* species.

In some crustose lichens as in genus *Rhizocarpon*, there is a prothallus which is a photobiont-free, white or dark brown to black zone, visible between the areolate and at the growing margins of the thallus.

A thallus is effigurate when the marginal lobes are prolonged and are radially arranged, as in genus *Caloplaca*, *Dimelaena* and *Acarospora*.

The squamulose thallus is an intermediate form between crustose and foliose forms. It generally has rounded to oblong minute lobules or squamules distinct from each other as they develop individually. The squamules are dorsiventral, attached to the substratum along an edge or by rhizoidal hyphae from the lower surface as in genera *Peltula*, *Psora* and *Toninia*. The flat scaled squamulose thalli, with more or less central attachment area on the lower surface, are called peltate as in *Peltula euploca*. Extremely inflated squamules in the lichen genus *Mobergia* are called bullate, while in some genera coralloid

Fig. 1.2 External morphology of the lichen thallus can be categorised into (**a**) fruticose, (**b**) foliose, (**c**) crustose and (**d**) dimorphic growth forms

tufted cushions termed as subfruticose are formed. In some species of *Caloplaca* and *Lecanora*, a thallus becomes radially striated with marginal partially raised lobes termed as lobate thallus.

The foliose or leafy thallus of lichens is typically flattened, dorsiventral, spreading and expanding horizontally outwards and usually attached to the substratum by rhizines arising from the lower surface. Foliose lichens develop a great range of thallus size and diversity.

Lacinate lichens are typical foliose lichens. They are lobate and vary considerably in size, which may be gelatinous–homoiomerous (e.g. *Collema*, *Leptogium*) or heteromerous. The lobes may be radially arranged (*Parmelia*) or overlapping like tiles on a roof (*Peltigera*). Sometimes thallus lobes can become inflated, having a hollow medullary centre (*Menegazzia* species). These foliose lichens are cosmopolitan in distribution.

Umbilicate foliose lichens have circular thalli, consisting either of one single, unbranched lobe or multilobate thalli with limited branching patterns, attached to the substrate by a central umbilicus from lower surface.

An ecologically interesting group of foliose lichens are vagrant lichens such as *Xanthomaculina convolute*, which do not remain adhered to the substratum and have hygroscopic movement.

In dry conditions, the thalli roll up and expose lower cortices. When they take up water, thalli unroll and expose their upper surface to the sunlight. In rolled condition they can be blown with the wind to longer distances.

Fruticose type of lichen is either erect or pendent shrubby growth attached to the substratum at the base by basal disc or holdfast (formed by mycobiont hyphae). Some groups have dorsiventrally arranged thalli (*Sphaerophorus melanocarpus*, *Evernia prunastri*) but majority possess radial symmetric thalli (*Usnea* species and *Ramalina* species).

Genera like *Baeomyces* or *Cladonia* develop a twofold, dimorphic thallus that is differentiated into a fruticose thallus verticalis and a crustose–squamulose to foliose thallus horizontalis.

Another peculiar type of anatomy is that of the thread-like growing genus *Usnea*; *Usnea longissima* is the world's longest lichen, which has a strong central strand of periclinally arranged, conglutinated hyphae that provide mechanical strength along the longitudinal axis.

Highly branched fruticose lichens have a high surface to volume ratio that results in more rapid drying and wetting pattern compared with lichens having lower surface to volume ratio. This phenomenon seems to be the main reason for attributing high sensitivity to this class of lichens towards slight change in the microenvironment. Fruticose growth forms can preferentially occur either in very wet, humid climates and in the dry urban climates its distribution is restricted.

Among the different growth forms of lichens in the evolutionary series, leprose are considered pioneers followed by crustose, placodioid, squamulose and foliose, dimorphic and fruticose being the latest. Leprose, crustose, some placodioid and squamulose lichens are called microlichens as they are smaller in size and mostly require a compound microscope for identification. Foliose, fruticose and dimorphic lichens on the other hand are called macrolichens. Macrolichens have a comparatively larger thallus and a hand lens and stereozoom microscope are sufficient for identification.

1.2.2 Anatomical Organisation of Algal and Hyphal Layer

1.2.2.1 Homoiomerous and Heteromerous (Stratified) Thallus

In homoiomerous thalli mycobiont and photobiont are evenly distributed, while in majority of lichens including many crustose species, internally stratified (heteromerous) thalli are found. The main subdivisions are into upper cortex, photobiont layer, medulla and lower cortex. These layers may include various tissue types, and their terminology follows the general mycological literature. Pseudoparenchymatous and/or prosoplectenchymatous tissue types are present where hyphae are conglutinated to the extent that single hyphae are indistinguishable.

1.2.2.2 Cortex, Epicortex and Epinecral Layer

In most of the lichens, the algal layer is covered by a cortical layer ranging from a few microns to several hundred microns. In many dark lichens, pigmentation (which is due to secondary metabolites' UV protective nature) is confined to fungal cell walls of cortical hyphae (Esslinger 1977; Timdal 1991) or the epinecral layer. In gelatinous lichens the secondary metabolites are primarily confined to the outer wall layers of the mycobiont (Büdel 1990). Pseudoparenchymatous or a prosoplectenchymatous fungal tissue mainly forms the cortex in foliose and fruticose lichens. Usually living or dead photobiont cells are completely excluded from the cortex, but in the so-called phenocortex collapsed photobiont cells are included (e.g. *Lecanora muralis*). In Parmeliaceae some species have a 0.6–1 mm thick epicortex, which is a noncellular layer secreted by the cortical hyphae. This epicortex can have pores, as in *Parmelina*, or have no pores, as in *Cetraria*. In a broad range of foliose to crustose lichens, an epinecral layer of variable thickness is often developed which is composed of dead, collapsed and often gelatinised hyphae and photobiont cells. Thalli often have a whitish, flour-like surface covering, the so-called pruina that consists primarily of superficial deposits, of which calcium oxalate is the most common as in *Dirinaria applanata*. The amount of calcium oxalate is probably dependent on ecological parameters, such as calcium content of the substrate and the aridity of the microhabitat (Syers et al. 1964).

Functions of the upper cortex and/or its pruina include mechanical protection, modification of energy budgets, antiherbivore defence and protection of the photobiont against excessive light (Büdel 1987). Light- and shade-adapted thalli of several species differ considerably in the anatomical organisation of their upper cortical strata (e.g. *Peltigera rufescens*). In a nearby fully sun-exposed habitat, the thickness of the cortex is reduced, but a thick, epinecral layer with numerous air spaces is formed giving the thallus a greyish-white surface due to a high percentage of light reflection (Dietz et al. 2000). This cortical organisation results in decreased transmission of incident light by 40 % in the sun-adapted thallus measured at the upper boundary of the algal layer. Epinecral layers of *Peltula* species are also known to contain airspaces that may also act as CO_2 diffusion paths under supersaturated conditions (Büdel and Lange 1991).

1.2.2.3 Photobiont Layer and Medulla

In most foliose and fruticose thalli, the medullary layer occupies the major part of the internal thalline volume. Usually it consists of long-celled, loosely interwoven hyphae forming a cottony layer with high intercellular spaces. The upper part of the medulla forms the photobiont layer. In many lichens, the hyphae of the photobiont layer are anticlinally arranged and may sometimes form short or globose cells. The hyphal cell walls of the algal and medullary layer are often encrusted with crystalline secondary products which make medullary hyphae hydrophobic (Honegger 1991).

1.2.2.4 Lower Cortex

In some foliose lichens such as *Heterodermia*, the medulla directly forms the outer, lower layer of the thallus. However, typical foliose lichens of the Parmeliaceae and many other groups have a well-developed lower cortex. As is the case with the upper cortex, it is either formed by pseudoparenchymatous or a prosoplectenchymatous tissue. But unlike the upper cortex, the lower cortex is often strongly pigmented. Its ability to absorb water directly is well documented. Only low water conductance has been found thus far. However, it may play a major role in retaining extrathalline, capillary water (Jahns 1984).

1.2.2.5 Attachment Organs and Appendages

In order to provide strong hold of the substratum for lichenisation, lichen develops a variety of attachment organs from the lower cortex and also occasionally from the thallus margin or the upper cortex to establish tight contacts to the substrate.

In foliose lichens attachment is mainly by simple to richly branched rhizines, mostly consisting of strongly conglutinated prosoplectenchymatous hyphae. Umbilicate lichens as well as *Usnea* and similarly structured fruticose lichens are attached to the substrate with a holdfast, from which hyphae may slightly penetrate into the substrate. Deeply penetrating rhizine strands are found in some squamulose, crustose or fruticose lichens growing in rock fissures, over loose sand.

In crustose lichens a prosoplectenchymatous prothallus is often formed around and below the lichenised thallus. It establishes contact with the substrate from where bundles of hyphae penetrate among soil particles. Members of various growth forms produce a loose web of deeply penetrating hyphae, growing outwards from the noncorticate lower surface of the thalli.

Cilia are fibrillar outgrowths from the margins or from the upper surface of the thallus. A velvety tomentum consisting of densely arranged short, hair-like hyphae may be formed on the upper or lower cortex. Tomentose surfaces are mainly reported from broad-lobed genera such as *Pseudocyphellaria*, *Lobaria*, and *Sticta* and few *Peltigera* species.

1.2.2.6 Cyphellae and Pseudocyphellae

Upper or lower cortical layers often bear regularly arranged pores or cracks. Pseudocyphellae, as found on the upper cortex of *Parmelia sulcata* or on the lower side of *Pseudocyphellaria*, are pores through the cortex with loosely packed medullary hyphae occurring to the interior. Cyphellae are bigger and anatomically more complex than pseudocyphellae and are only known from the genus *Sticta*. In the interior portions of the cyphellae, hyphae form conglutinated, globular terminal cells, and this is the main difference from pseudocyphellae.

Pseudocyphellae and young cyphellae are the regions in the cortex having lower gas diffusion resistance. The pseudocyphellae and cyphellae are hydrophobic structures of lichens and may act as pathways for gas diffusion into thalli only when there is drying of the thallus (Lange et al. 1993).

1.2.2.7 Cephalodia (Photosymbiodemes)

Representatives of the foliose Peltigerales with green algae as the primary photobiont, and members of such genera as *Stereocaulon*, *Chaenotheca*, and *Placopsis* usually possess an additional cyanobacterial photobiont. In *Solorina*, the secondary photobiont may be formed as a second photobiont layer underneath the green-algal layer, but usually it is restricted to minute to several millimetres wide cephalodia. Cephalodial morphology is often characteristic on a species level and ranges from internal verrucae to external warty, globose, squamulose or shrubby structures on the upper or lower thallus surfaces. Cephalodial morphology usually differs completely from the green-algal thallus, and this emphasises the potential morphogenetical influence of the photobiont on the growth form of the mycobiont–photobiont association. Because many cyanobacterial photobionts are nitrogen-fixing, these lichens may considerably benefit from cephalodia, especially in extremely oligotrophic habitats.

1.2.2.8 Differentiating Lichens from Other Groups of Plants

Along with lichens, non-lichenised fungi, algae, moss, liverworts (bryophytes) are the plants which grow on rocks, bark and soil and may create confusion for beginners in the field. However, lichens can be easily be differentiated from these group of plants as lichens are never green as algae, liverworts and mosses. Foliose lichens growing in moist places or in wet condition may look greener but have thick, leathery thallus, while in contrast liverworts have non-leathery and slimy thallus. The dimorphic forms of lichens such as *Cladonia* may be confused with the leaf liverworts and mosses in the field. Leafy liverworts and mosses have dense small leaf-like structures throughout the central axis of the plant, while in the case of dimorphic lichens, the squamules of semicircular shape usually present at the base of the central axis or sparse throughout. Dried algal mat on rocks and bark may look like lichens, but it may be checked by spraying some water on these mats as upon hydration algae regain its colour and texture which may be distinguished clearly.

The non-lichenised fungi are the most confusing ones with crustose lichens in the field. Such fungus usually forms patches with loosely woven hyphae, while lichens form smooth, perfect thallus which can be distinguished under the microscope. The fungus are usually whitish in colour and lichens are usually greyish, off-white, but sometimes yellowish, yellowish-green or bright yellow or yellow-orange in colour which is attributed to the presence of lichen substances in the cortex of the thallus as in the case of *Xanthoria parietina* (orange) due to the presence of parietin. The greenish colour of the thallus is mainly due to the presence of algal cells in the thallus. The lichen thallus usually bears cup-shaped structures called apothecia or bulged, globular structures called perithecia. Some crustose lichens belonging to family Graphidaceae bear worm-like structures, lerielle, which are modified apothecia (Fig. 1.3). Asexually reproducing fruiting bodies present on the thallus surface may be finger-like projections called isidia or granular, powder-like structure called soredia (Fig. 1.4). While collecting lichens, it is necessary to look for such structures with the help of the hand lens. When a lichen thallus does not have any such structures, it makes it difficult to differentiate it from fungus.

In any case it is observed that a beginner usually collects fungus and other plants in place of lichens. Usually fungus of various colour (mostly appearing like mushrooms) are confused for lichens and collected by beginners. Such specimens can be identified by taking a thin section of the thallus and studying them under a

Fig. 1.3 Sexually reproducing fruiting bodies, (**a**) apothecia, (**b**) perithecia and (**c**) lerielle, which are exclusively produced by fungal partner

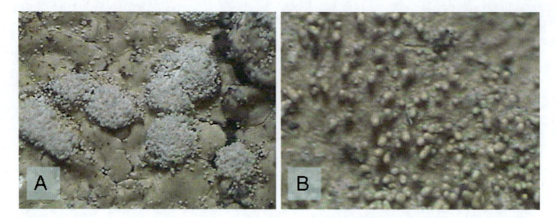

Fig. 1.4 Asexually reproducing bodies. (**a**) Soredia. (**b**) Isidia containing both fungal hyphae and algal cells

microscope. If the section contains both fungal tissue and algal cells, then the specimen is lichen; otherwise, it may be some other plant group. In India basidiolichens (looking like mushrooms) are rare or absent (Nayaka 2011).

1.3 Development and Establishment

Development and establishment of lichen thallus, termed as morphogenesis, involves the process of lichenisation (spore/propagule dispersal), acquisition of compatible partners, competition and growth of lichen thallus resulting in its establishment on the substratum.

Lichen-forming fungi express their symbiotic phenotype (produce thalli with species-specific features) only in association with a compatible photobiont. About 85 % of lichen mycobionts are symbiotic with green algae, about 10 % with cyanobacteria ('blue-green algae') and about 3–4 %, the so-called cephalodiate species, simultaneously with both green algae and cyanobacteria (Tschermak-Woess 1988; Peršoh 2004).

Majority of lichen-forming fungi reproduce sexually and thus have to re-establish the symbiotic state at each reproductive cycle. Compatible photobiont cells are not normally dispersed together with ascospores; exceptions are found in a few species of Verrucariales with hymenial photobionts (e.g. *Endocarpon pusillum*). Many tropical lichen-forming fungi associate with green-algal taxa that are widespread and common in the free-living state (e.g. *Cephaleuros*, *Phycopeltis*, *Trentepohlia*) (Honegger 2008).

1.3.1 Colonisation

The ability of lichens as primary colonisers on barren rock is evident by distribution of *Rhizocarpon geographicum* on rocks and moraines exposed by glacier retreat phenomenon and is used worldwide in lichenometric studies to study paleoclimatic condition. Resistance to drought condition and extreme temperatures and presence of lichen substances enable lichens to colonise on rock surface.

Colonisation on bark surface is governed by water relation, pH, light and nutrient status which determine lichen diversity on the bark surface, while on soil surface colonisation by lichens is limited by their poor competitive abilities compared to the higher plants, and because of their small structures, they are more prone to trampling and trekking activities. Most of the soil-inhabiting lichens fix atmospheric nitrogen like species of *Stereocaulon*, *Peltigera* and *Collema* due to the presence of cyanobacteria *Nostoc*. The ability to accumulate nutrients from rain or runoff (in which nutrients are present in small quantities) enables them to colonise on varied substratum (Brodo 1973).

1.3.2 Growth

Growth rate has a pronounced effect on colonisation and competitive ability of the new thalli resulting in establishment of the thalli. The balance between growth rate and reproductive activity varies with environmental changes (Seaward 1976). Growth varies over the geographical range of a species, and the size ratio of equal-aged thalli of two species may differ in different regions. Rainfall is important for many species as in *Xanthoria elegans* growth rate correlated linearly with the annual rainfall (Beschel 1961).

In species *Parmelia caperata* occurring in colder climate shows frost damage, while alternations in regimes of incident light and humidity are essential for growth of *Parmelia sulcata* and *Hypogymnia physodes*.

Nutrient enrichment also affects growth rate; moderate levels of nutrients are found to favour growth, while excessive levels of nutrients (pollutants) result in slow growth rate depending on the species.

Lichens colonising on non-vertical surfaces of the substratum show good growth rate in comparison to the vertical surfaced substratum, as non-vertical surface retains moisture content for longer periods of time.

Growth rate also depends on the age of the thalli. The life cycle of fruticose lichen *Cladonia* has been divided into three periods: generation, renovation and decline. In generation, phase growth is continuous (5–25 years); in the second podetia decay from below, while growth continues at the tips (80–100 years); and in the final period (20 years) decay predominates. But in the case of crustose lichen especially *Rhizocarpon geographicum*, growth rate is approximately constant with time and this feature is exploited in lichenometric studies. Growth rate affects biomass, i.e. productivity and carbon assimilation potential of the species (Seaward 1976).

1.3.3 Succession

Replacement of one growth form or community structure with another in the course of colonisation and establishment refers to succession. Succession, in general, begins with crustose continuing with foliose and concluding with fruticose. But in the case of epiphytic lichens, foliose lichens may colonise first on tree bases and on twigs (Kalgutkar and Bird 1969).

In successions on soil, colonising species vary; it has been observed that after forest fires *Lecidea* species colonises first; same is the case in the Arctic where *Stereocaulon*, *Cetraria* and *Peltigera* are primary successors. Many successions, especially in mesic or humid environments, do not involve lichens and in xeric environments, mosses compete with lichens by colonising on similar niches. In several studies conducted involving rock substratum like granite, rock ledges and sloping cliffs, mosses are found to be the primary coloniser (Topham 1977).

The prominent role of lichens in succession indicates their contribution to community structure. Lichens increase the intrabiotic fraction of inorganic nutrients, as corticolous species trap nutrients from the phorophytes and from rainwater, while saxicolous species mobilise and extract mineral nutrients from substrates, rainwater and airborne dust.

In the process of succession, lichen diversity increases as do the vascular plants. Species diversity is found to increase in early stages of succession until a closed community is formed (Degelius 1964).

Pioneer species in diverse habitat are morphologically adapted with reduced thalli and abundant production of small ascospores which are soon replaced by taller growing life-forms with larger reproductive bodies with spore, soredia, isidia, thalline lobule and fragments. In general as the community becomes closed and ecosystem more matured, diaspore size increases, few in number with longer time to reproduce.

Reindeer lichens and Lobarion lichens show the trend of increasing thallus size in climax communities. Thus, succession by lichens is highly influenced by generation time, diaspore size and thallus size, which ultimately decide its position in the developing ecosystem.

1.3.4 Competition

Being pioneers lichens usually lack competitive ability in comparison to other plants especially vascular plants. In the process of succession, an organism competes successfully with its immediate predecessors and succumbs to its immediate successor. In general crustose lichens are less competitive than foliose, which are less aggressive than fruticose ones (Brightman and Seaward 1977). Competitive ability corresponds to stages in succession.

Allelopathic nature of lichen substances also enables certain species to predominate in the area by the creation of an 'inhibition zone'. The 'inhibition zones' are zones of 1–5 cm where concentrations of lichen substances produced by lichens are maximum. This phenomenon is quite pronounced in *Rhizocarpon* species.

The study of competition is also closely linked to autecology of the individual species, since successful species in a competitive condition is usually best adapted to its environment (Topham 1977).

1.4 Role of Lichens in Biodeterioration and Soil Establishment

There is no distinct demarcation between the terms biodeterioration and soil formation (pedogenesis), and they may be treated as two distinct phases of a process in which primarily physical and chemical weathering results in loosening of rock and later 'windblown material' soil is formed (Joffe 1949). Biodeterioration (rock weathering) may be defined as any undesirable change in the properties of a material caused by vital activities of living organisms (Saiz-Jimenez 1995, 2001, 1994; Wilson et al. 1981). The development of specific biological species on a particular stone surface is determined by the nature and properties of the stone. Many building materials are prone to colonisation by living organisms. The invasion results in changes in colour and in the chemical or physical properties of the materials (Saiz-Jimenez 1981). In order to assess the impact of colonisation on the material, 'bioreceptivity' of the material is estimated. Bioreceptivity is defined as the overall properties of the material that contribute to the establishment, anchorage and development of flora and or fauna (Guillitte 1995). Other than bioreceptivity of the material, the response of living organisms to a potentially colonisable surface depends on the ecological and physiological requirements of the biological species involved (Fig. 1.5) (Caneva and Salvadori 1988; Seaward and Giacobini 1991). The predominant role of lichens as pedogenetic agent is also related to physical and chemical characteristics which makes them potential biodeteriorating agents. Pedogenesis (soil formation) is thus defined as the transformation of rock material into soil. Linnaeus in 1762 first discussed the ability of crustose lichens to colonise on unweathered rocks and its role in soil formation and plant succession (Bajpai 2009).

1.4.1 Biodeterioration (Rock Weathering)

Lichens are known to occur on various substrates including rocks, which is mainly due to their resistance to desiccation in extreme temperature and efficiency in accumulating nutrients (Martin 1985; Chaffer 1972; Seaward 1979, 1988). Anchoring structures present in lichens which facilitate attachment to the rock substrate are rhizoids and hyphae (Chen et al. 2000). The

1.4 Role of Lichens in Biodeterioration and Soil Establishment

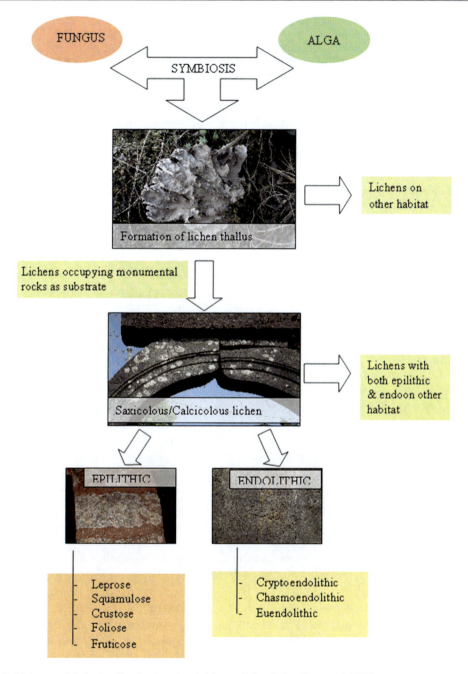

Fig. 1.5 Lichens and their classification based on habitat variation (After Chen et al. 2000)

growth of lichens may either be over the rock surface, epilithic, or it may be entirely living beneath a stone surface, endolithic (Danin et al. 1982, 1983).

Various factors (Table 1.1) are responsible for the establishment of lichens on rock which include pH range, humidity, incident light and nitrogen supply (Erick et al. 1996). Lichens which establish and survive in very narrow range of factors are stenoic, while other species which grow over a wider interval of conditions are euroic. pH plays a predominant role in the development and estab-

Table 1.1 Factors influencing colonisation of lichens and its biodeterioration potential

	Factors	
1	Environmental factors	Incident light (exposed to Sun or shady), humidity, temperature
2	Inorganic and organic pollutants	As carbon, nitrogen and sulphur source which act as growth inhibitors
3	Surface bioreceptivity	Nature of material, pH range, conservation measures, length of exposition
4	Colonising species	Ecological and physiological requirement of the particular species
5	Treatment	Biocides, surfactants/hydrophobic compounds

Modified from Urzi and Krumbein (1994)

lishment of lichens on rocks. Calcicolous species develop on neutral and alkaline substrates, e.g. Lecanora calcarea grow on basic substrates, while silicicolous species grow on acidic substrate, e.g. Parmelia saxatilis (Lisci et al. 2003).

Lichen establishes more readily on a substrate after it has been partly transformed by chemical substances in the air and then by bacteria. This fact is evident from the findings that lichens grow more readily on archaeological ruins, on which mould and porous surfaces facilitate bacterial growth. Biodeterioration caused by lichens has been reviewed by several workers (Hale 1983; Seaward 1988; Monte 1991; Clair and Seaward 2004).

1.4.1.1 Biogeophysical Deterioration

Physical weathering of the rock by saxicolous lichens is mainly brought about by rhizinae penetration and thallus expansion/contraction. Mechanical damage is mainly due to penetration of rhizinae deep into the substrate. The depth to which thallus penetrates depends on the lichen species and nature of the substrate. According to Syers (1964) penetration ranges from 0.3 to 16 mm. Frequent dying and saturation events of the thallus result in loss of cohesion. Porous and calcareous sedimentary rocks have been found particularly susceptible to physical rhizinae penetration by lichens. Among the different growth forms of lichens, foliose and crustose lichens are the most harmful biodeteriogens (Singh and Upreti 1991).

Expansion and contraction of the lichen thallus result in formation of cleavages on the rock surface. Epilithic crustose lichens growing on rocks have a medullary zone that is attached directly to the substratum, and when the cortical tissue at the marginal fringe of the thallus contracts during drying, it creates a pulling strain which may tear the thallus and leave the extreme margins attached to the substratum. In species with thicker thalli, the hyphal tissue may be torn from the substratum detaching rock fragments (Syers and Iskandar 1973; Fry 1922, 1924, 1927).

The squamulose form of lichens has peculiar thallus morphology as they have slightly upward curled margins in dry conditions. Numerous soil particles of the substratum adhered on their lower surface. The curling of margins in dry season thus seems to be the result of the well-documented pulling action of lichen thallus when it dries up. On receiving moisture it expands and squamules margins flatten to attach themselves to the substratum once more; these conditions may accelerate the physical deterioration of monuments (Griffin et al. 1991).

The extent of biogeophysical weathering inhabited by lichens is influenced mainly by the nature of the thallus and by chemical and physical composition of the rock substratum. Because of the nature of attachment, foliose species are probably more effective in biogeophysical weathering. As rhizinae and haptera are more abundant in peripheral region of the thallus, therefore damage caused by expansion and contraction of the thallus is more pronounced in the rocks covering peripheral region (Fry 1924).

These biogeophysical processes result in increase of the surface area of the mineral or rock, making it more susceptible to biogeochemical weathering. The mechanical disruption of crystals due to penetration of the rhizinae accelerates chemical decomposition.

1.4.1.2 Biogeochemical Weathering

The biocorrosive activity by lichens is mainly due to the presence of substances like carbonic acid, lichenic acid and oxalic acid which by acidolysis release proton capable of corroding the material via proton exchange phenomenon (Salter 1856; Salvadori and Zittelli 1981). These weak acidic lichen substances cause chemical degradation of the substrate (Warscheid and Braams 2000; Jackson and Keller 1970).

Water is an essential component in any chemical reaction (being the source of –OH and –H). Lichens, especially crustose lichens, have a tendency to absorb water up to 100–300 % of the dry weight in order to withstand extreme conditions of desiccation. Water is absorbed in the medulla of the thallus, and since the medulla of crustose lichens is in direct contact with the rock surface, the substratum becomes more prone to chemical weathering in the presence of a proton donor (from water).

CO_2, produced during the metabolic process in the thallus, in presence of water forms carbonic acid which being unstable dissociates into H^+ and HCO_3^-. The ability of H+ to accelerate chemical decomposition is a well-established fact (Keller and Frederickson 1957).

Oxalic acid has significant contribution in rock weathering process (Beeh-Andersen 1986). Salter (1856) first observed the role of oxalic acid produced by lichens in disintegration of rocks. Affinity of oxalic acid for calcium (limestone as substratum) may be attributed to their electronic configuration, because of which they form insoluble calcium oxalate. The insoluble calcium oxalate usually deposits on the outer surface of the fungal hyphae in the lichen thallus or on the surface of the upper cortex (Smith 1962). Concentration of calcium oxalate varies from species to species; it may range up to 66 % of the thallus dry weight as in the case of Lecanora esculenta (Euler 1939).

Biocorrosive activity of lichens is mainly characterised by the excretion of organic acids complexation. These acids are capable of chelating cations such as calcium and magnesium from mineral forming stable complexes. It has been shown that biogenic organic acids are considerably more effective in mineral mobilisation than inorganic acids and are considered as one of the major damaging agents affecting monument deterioration. The higher concentration of calcium and magnesium is reported in crustose forms of lichens (*Caloplaca subsoluta*, *Diploschistes candidissimum*) followed by squamulose forms (*Peltula euploca*, *Phylliscum indicum*). The lichen taxa tightly adpressed to the substratum exhibit higher concentration of Ca and Mg than the lichen having loose thallus with free lobe margins.

As secondary substances of lichens including various organic acids, which actively chelate substrate cations and thus modify the chemical and physical structure of mineral substrata. The corrosion of rocks in the study area cannot be ruled out as it has drier and warmer climate coupled with luxuriant growth of oxalic acids and chelating substance producing lichens. Thus, enumeration of lichens and their type of biodeterioration capacity data will be helpful in conducting future biomonitoring studies and to the conservators for adopting conservation practices for the monuments (Bravery 1981; Charola 1993; Koestler et al. 1994, 1997).

Lichen species growing on the monuments produce different secondary metabolites. These chemicals secreted by the mycobiont of many lichens are commonly considered to play a crucially important role in the chemical weathering of monuments (Pinna 1993). In the present study, out of 95 species, 65 have the capability to produce secondary substances. A total of 26 different secondary metabolites are reported, of which atranorin, zeorin, parietin and norstictic, lecanoric and usnic acid are the most common.

Metal chelation ability of the lichen substances is another aspect which contributes towards rock weathering by formation of complexes of the mineral component of the rock with electron-rich lichen substances (Neaman et al. 2005). Lichen substances frequently contain polar donor groups in ortho position which favour the complex formation of cations (Syers 1969). According to Culberson (1969), apart from the presence of electron-rich moieties, water-soluble phenolic groups have an important role to play in metal chelation.

The effect of biogeochemical weathering is pronounced in limestone inhabited by lichens due to the slight solubility of calcium carbonate in water. Many calcicolous species are known to immerse in substratum and form endolithic thallus. Formation of perithecial pits (foveolae) is an evidence of the effect of chemical weathering by lichens on the substratum. When the perithecium dies, a hemispherical pit is left exposed to further weathering. Jackson and Keller (1970) found that lichen-covered rock surface weathered more in comparison to the uncovered surface.

The formation of primitive or lithomorphic soil below saxicolous lichens is well documented (Emerson 1947). Microorganisms and insects feed on living and dead lichen thallus, and organic acids produced by the microorganisms probably decompose minerals of the substratum, thus accelerating biogeochemical weathering (Golubic et al. 1981; Lewis et al. 1985). Due to the thallus morphology, lichens not only accumulate nutrients but also trap dust particle. The trapped dust particle gets mixed with the organic matter produced by the decomposing thalli and with detached particles of the underlying rock altered by weathering processes. Polynov (1945) first discussed the formation of organomineral particle below lichens, which may be considered the first manifestation of the unique and most characteristic feature of soil formation, i.e. lithomorphic soil (Ascaso et al. 1976, 1982).

The formation of lithomorphic soil is significant in plant succession as nutrients particularly P, S, Mg, Ca and K which are essential for plant growth are potentially present in bioavailable form in lichens. Together with cyanobacteria, the lichens play an important role as pioneer organisms in colonising rocks as few cyanolichens have an ability of nitrogen fixation. Water-holding capacity increases with the accumulation of organomineral material which provides favourable habitat for the development of plants especially mosses. Lichens provide habitat for a group of oligonitrophile microorganisms involved in the nitrogen cycle in primitive soils (Evdokimova 1957; Stebaev 1963). Lichens are known to convert the primary calcium phosphate, apatite, into bioavailable form for the growth of species growing in succession (Bobritskaya 1950; Syers and Iskandar 1973; Schatz 1962, 1963).

As lichens have slow growth rate, soil formation induced by lichens is a rather slow process which restricts its predominant role in soil formation, but on plane rock surfaces, lichens are undoubted pioneers (Hale 1961). Slow growth rate in no way hinders the role of lichens as biodeteriorating agents (both physical and chemical), being reservoir of nutrients necessary for plant succession and precursors of organomineral material which results in formation of lithomorphic soil.

Thus, the knowledge of the type of lichen growth form, frequency, density and abundance of lichens and the type of lichen substance produced together with the extent of mycobiont penetration into the rock can provide information about the kind of damage caused by lichens.

Environmental pollution plays a significant role in eliminating a large number of lichen species as they cannot tolerate soot and sulphate and thus have no chance to invade monuments in such areas. A few resistant species that perhaps could not compete with other lichens in earlier unpolluted atmospheric condition find a competition-free field to thrive, or even some species get metabolic stimulation by certain pollutants; they thus spread rapidly. Such lichens are highly virulent as monument biodeteriorating agents (Ayub 2005). Some poliotolerant species, like *Phaeophyscia hispidula*, *Phylliscum indicum*, *Rinodina* sp. and *Buellia* sp. that exhibited aggressive behaviour, spreading rapidly, covering a variety of substrates replacing the disappearing harmless or less harmful species that are susceptible to air pollution, have been reported. Under these conditions, it becomes necessary to determine the interaction between species of lichens and monument stone surface. Such increased lichen activity coupled with direct action of pollutants clearly explains why the monuments apparently unchanged for centuries in the past appear now vulnerable to deterioration by lichen attacks (Singh and Upreti 1991).

Association of lichens and monuments in atmospherically unpolluted areas is a common sight throughout the world (Fig. 1.6). The growth of lichens on monuments and buildings is variously

1.4 Role of Lichens in Biodeterioration and Soil Establishment

Fig. 1.6 An ancient monument (**a**) Panch Ganga Temple in Old Mahabaleshwar (**b**) Lichen species growing luxuriantly on the pillar colonised by lichens (**c**) *Dirinaria* species and (**d**) *Diploschistis* species

interpreted. Some lay greater emphasis to the protective role of lichens, while others consider them as harmful agents. The multicoloured mosaic lichens on monuments' surface has an aesthetic appeal to the viewers, and it also forms a protective cover against external weathering agents. The lichens' cover provides a protective plastering over the substrate. After the removal of lichens, the rock surface under lichen thallus may be more prone to abiotic and biotic factors such as wind and moisture and insects, respectively (Hawksworth 2001; Hyvert 1978).

Damage caused by lichens to monuments has drawn worldwide attention. The extent of damage caused by lichens varies with species (Gayathri 1980; Tiano 1993). Some species are highly corrosive to rocks due to chemical substances widely known as lichen substances. In a study carried out by Chatterjee et al. (1996), different monuments in culturally rich heritage of Karnataka, Orrisa and Uttar Pradesh were studied (Upreti 2002; Upreti et al. 2004; Jain 2001). It was observed that architectural patterns creating many microclimatic conditions also seem to control the distribution and frequency of different lichen species, at different niches. In Karnataka monuments are generally topped by flat roof covered with plaster of lime. It accumulates dust, dirt and plant remains and is exposed to direct sun rays during greater part of the day, producing xeric conditions. In such habitats, *Endocarpon*, *Peltula* and *Phylliscum* are common, while in vertical walls lichen taxa occupy different niches of walls depending upon light exposure and moisture content of the wall. Basal portions of walls are more shady and humid which is preferred by *Lepraria* species, while higher locations are occupied by *Caloplaca*, *Buellia* and *Lecanora* species. Higher and more lightened walls are inhabited by *Pyxine cocoes*, *P. petricola* and *Candelaria concolor*; other species are *Physcia*, *Dirinaria*, *Heterodermia*, *Leptogium*, *Coccocarpia* and *Parmelia* (Singh and Dhawan 1991; Singh et al. 1999). The taxonomic identification and ecological properties of each species growing on monuments are necessary since different species contribute to the process of degradation in a different way. Lichen communities growing on rocks undergo regular patterns of successional change; one assemble of species may occupy a given rock surface for several years, steadily altering the substratum in ways that eventually better accommodate a new combination of species. Thus, over time, the changing lichen community relentlessly changes the rock surfaces. The correct determination of lichens species together with their frequency, abundance and density can provide a clue to the conservators to know the nature of the damage caused by lichens (Singh and Upreti 1991). In order to understand the growth and activity of lichens on different monuments, monuments and rocks representing seven districts of Madhya Pradesh exhibited occurrence of a total number of 95 species belonging 34 genera and 17 families of lichens. Most of the species were recorded from monuments of Raisen district followed by Hoshangabad, Dhar and Anuppur district as 43, 36, 29 and 28 species, respectively. Among these lichen families, Physciaceae, Peltulaceae and Verrucariaceae were the most common in most of the monuments, while Collemataceae, Teloschistaceae and Bacidiaceae were restricted in some localities. Among the different substrates, the sandstone bears the occurrence of maximum species followed by calcareous and bauxite rocks. The reason for more growth of lichens on cement plaster and sandstone is due to their excellent water-holding capacity, while siliceous and bauxite rocks are less porous and exhibit poor growth of lichens (Bajpai 2007).

All the recorded species of lichens from the study area can be grouped into four growth forms. Among the different growth forms, the crustose forms exhibit the dominance of 67 species followed by foliose, squamulose and leprose as 54, 44 and 7 species, respectively, in all the 7 districts. Hoshangabad district exhibits the maximum diversity of crustose form, while Anuppur district is dominated by foliose form of lichens. Dhar district showed luxuriant growth of leprose lichens on the monuments, while Raisen has the dominance of squamulose forms.

The negative role of lichens on monuments has greater acceptance. The monuments of the state of Madhya Pradesh bear luxuriant growth of both more effective biodeteriorating and less effective biodeteriorating species. The information presented about the lichens, substrata and chemical substances will be helpful to the conservators of the monuments to clearly distinguish between the aggressive and less harmful groups and to form appropriate strategies for conservation of monuments in the different states of the country.

1.5 Effect of Microclimatic Condition on Lichen Diversity

The microclimate plays an important role in colonisation of lichens on the substrata. Environmental factors including precipitation, light and shady conditions, humidity and dryness, air quality and wind currents show pronounced impact on the lichen diversity of an area.

Light is a vital ecological factor for a lichen-forming fungus being dependent on a close symbiotic association with its autotrophic photobiont partner. The amount of light received by the photobiont during periods of thallus hydration may determine lichen growth (Dahlman and Palmqvist 2003).

Variation in the incident sunlight gives rise to light and shady conditions. Bright sunlight and high temperature have an inhibiting effect on the lichen growth. In order to protect the thallus especially the algal partner from high irradiance, lichens develop cortical pigments such as parietin or melanin which act as UV filters. Such conditions are quite prevalent in alpine cold deserts and urban centres.

The most suitable temperature for the growth of lichens is 20–25 °C, which is evident by the high diversity in the temperate regions of the world. Many lichens are able to withstand high temperatures of the tropics because of high moisture content due to heavy rainfall. Lichen diversity in tropical areas is mostly corticolous. Certain alpine lichens such as *Rhizocarpon* and *Acarospora* can tolerate very low temperature in high altitudes in the Arctic and Antarctic regions. Lichen species which withstand high temperature are termed as thermophilous lichens.

Moisture content reflected by humidity and dryness affects lichen diversity. Semi-moist tropical areas which have seasonal rains provide favourable conditions for members of Lichniaceae which retain moisture for longer periods of time. In temperate regions, rains are intermittent, so foliose and fruticose lichens grow luxuriantly, while in desert conditions having no rains and dry conditions, they exhibit poor growth of lichens. Only crustose species and some vagrant lichen species compose the flora of the area.

Assessment of air quality is an important parameter for lichen diversity studies. Lichen diversity has been found to be highly influenced by the air quality of an area as lichens are sensitive to phytotoxic gases especially sulphur dioxide which impairs the photosynthetic apparatus by irreversible conversion of chlorophyll a in phaeophytin. For details of the effect of air quality on lichen diversity, see section on Ecosystem Monitoring.

The intensity of wind currents influences lichen diversity. Areas with intense wind have dominance of crustose lichens as they tightly adhered to the substratum. Species of *Usnea* and *Ramalina*, having thin strands, aerodynamically allow wind to pass through the strands without being uprooted by the strong wind current.

Hence, changes in the biotic factors contribute towards varying lichen diversity across the globe. Lichen diversity under natural conditions is dependent on the microclimatic condition and changes in the lichen community structure clearly indicates the changes in the surrounding environment.

References

Ahmadjian V (1993) The lichen symbiosis. Wiley, New York

Ascaso C, Galvan J, Ortega C (1976) The pedogenic action of *Parmelia conspersa, Rhizocarpon geographicum* and *Umbilicaria pustulata*. Lichenologist 8:151–171

Ascaso C, Galvan J, Rodriguez P (1982) The weathering of calcareous rocks by lichens. Pedobiologia 24:219–229

Awasthi DD (2000) A hand book of lichens. Bishan Singh Mahendra Pal Singh, Dehradun

Ayub A (2005) Lichen flora of some major historical monuments & buildings of Uttar Pradesh. PhD thesis, Dr. R. M. L. Avadh University, Faizabad

Bajpai A (2007) Taxonomic & ecological studies on lichens of Bhimbetka world Heritage zone, Raisen district, M. P. PhD thesis, Lucknow University, Lucknow

Bajpai R (2009) Studies on lichens of some monuments of Madhya Pradesh with reference to Biodeterioration and Biomonitoring. PhD thesis, Babasaheb Bhimrao Ambedkar University, Lucknow

Bech-Andersen J (1987) Oxalic acid production by lichens causing deterioration of natural and artificial stones. In: Morthor LHG (ed) Proceedings of the biodeterioration society meeting on biodeterioration of constructional materials, Delft, pp 9–13

Beeh-Andersen J (1986) Oxalic acid production by lichens causing deterioration of natural and artificial stones. In: Prceedings of the biodeterioration Society Meeting on Biodeterioration of constructional materilas (L.H.G. Morthon, ed.), Delft, pp 9–13

Beschel RE (1961) Dating rock surfaces by lichen growth and its application to glaciology and physiography (lichenometry). In: Raasch GO (ed) Geology of the Arctic, vol 2. University of Toronto Press, Toronto, pp 1044–1062

Bobritskaya MA (1950) Absorbtion of mineral elements by lithophilic vegetation on massive crystalline rock. Tr Pochv Inst Akand Nauk SSSR 34:5–27

Bravery AF (1981) Preservation in the conservation industry. In: Russell AD, Hugo WB, Ayliffe AJ (eds) Principles & practices of disinfection, preservation & sterilization. Blackwell Scientific, Oxford, pp 370–402

Brightman FH, Seaward MRD (1977) Lichens of man-made substrates. In: Seaward MRD (ed) Lichen ecology. Academic Press, London/New York/San Francisco, pp 253–293

Brodo IM (1973) Substrate ecology. In: Ahmadjian V, Hale ME (eds) The Lichens. Academic, New York, pp 401–441

Büdel B (1987) Zur Biologie und Systematik der Flechtengattungen Heppia und Peltula im sü dlichen Afrika. Bibliotheca Lichenologica 23:1–105

Büdel B (1990) Anatomical adaptations to the semiarid/arid environment in the lichen genus *Peltula*. Bibliotheca Lichenologica 38:47–61

Büdel B, Lange OL (1991) Water status of green and blue green algal photobiont in lichen thalli after hydration by water vapour uptake: do they become turgid. Botanica Acta 104:361–366

Büdel B, Scheidegger DC (2008) Thallus morphology and anatomy. In: Nash TH III (ed) Lichen biology, 2nd edn. Cambridge University Press, Cambridge, pp 40–68

Caneva G, Salvadori O (1988) Biodeterioration of stone. In: Lazzarini L, Pieper R (eds) Deterioration and conservation of stone, Studies and documents on the cultural heritage, no 16. UNESCO, Paris, pp 182–234

Chaffer RJ (1972) The weathering of natural stones. DSIR building research special report No. 18, London

Charola AE (1993) General report on prevention & treatment, cleaning, biocides & mortars. In: Actes of the congreee International Sur la conservation de la Pierse at autres Materiauxe, UNESCO, 29-6-1-07,1993, Paris, pp 65–68

Chatterjee S, Sinha GP, Upreti DK, Singh A (1996) Preliminary observations on lichens growing over some Indian monuments. Flora Fauna 2(1):1–4

Chen J, Blume H, Bayer L (2000) Weathering of rocks induced by lichens colonization – a review. Catena 39:121–146

Culberson CF (1969) Chemical and botanical guide to lichen products. University of North Carolina Press, Chapel Hill

Dahlman L, Palmqvist K (2003) Growth in two foliose tripartite lichens, *Nephroma arcticum* and *Peltigera aphthosa*: empirical modelling of external vs internal factors. Funct Ecol 17:821–831

Danin A, Gerson R, Marton K, Garty J (1982) Pattern of lime stone and dolomite weathering by lichens and blue green algae and their palaeoclimatic significance. Palaeogeogr Palaeoclimatol Palaeoecol 37: 221–233

Danin A, Gerson R, Garty J (1983) Weathering patterns on hard limestone and dolomite by endolithic lichens and cyanobacteria: supporting evidence for eolian contribution to Terra Rossa soil. Soil Sci 136:213–217

Degelius G (1964) Biological studies of the epiphytic vegetation on twigs of *Fraxinus excelsior*. Acta Horti Gotoburg 27:11–55

Dietz S, Büdel B, Lange OL, Bilger W (2000) Transmittance of light through the cortex of lichens from contrasting habitats. In: Schroeter B, Schlensog M, Green TGA (eds) Aspects in cryptogamic research. Contributions in honour of Ludger Kappen. Gebrü der Borntraeger Verlagsbuchhandlung, Berlin, pp 171–182

Emerson FW (1947) Basic botany. McGraw-Hill (Blakiston), New York

Erick MJ, Grossl PR, Golden DC, Sparks DL, Ming DW (1996) Dissolution kinetics of a lunar glass simultant at 298K. The effect of Ph and organic acids. Geochim Cosmochim Acta 58:4259–4279

Euler WD (ed) (1939) Effect of sulphur dioxide on vegetation. National Research Council Canada, Ottawa

Esslinger TL (1977) A chemosystematic revision of the brown Parmeliae. J Hattori Botanical Lab 42: 1–211

Evdokimova TI (1957) Soil formation processes on metamorphic rocks of Karelia. Pochvovedenie 9:60–69

Fry EJ (1922) Some types of endolithic limestone lichens. Ann Bot 36:541–562

Fry EJ (1924) A suggested explanation of the mechanical action of lithophytes lichens on rock (shale.). Ann Bot 38:175–196

Fry EJ (1927) The mechanical action of crustaceous lichens on substrata of shale, schist, gneiss, limestone, and obsidian. Ann Bot 41:437–460

Gayathri P (1980) Effect of lichens on granite statues. Birla Archaeol Cult Res Inst Res Bull 2:41–52

Golubic S, Friedmann EI, Schneier J (1981) The lithobiotic ecological niche, with special reference to microorganism. J Sediment Petrol 51:475–478

Griffin PS, Indicator N, Kostler RJ (1991) The biodeterioration of stone: a review of deterioration mechanism, conservation case histories & treatment. Int Biodeterior Spec Issues Biodeterior Cult Prop 28:187–207

Guillitte O (1995) Bioreceptivity: a new concept for building ecology studies. Sci Total Environ 167:215–220

Hale ME (1961) Lichen handbook. Smithsonian Institute, Washington, DC

Hale ME (1983) The biology of lichens, 3rd edn. Edward Arnold, London

Hawksworth DL (1988) The fungal partner. In: Galun M (ed) CRC handbook of lichenology, vol 1. CRC Press, Boca Raton, pp 35–38

Hawksworth DL (2001) Do lichens protect or damage stonework? Mycol Res 105:386

Honegger R (1991) Haustoria-like structures and hydrophobic cell wall surface layers in lichens. In: Mendgen K, Lesemann DE (eds) Electron microscopy of plant pathogens. Springer, Berlin, pp 277–290

Honegger R (2008) Morphogenesis. In: Nash TH (ed) Lichen biology, 2nd edn. Cambridge University Press, Cambridge, pp 69–93

Hyvert G (1978) Weathering & restoration of Borobudur Temple, Indonesia. In: Winkler EM (ed) Decay & preservation of stone, Engineering geology case histories No.11. Geological Society of America, Boulder, pp 95–100

Jackson TA, Keller WD (1970) A comparative study of the role of lichens and "inorganic" processes in the chemical weathering of recent Hawaiian lava flows. Am J Sci 269:446–466

Jahns HM (1984) Morphology, reproduction and water relations – a system of morphogenetic interactions in Parmelia saxatilis. Nova Hedwigia 79:715–737

Jain AK (2001) Biodeterioration of rock cut images at Gwalior fort. In: Agarwal OP, Dhawan S, Pathak N (eds) Studies in biodeterioration of material. ICCI & ICBCP, Lucknow, pp 59–68

Joffe JS (1949) Pedology. Pedology Publications, New Brunswick

Kalgutkar RM, Bird CD (1969) Lichens found on Larix lyallii and Pinus albicaulis in South Western Alberta, Canada. Can J Bot 47:627–648

Keller ND, Frederickson AF (1957) Role of plants & colloidal acids in the mechanism of weathering. Am J Sci 250:594–608

Koestler RJ, Brimblecomb P, Camuffo D, Ginell W, Graedel T, Leavengood P, Petuskhova J, Steiger M, Urzi C, Verges-Belmin Warscheid T (1994) Group report: how do external environmental factors accelerate change? In: Krumbien WK, Brimblecombe P, Cosgrove DE, Stauforth S (eds) Durability & change. Wiley, Chichester, pp 149–163

Koestler RJ, Warscheid T, Nieto F (1997) Biodeterioration risk factor & their management. In: Baer NS, Snethlage R (eds) Saving our cultural heritage. The conversation of historic stone structure. Wiley, New York, pp 25–36

Lange OL, Büdel B, Meyer A, Kilian E (1993) Further evidence that activation of net photosynthesis by dry cyanobacterial lichens requires liquid water. Lichenologist 25:175–189

Lewis FJ, May E, Bravery AF (1985) Isolation and enumeration of autotrophic and heterotrophic bacteria from decayed stone. In: Vth international congress on deterioration & conservation of stone, vol 2. Press Polytechnique Romandes, Lausanne, pp 633–642

Lisci M, Monte M, Pacini E (2003) Lichens and higher plants on stone: a review. Int Biodeter Biodegr 51:1–17

Martin MAE (ed) (1985) Concise dictionary of biology. Oxford University Press, Oxford

Monte M (1991) Multivariate analysis applied to the conservation of monuments: lichens on the Roman Aqueduct Anio Vetus in S. Gregorio. Int Biodeterior 28:151–163

Nayaka S (2011) Collection, identification and preservation of lichens. In: Workshop on methods and approaches in plant systematics, 5–14 December 2011. CSIR National Botanical Research Institute, pp 93–110

Neaman A, Chorover J, Brantley SL (2005) Implication of the evolution of organic acid moieties for basalt weathering over ecological time. Am J Sci 305:147–185

Peršoh D (2004) Diversity of lichen inhabiting fungi in the *Letharietum vulpinae*. In: Randlane T, Saag A (eds) Lichens in focus. Tartu University Press, Tartu, p 34

Pinna D (1993) Fungal physiology & the formation of calcium oxalate film on stone monuments. Aerobiologia 9:157–167

Polynov BB (1945) The first stages of soil formation on massive crystalline rocks. Pochcowdenie 7, 327–339 (Israel program for technical translations Cat. No. 1350)

Saiz-Jimenez C (1981) Weathering of building material of the Giralda (Seville, Spain) by Lichens. In: Proceedings of the 6th triennial meeting ICOM committee for conservation, Ottawa, 4 October 1981

Saiz-Jimenez C (1994) Biodeterioration of stone in historic buildings & monuments. In: Liewellyn GC, Dashek WW, Rear CEO (eds) Biodeterioration research 4: Mycotoxins, wood decay, plant stress, Biocorrosion and general biodeterioration. Plenum, New York, pp 587–603

Saiz-Jimenez C (1995) Deposition of anthropogenic compounds on monuments & their effect on airborne microorganism. Aerobiologia 11:161–175

Saiz-Jimenez C (2001) The biodeterioration of building materials. In: Stoecker J II (ed) A practical manual on microbiologically influenced corrosion, vol 2. NACE, Houston, pp 4.1–4.20

Salter JW (1856) On some reaction of oxalic acid. Chem Gaz 14:130–131

Salvadori O, Zitelli A (1981) Monohydrate and dihidrate calcium oxalate in living lichen incrustation biodeteriorating marble columns of the basilica of S. Maria Assunta on the island of Torecello (Venice). In: Rossi Manaresi R (ed) Proceeding of IInd international symposium the conservation of stone II. Centro per la Conservazione delle Sculture all Aperto, Bologna, pp 759–767

Saxena S, Upreti DK, Singh A, Singh KP (2004) Observation on lichens growing on artifacts in the Indian subcontinents. In: St. Clair L, Seaward M (eds) Biodeterioration of stone surface. Kluwer Academic Publishers, Dordrecht, pp 181–193

Schatz A (1962) Pedogenic (soil-forming) activity of lichens acids. Naturwissenchaften 59:518–522

Schatz A (1963) Soil microorganisms & soil chelation. The pedogenic action of lichens & lichen acids. Agric Food Chem 11:112–118

Seaward MRD (1976) Performance of *Lecanora muralis* in an urban environment. In: Brown DH, Hawksworth DL, Bailey RH (eds) Lichenology: progress and problems. Academic, London, pp 323–357

Seaward MRD (1979) Lower plants & the urban landscape. Urban Ecol 4:217–225

Seaward MRD (1988) Lichens damage to ancient monuments: a case study. Lichenologist 20(3):291–295

Seaward MRD, Giacobini C (1991) Lichens as biodeteriorators of archaeological materials with particular reference to Italy. In: Agarwal OP, Dhawan S (eds) Biodeterioration of cultural property. Mac Millan India Ltd., New Delhi, pp 195–206

Singh A, Dhawan S (1991) Interesting observation on stone weathering of an Indian monument by lichens. Geophytology 21:119–123

Singh A, Upreti DK (1991) Lichen flora of Lucknow with special reference to its historical monuments. In: Agrawal OP, Dhawan S (eds) Biodeterioration of cultural property: proceedings of the international conference on biodeterioration of cultural property, 20–25 February 1989, held at National Research Laboratory for Conservation of Cultural Property, in collaboration with ICCROM and INTACH. Macmillan India, New Delhi, pp 219–231

Singh A, Chatterjee S, Sinha GP (1999) Lichens of Indian monuments. In: Mukerjee KG, Chamola BP, Upreti DK, Upadhyaya RK (eds) Biology of lichens. Aravali Book International, New Delhi, pp 115–151

Smith DC (1962) The biology of lichen thalli. Biol Rev 37:537–570

St. Clair LL, Seaward MRD (2004) Biodeterioration of rock substrata by lichens: progress and problems. In: St. Clair LL, Seaward MRD (eds) Biodeterioration of stone surfaces. Kluwer, Dordrecht, pp 1–8

Stebaev IV (1963) Die Veranderung der Tierbevolkerung der Boden im Laufe der Bodenentwicklung auf Felsen und auf Verwitterungsprodukten im Wald-Wiesenlandschaften des Sud-Urals. Pedobiologia 2:265–309 (in Russian)

Syers JK (1964) A study of soil formation on carboniferous limestone with particular reference to lichens as pedogenic agents. PhD thesis, University of Durham England

Syers JK (1969) Chelating ability of fumarprotocetraric acid and *Parmelia conspersa*. Plant Soil 31:205–208

Syers JK, Iskandar IK (1973) Pedogenic significance of lichens. In: Ahamadjian V, Hale ME (eds) The lichens. Academic, New York, pp 225–248

Syers JK, Birine AC, Mitchell BD (1964) The calcium oxalate content of some lichens growing on lime stone. Lichenologist 3:409–424

Tiano P (1993) Biodeterioration of stone monuments: a critical review. In: Garg KL, Garg N, Mukerji KG (eds) Recent advance in biodterioration and biodegradation, vol 1. Naya Prakashan, Calcutta, pp 301–321

Timdal E (1991) A monograph of the genus Toninia (Lecideaceae, Ascomycetes). Opera Bot 110:1–137

Topham PB (1977) Colonization, growth succession and competition. In: Seaward MRD (ed) Lichen ecology. Academic Press Inc, London, pp 31–68

Tschermak-Woess E (1988) The algal partner. In: Galun M (ed) CRC handbook of lichenology, vol 1. CRC Press, Boca Raton, pp 39–92

Upreti DK (2002) Lichens of Khajuraho temple and nearby area of Mahoba and Chattarpur district. In: Srivastava RB, Mathur GN, Agarwal OP (eds) Proceedings of national seminar on biodeterioration of material- 2. DMSRDE, Kanpur, pp 123–127

Upreti DK, Nayaka S, Joshi Y (2004) Lichen activity over rock shelter of Bhimbetka world Heritage Zone, Madhya Pradesh. Rock Art research: changing paradigms. In: The 10th congress of the International Federation of Rock Art Organization (IFRAO) & federation of Rock Art, 28th Nov–2nd Dec 2004. Rock Art Society of India, pp 30–31

Urzi C, Krumbein WE (1994) Microbiological impact on cultural heritage. In: Krumbein WE, Brimblecobe P, Consgrove DE, Staniforth S (eds) Durability and change: the Science, responsibility and cost of sustaining cultural heritage. Wiley, New York, pp 107–135

Warscheid TH, Braams J (2000) Biodeterioration of stone: a review. Int Biodeter Biodegr 46:343–368

Wilson MJ, Jones D, McHardy WJ (1981) The weathering of serpentines by *Lecanora atra*. Lichenologist 13:167–176

Secondary Metabolites and Its Isolation and Characterisation

2

Secondary metabolites are known to protect lichens against increasing environmental stresses such as light exposure, water potential changes, microbial and herbivore interactions and other changes associated with changes in environmental conditions. Toxitolerant lichens show resistance to ambient environmental levels of pollutants which may be phenotypic or genotypic. In recent years, more attention is being paid to the chemical characterisation of the phenotype. The metabolome consists of two types of compounds, the primary metabolites and the secondary metabolites. The primary metabolites are compounds involved in the basic functions of the living cell, such as respiration and biosynthesis of compounds needed for a living cell, while some secondary metabolites are species specific, which play a role in the interaction of a cell with its environment and may be used as tool to protect lichens from external biotic and abiotic factors, including its defence against elevated pollutant concentration. As metabolic profiles in lichens, may be used as pollution indicators (biosensors), thus lichens have great potential for the risk assessment of ecosystem. Therefore, isolation and characterisation of metabolites may provide direct evidence about the air quality-induced metabolomic changes in lichens. This chapter provides insight into various chromatographic techniques and modern spectroscopic techniques involved in characterisation of lichen substances.

The use of plant bioindicators has proved to be a complementary method of investigation for pollutant analysis, as they constitute real biological integrators, capable of providing a basis for assessment of environmental quality and/or contamination (Hawksworth and Rose 1970; Howell 1970; Seaward 1993; Monna et al. 1999; Alaimo et al. 2000; Conti and Cecchetti 2001). Over the past few years, research has focused on the measurement of chemical compounds in plants, as indicators of a particular environment state. This provides a basis for determining the long-term impact of even low levels of pollution and, because these physiological changes appear before morphological and anatomical symptoms, they provide an early warning signal of modifications in environmental quality. Among the chemical compounds in plants, secondary metabolites are of great importance in plant–environment relationships (Beart et al. 1985; Hagerman and Robbins 1987; Haslam 1989; Rhodes 1994; Waterman and Mole 1994; White 1994; Macheix 1996; Spaink 1998). Among these secondary metabolites, phenolic compounds are of particular interest because of (1) their major ecological role in allelopathic processes (Haslam 1989; Rhodes 1994; Cooper-Driver and Bhattacharya 1998), (2) their role in the protection of plants against herbivores (Pisani and Distel 1998) and (3) their involvement in the response of plants to environmental stress such as intra- and/or interspecific competition or atmospheric pollution (Muzika 1993; Karolewski and Giertych 1994; Penuelas et al. 1996).

2.1 Environmental Role of Lichen Substances

The unique combination of alga and fungus produces varieties of chemicals which can be categorised in to two major classes: (1) primary metabolites and (2) secondary metabolites. Primary metabolites are intracellular products which are directly involved in the metabolic activities of the lichens such as growth, development and reproduction. They include proteins, amino acids, polyols, carotenoids, polysaccharides and vitamins, which are bound to the cell walls and protoplasts. They are often water soluble and can be extracted with boiling water. The primary metabolites are either of fungal or algal origin or both. They are also nonspecific and present in free-living alga, fungus, higher plants and other organisms (Culberson and Elix 1989; Ahmadjian and Jacobs 1981).

In lichen, all the secondary metabolites are of fungal origin. They are produced by utilising the primary metabolites and are not involved in the direct metabolism of lichen. Secondary metabolites are the byproduct of primary metabolism and biosynthetic pathways act as storage substances, few have an important ecological role. The secondary metabolites are deposited on the surface of the hyphae rather than within the cells; hence, they are called as extracellular compounds. Popularly they are known as lichen acids or lichen substances. Secondary metabolites are insoluble in water but can be extracted using organic solvents. The lichens produce more than 800 secondary metabolites, most of them are unique to lichens and only small portion of about 50–60 occur in other fungi or higher plants. For example, the anthraquinone, parietin, the orange pigment that is common in most Teloschistales, occurs in non-lichenised fungal genera *Achaetomium, Alternaria, Aspergillus, Dermocybe, Penicillium* as well as in the vascular plants *Rheum, Rumex* and *Ventilago*. The lecanoric acid also occurs in fungus *Pyricularia*, while the sterol, brassicasterol of higher plant, is also available in the lichens (Boustie and Grube 2005; Shukla et al. 2004, 2010).

Probable hypothesis for the production of secondary metabolites postulates that a fungus undergoes maximum growth when all required nutrients are available in optimal quantities and proportions. If one nutrient becomes altered, then primary metabolism is affected and fungal growth is slowed, which may be due to abiotic factors including environmental changes (Deduke et al. 2012). The intermediates of primary metabolism that are no longer needed in the quantity in which they are produced may be shifted to another pathway. It is thought that the intermediates may be used in the secondary metabolic pathways (Moore 1998) serving as an alternative sink for the extra products of primary metabolism while allowing nutrient uptake mechanisms to continue to operate. Thus secondary metabolites are mainly products of an unbalanced primary metabolism resulting from slowed growth, including metabolites that are no longer needed for growth.

A shift in secondary metabolism to allow survival in a particular habitat may promote changes in species and therefore functional attributes of phenotype. Environmental changes influence many cellular activities and also serve as triggers for a change in mode of reproduction, influencing the entire biology of the species. Since most species have diagnostic compounds that are consistently produced because of genetic inheritance and species adaptation to particular niches, chemical diversity can be correlated with taxonomy (Lawrey 1977). The chemical correlation with taxonomy is referred to as chemotaxonomy (Hawksworth 1976; Frisvad et al. 2008). Knowledge of species taxonomic diversity is a first clue to understanding the polyketide diversity in any habitat. *Ramalina americana* was split into two different species (*R. culbersoniorum* and *R. americana*) based on secondary metabolite and nucleotide sequence divergence (LaGreca 1999). The *Cladonia chlorophaea* complex contains at least five chemospecies, which are named and determined by the secondary metabolite produced (Culberson et al. 1977). Other examples exist to show variability among individuals within the same geographic area. Secondary metabolites may also vary even within chemospecies. For example, the diagnostic metabolite for *Cladonia grayi* is grayanic acid, and for *C. merochlorophaea* is merochlorophaeic acid. However, these species may or may not produce fumarprotocetraric acid, a polyketide that is considered to be an accessory compound since it is not consistently produced among individuals within a species which may be attributed to changes in the environment (Culberson et al. 1977) affecting regulatory pathways that depend on fungal developmental and environmental changes.

It has been observed that polyketide-producing organisms derive benefit from these compounds, which allow them to survive in discrete ecological niches by reacting to environmental variables such as light or drought or protecting themselves from predators and parasites (Armaleo et al. 2008; Huneck 1999). Secondary metabolites have also been hypothesised to play a role in herbivory defence, antibiotics or as metal chelators for nutrient acquisition (Gauslaa 2005; Lawrey 1986; Huneck 1999). Recently it was hypothesised that polyketides play a role in protection against oxidative stress in fungi (Luo et al. 2009; Reverberi et al. 2010; Bennett and Ciegler 1983) and that some metabolites such as fumarprotocetraric acid, perlatolic and thamnolic acids contribute to the ability of lichens to tolerate acid rain events and consequences (Hauck 2008; Hauck et al. 2009a, b).

One explanation for high levels and numbers of secondary metabolites in lichen fungi is the slow growth of the lichen. It is known that lysergic acids are produced in the slow-growing overwintering structures (ergot) of the non-lichenised fungus *Claviceps purpurea*. The ergot in *C. purpurea* represents the slow-growing overwintering stage of the fungus following the fast-growing mycelial stage during the summer season where infection of the host occurs. However, lichens have no fast-growing stage in comparison with *C. purpurea*, and there appears to be no limitation to production of polyketides. The detoxification of primary metabolites is another hypothesis that has been proposed to explain the production of secondary metabolites. If growth of the fungus slows down, but metabolism is still very active, toxic products of primary metabolism may accumulate. The transformation of these into secondary metabolites may be one method to prevent toxic accumulation of byproducts. This hypothesis may be integrated within the first hypothesis on slow growth rates to explain the production of secondary compounds by fungi.

Regardless of the reason for secondary metabolite production (byproduct, detoxification of primary metabolism or leftover products after growth slows), they often elicit a function that is advantageous to survival of the lichen within its ecological niche. The advantage(s) may in part be understood by the location of the compounds within the thallus such as atranorin and usnic acid occurring more frequently in the cortical hyphae than the medullary hyphae and having a function related to photoprotection. These chemical characters are thought to be adaptive features because of their perceived ecological role (Bjerke et al. 2004).

The most widely studied secondary metabolite produced by lichen-forming fungi is usnic acid, a cortical compound that absorbs UV light (Cocchietto et al. 2002). Seasonal and geographic variation has been shown to occur in populations of the usnic acid-producing lichens *Flavocetraria nivalis* and *Nephroma arcticum* in Arctic and Antarctic regions (Bjerke et al. 2004, 2005; McEvoy et al. 2007a, b). These are regions that are highly exposed to strong UV light, desiccating winds and harsh temperature changes (Hamada 1982, 1991). Other secondary metabolites examined on large geographic scales include alectoronic acid, a-collatolic acid and atranorin produced by *Tephromela atra*, a crustose lichen that grows on tree bark. That study showed a significant variation between localities (Hesbacher et al. 1996), but no relationship with tissue age, grazing or reproductive strategy. In a study on the *Cladonia chlorophaea* complex, the levels of fumarprotocetraric acid increased from coastal North Carolina to the Appalachian Mountains in the interior of the state (Culberson et al. 1977). The authors interpreted this geographical gradient of higher levels of fumarprotocetraric acid in mountain populations, as providing protection against harsher environmental conditions in the mountains than in the coastal area. If environment influences secondary metabolite production, then changes should be observed along a gradient of environmental conditions over a species distribution. Although Hesbacher et al. (1996) showed that thallus age has no affect on secondary compound concentrations for atranorin and alectoronic acid, Golojuch and Lawrey (1988) showed that concentrations of vulpinic and pinastric acids are higher in younger lichens.

Bjerke et al. (2002) showed that the most exposed sections of the thallus (such as the tips of C. mitis) accumulate greater concentrations of secondary compounds than less exposed sections of the thallus. However, it is not known if the metabolites are actively produced in the exposed and younger tips or if the metabolites are lost in the older parts of the thallus as the thallus ages and the fungal tissue degrades, giving the appearance that the tips have more metabolites. High concentrations of secondary metabolites were reported in the sexual and asexual reproductive bodies rather than the somatic (vegetative) lichen tissue (Liao et al. 2010; Culberson et al. 1993). Geographic and intrathalline variation suggests a functional role for these metabolites that has been described in a theory called optimal defence theory (ODT). The theory states that plants and fungi will allocate secondary compounds where they are most beneficial to the organism (Hyvärinen et al. 2000), implying an active production of secondary metabolites, which is contrary to the current theories of secondary metabolite production. The inconsistency in findings to explain geographic trends and the intrathalline variation in secondary metabolite production may be addressed by increasing sample size and geographic distance to capture the population variation and prevent saturation of larger scale geographic variation. Relationships between metabolite production and geographic location should be evident in a north–south direction because of differences in climate. It would also be expected that the production of intrathalline metabolites would be coordinated because of their hypothesised function regarding environmental changes.

A number of environmental predictions of future global climate conditions are predicted in the fourth assessment of the United Nations Intergovernmental Panel on Climate Change. The outlook included an increase in average temperature; an increase in intensity and length of droughts; an increase in global water vapour, evaporation and precipitation rates which will cause increasing tropical precipitation and decreasing subtropic precipitation; an increase in sea levels from glacial melt; and anthropogenic carbon dioxide production will further increase atmospheric carbon dioxide levels (Meehl et al. 2007). Most of these changes will have implications on the future adaptability and secondary metabolite production of lichen species. These secondary metabolites protect against increasing environmental stresses such as light exposure, water potential changes, microbial and herbivore interactions and other changes associated with changes in environmental conditions (MacGillvray and Helleur 2001).

Increases in temperature may require the increase of secondary metabolites such as salazinic acid to mitigate the effects of higher temperatures on lichen biology. The relationship between temperature and production of salazinic acid is thought to be related to the effect of hydrophobic properties of the metabolite. The metabolite, being produced by medullary hyphae, would ensure a hydrophobic environment to optimise carbon dioxide transfer to the algal cells. A higher temperature increases the water potential of the thallus and more need for hydrophobic conditions to allow optimal carbon dioxide exchange between air spaces and algal cells. However, a higher thallus temperature may also promote the initiation of transferring one algal partner for another partner. Depending on the taxonomic extent of different algal partners, this may invoke different carbohydrate starting units or trigger a different biosynthetic pathway for secondary metabolite production. The predicted increases in average annual temperature in northern geographic areas may also promote temperate species of lichens to move further north into previously uninhabitable environments. Simultaneously, this may cause a more northerly movement of lichens that are adapted to or can tolerate cooler environments. The effects on epiphytic lichens will also be significant based on the availability of host tree species and how well the host trees adapt to climate change. Cool temperature plant species

2.1 Environmental Role of Lichen Substances

that do not adapt well to warmer temperatures may become fewer in number in northern regions. Fewer plant species may reduce the availability of suitable habitat for lichens specialised to growing on the bark of specific tree species. Species of lichens that are generalists, colonising a number of different tree species or other substrata, will be better adapted to environmental changes than specialist species, because previously lost tree hosts may be replaced by succeeding species of plant host. Pollution is also thought to be responsible for the increased levels of ultraviolet light caused by the loss of atmospheric ozone. Cortical compounds and other compounds within the thallus that offer protection to the sexual and asexual reproductive structures and photobionts may ensure that those lichen species will have some protection from increase ultraviolet light. Species lacking those photoprotective compounds may endure degradation of photobionts and an increased frequency of mutations due to ultraviolet light exposure. Environmental stress may stimulate the production of cortical compounds in species that normally do not produce them, in species that do not produce them frequently and in increased quantities for the species that already produce them.

Multiple parameters are responsible for conferring stress tolerance especially to acidic conditions in a lichen species, and presence of secondary metabolites has been shown to be a key feature in enabling certain species to withstand harmful levels of pollutants by either metal homeostasis or having low pK_{a1} values (first dissociation constant). It has been observed that lichen species producing depsides and depsidones (fumarprotocetraric, perlatolic or thamnolic acids) have high acidity tolerance (Hauck et al. 2009a, b).

Hydrophobicity of the thallus surface determines SO_2 tolerance in lichens. Hydrophobicity in lichen thalli is produced both by the structure of the surface at nano- and micrometre scale and by the covering of the cell walls with lichen substances. Lichen species with a strongly hydrophobic surface, which has been originally evolved to facilitate the uptake of CO_2 into the thallus

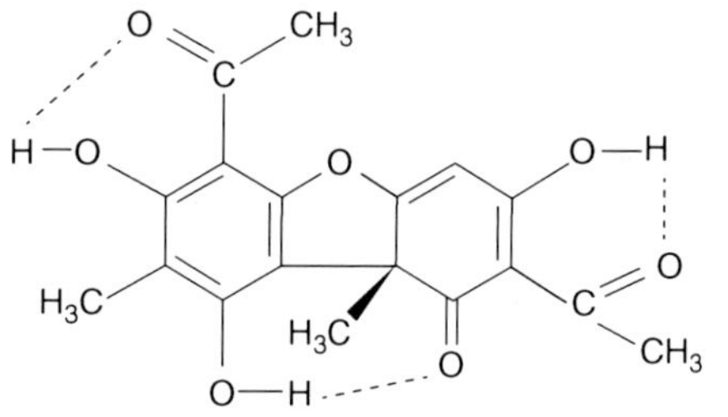

Fig. 2.1 Chemical structure of usnic acid, a potential broad-spectrum antibiotic compound

even under wet conditions (Shirtcliffe et al. 2006), are tolerant to SO_2, because their hydrophobic surfaces inhibit the diffusion of aqueous solutions containing SO_2 and its derivatives. Lichens with hydrophilic surfaces are generally sensitive to SO_2 (Hauck et al. 2008). Most of the lichen substances are hydrophobic in nature; a common hydrophobic compound, usnic acid (dibenzofuran), is widespread in the cortical regions of the thallus. Usnic acid (Fig. 2.1) is known to play an important role in protecting lichen thallus against biotic and abiotic stresses. Under specific pH range of 3.5–5.5, usnic acid plays a crucial role in the acidity tolerance of lichens depending on its pK_{a1} value of 4.4 (Ingólfdóttir 2002; Hauck et al. 2009a, b; Hauck and Jürgens 2008).

Formation of metal complexes by lichen substances (commonly termed as *chelation*) is another lichen substance-mediated control of metal uptake. Usnic acid is known to form complexes under acidic conditions, while parietin and most compounds of the pulvinic acid group form complexes under alkaline conditions. While rhizocarpic acid is known to form metal complexes in both acidic and alkaline pH range, these pH preferences in complex formation provide explanation of the strong preference of lichens having usnic or rhizocarpic acids towards acidic substrata and preference of lichens with parietin and some lichens with compounds of the pulvinic acid group either for nutrient-rich substrata at low pH or for calcareous substrata (Hauck et al. 2008).

Heavy metal uptake in lichens is also known to control the extracellular adsorption of metal ion by lichen substances, termed as *metal homeostasis*.

Lichen substances mainly depsides and depsidones having lone pairs of electrons and number of hydroxyl groups provide an inherent characteristic chelating property. Hauck et al. (2008) showed that lichen substances present in *Hypogymnia physodes* control the uptake of Cu and Mn. Adsorption characteristics are dependent on the chemical nature of lichen substance and the electrons available in the metal ion for binding (Hauck and Huneck 2007a, b).

Pyxine subcinerea grows luxuriantly in the metal-enriched urban environment in the tropical and temperate regions of India. In *P. subcinerea* metal concentrations were correlated with its corresponding chemical profile which indicates that lichaxanthone had significant negative correlation with lead (−0.7212) and zinc (−0.6443) while with Cd and Cr had significant positive correlation 0.8466 and 0.8599, respectively. Cadmium and chromium having variable valency are known to have higher affinity to form coordinate compounds in comparison to lead and zinc (Shukla 2012).

Correlation between metal concentration (Cd) and lichen substance of *Phaeophyscia hispidula* and *P. subcinerea* was affirmed by carrying out experimental analysis (Fig. 2.2) in which filter paper impregnated with lichen substances following the procedure of Hauck et al. (2008) inhibited metal absorption (Cd) on filter paper. Result revealed that *P. subcinerea* inhibited metal absorption more than *P. hispidula*, thus establishing the role of lichen substances in metal homeostasis (Shukla 2012).

Lichen substances play an important role in photoprotection, enabled by the presence of coloured lichen substances like parietin, usnic acid, melanins and calcium oxalate crystals (Table 2.1) (Bjerke and Dahl 2002; Nybakken et al. 2004; Hauck et al. 2009b). Majority of the PAR (Photosynthetically Active Radiation)-absorbing lichen substances are present in the cortical region of the thallus. The coloured pigments mainly belong to the dibenzofurans, anthraquinones, xanthones (faintly yellow) and pulvinic acid derivatives (Huneck and Yoshimura 1996).

The anthraquinones include several orange substances that are synthesised by lichens, such as chrysophanol, emodin, fallacinol, parietin or parietinic acid (Huneck and Yoshimura 1996). Parietin is the most common anthraquinone in lichens, which occurs in species of the families Teloschistaceae, Brigantiaceae, Letroutiaceae and Psoraceae (Solhaug and Gauslaa 1996; Hafellner 1997; Johansson et al. 2005). Pulvinic acid and its derivatives, such as calycin, epanorin, pulvinic acid lactone as well as leprapinic, pinastric, rhizocarpic and vulpinic acids, form a group of mostly yellow pigments (Huneck and Yoshimura 1996).

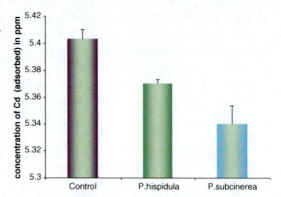

Fig. 2.2 Experimental analysis showing role of lichen substances in metal homeostasis

Table 2.1 Chemical constituents present in lichens playing probable role in photo protection of lichen thallus

S. no.	Chemical constituents	Environmental role	References
1	Usnic acid, diffractic acid and barbatic acid	UV-C stress and cold temperature stress	Hager et al. (2008)
2	Depsidones and melanins	Light-screening pigments	McEvoy et al. (2007a)
3	Phenolics, depsides and triterpenes		Cuellar et al. (2008)
4	Usnic acid	UV Screening	McEvoy et al. (2007b)
5	Parietin	UV-protecting agent	Solhaug et al. (2003)
6	Secondary metabolites	Low temperature stress	Stocker-Wörgötter (2001)
7	Gyrophoric acid, lecanoric acid and methyl orsellinate	Inhibition of PTP1B	Seo et al. (2009)

Lichen is also known to synthesise UV-B-absorbing cortical pigments, which are produced by the mycobiont and induced by UV-B radiation such as brown melanic compounds in *Lobaria pulmonaria* and orange parietin in *Xanthoria* species. The UV-B-absorbing compounds act as UV filter and protects the inner algal layer from harmful effects of UV radiation. UV-B has been reported to induce production of usnic acid (Solhaug et al. 2003; McEvoy et al. 2006; Nybakken and Julkunen-Tiitto 2006). Low temperature also can set off an increased production of secondary metabolites, as reported by Stocker-Wörgötter (2001). According to Swanson and Fahselt (1997) and Begora and Fahselt (2001), UV-A irradiation alone causes the production of higher amount of phenolic compounds in lichens than exposed to both UV-A and UV-B radiations.

Thus, lichen substances are an important component in high success of lichens in extreme environmental conditions.

2.2 Biosynthetic Pathways of Lichen Secondary Metabolites

The chemical substances present in lichen, depsides, depsidones and dibenzofuran (Asahina and Shibata 1954; Culberson 1969, 1970; Huneck and Yoshimura 1996) are unique as most of such chemicals are not known in other organisms. Lichens produce wide array of chemicals varying in their structures and biological activity. The lichen substances are synthesised via three metabolic pathways mainly polymalonate, shikimic acid and mevalonic acid (Boustie and Grube 2005). Based on the chemical structures, most lichen substances are phenolic compounds (orcinol and β-orcinol derivatives), dibenzofurans (e.g. usnic acid), depsides (e.g. barbatic acid), depsidones (e.g. salazinic acid), depsones (e.g. picrolichenic acid), lactones (e.g. protolichesterinic acid, nephrosterinic acid), quinones (e.g. parietin) and pulvinic acid derivatives (e.g. vulpinic acid) (Shukla et al. 2010).

Lichen substances may comprise of up to 20 % of the dry weight of thallus (5–10 % is common). The physiological cost in energy and carbon used to produce chemical compounds suggests that they have important role in defence mechanism (Dayan and Romagini 2001, 2002). More than 50 % of lichen substances are synthesised to sustain and protect the symbiotic association from various abiotic and biotic factors. The orcinol and β-orcinol derivatives (depsides, depsidones and dibenzofuran) are especially interesting as they presumably play a role in the establishment of the lichen symbiosis and in the interaction of the symbionts with their environments (Armaleo 1995). The multitude of mechanisms is involved in the air pollution protection (1) induced or stimulated by exposure to pollutants and (2) constitutive defence mechanism (Coley et al. 1985).

In the symbiotic association between fungus and alga, the fungus entraps or absorbs moisture and nutrients from the atmosphere, and it is supplied to the alga; the alga by having photosynthetic apparatus produces carbohydrates with the help of sunlight and the nutrients, which is then supplied to the fungus. The fungus by utilising this carbohydrate produces variety of secondary metabolites. Mosbach (1969) summarised the overall carbon metabolic sequence in alga, carbohydrate synthesis, its transportation to fungus and the subsequent synthesis of secondary metabolites. The type of carbohydrate released to the fungus and the final secondary metabolite produced are depended on the type of algal partner. The cyanobacteria in lichen produce simple glucose, while green alga produces polyol (ribitol, erythritol, sorbitol). The secondary metabolites in lichens are produced through three major pathways: (1) acetyl-polymalonyl pathway, (2) mevalonic acid pathway and (3) shikimic acid pathway.

Acetyl-Polymalonyl Pathway includes most common lichen substances and about 80 % of the secondary metabolites are synthesised through this pathway. They are aromatic compounds formed by bonding of 2 or 3 orcinol or β-orcinol-type phenolic units through ester, ether and C–C linkage (e.g. depsides, depsidones, dibenzofurans, usnic acids, depsones). All the metabolites produced through this mechanism are peculiar to lichens. The other aromatic compound produced by acetyl-polymalonyl pathway is by internal cyclisation of single or more polyketide chain (chromones, xanthones and anthraquinones), which are identical or analogous to products of other fungi or higher plants. The structure of many compounds produced by acetyl-polymalonyl pathway is yet to be elucidated and but is given common names and assigned compound classes, because they are frequently encountered and easily recognised by microchemical methods. The common products of *mevalonic acid pathway* include steroids and triterpenes. Only about 17 % of the total compounds are produced by this pathway.

Very small portion of the secondary metabolites are synthesised by *shikimic acid pathway*; they include terphenylquinones and pulvinic acid derivatives. They are the most common yellow pigments in lichens.

The polyketides are the building blocks of secondary metabolites. Polyketides are a class of naturally occurring metabolites found in bacteria, fungi, algae and higher plants, as well as in animal kingdom. Polyketides are biosynthesised by sequential reactions catalysed by an array of polyketide synthase (PKS) enzymes. PKSs are large multienzyme protein complexes that contain a typical core of coordinated active sites. An immensely rich diversity of polyketide structural moieties has been detected and structurally elucidated and more await discovery.

2.3 Isolation and Modern Spectroscopic Techniques Involved in Characterisation of Lichen Substances

Since secondary metabolism is not required for survival and its products are dispensable, primary metabolism is essential for survival with anabolic and catabolic activities to maintain life, which points at the adaptive significance of secondary metabolites produced by lichen fungi including its environmental role (Fox and Howlett 2008; Lawrey 1977). Toxitolerant lichens show resistance to ambient environmental levels of pollutants which may be phenotypic or genotypic. A phenotype can be defined by morphological characteristics, by physiological measurements as well as by biochemical (transcriptomics or proteomics) and chemical (metabolomics) analysis. In recent years, more and more attention is being paid to the chemical characterisation of the phenotype. Chemical characterisation can be done on the level of macromolecules (e.g. proteomics and characterisation of polysaccharides and lignins) or low molecular weight compounds (the metabolome). The metabolome consists of two types of compounds, the primary metabolites and the secondary metabolites. The primary metabolites are compounds involved in the basic functions of the living cell, such as respiration and biosynthesis of the amino acids and other compounds needed for a living cell. Basically all organisms share the same type of primary metabolites, though per class of organisms there may be large differences in the biosynthetic pathways present, e.g. polyketide pathway exceptional to lichens. The some secondary metabolites are species specific; they play a role in the interaction of a cell with its environment, which can be used as tool to protect from external biotic and abiotic factors, including its defence against elevated pollutant concentration. Therefore, primary metabolites in plants are important for the growth and agricultural yields, whereas the secondary metabolites are concerned with resistance of species against external environment (Verpoorte et al. 2009).

The distribution of secondary metabolites in the lichen thallus varies according to their function. Some are located in the cortex while most of them are in medullary region. The secondary metabolites in lichen can be identified using 'spot test', microcrystallography, Thin Layer Chromatography (TLC) and High Performance Liquid Chromatography (HPLC). These tests are routinely or more frequently used for identification of secondary metabolites and hence lichen taxa.

Metabolomics is generally defined as both the qualitative and quantitative analysis of all metabolites in an organism. In analysing a metabolome, both the primary and secondary metabolites will be detected. Plants share the primary metabolites;

therefore, methods that can easily identify and quantify the primary metabolites with high reproducibility are required. Secondary metabolites such a database is needed, though in that case the occurrence of most compounds is restricted to one or a few species only. The reproducibility is also required for studying the effect of any external condition on the plant metabolome, as one needs to know the biological variation that requires large numbers of analyses to determine the metabolomic changes as function of daily, seasonal and developmental variation. The results should function as the control samples for all future analyses.

Thus, reproducibility is the most important criteria for developing a metabolomics technology platform. The other criteria concern the ease of quantitation and identification, the number of metabolites that can be measured and the time needed for an analysis, including the sample preparation. Obviously a high-throughput methodology is preferred. Based on the underlying principle, the methods which one may consider for a metabolomic platform can be grouped as follows.

2.3.1 Spot Test

This test has been used universally as rapid, nonspecific means for detecting the presence of certain unspecified lichen substances. This test is most convenient and simple to perform, even under field conditions. However, this is only a preliminary step in the process of identification of lichens or its substances. In order to identify accurately the secondary metabolite present in the lichen thallus, one has to perform more sensitive tests such as TLC or HPLC. Spot test is carried out by placing a small drop of reagent on the lichen thallus, either directly on the upper surface (cortex) or on the medulla. In the later case cortex is scraped or superficially cut with the help of blade. The reagents used are 10 % aqueous KOH solution (K), saturated aqueous solution of bleaching powder ($NaOCl_2$) or calcium hypochlorite ($Ca(OCl)_2$) (called as C) and 5 % alcoholic p-phenylenediamine solution (PD).

The colour changes at the reagent application point of the thallus are noted as + or −. These colour changes take place due to presence of particular secondary metabolite in the thallus, which is termed as spotting.

2.3.2 Microcrystallography

Some of the secondary metabolites form characteristic crystals when a crystallising reagent is added and gently warmed. The test is conducted on the glass slide. The lichen compound is extracted using acetone. Glycerol, ethanol and glacial acetic acids are some of the chemicals used in different combination to make the reagent. Microcrystallography has been largely superseded by the more sensitive and reliable method such as TLC. But the technique is still useful for a number of lichen compounds which are difficult to identify in TLC due to the same R_f class (or value) or spot colour (e.g. lecanoric and gyrophoric acids, barbatic and diffractaic acids). However, mixture of substances may be difficult to identify with this method and also minor substances may be undetectable.

2.3.3 Chromatography

Chromatography is essentially a group of techniques for the separation of the compounds from mixtures by their continuous distribution between two phases, one of which is moving past the other. The systems associated with this definition are:
1. A solid stationary phase and a liquid or gaseous mobile phase (adsorption chromatography)
2. A liquid stationary and a liquid or gaseous mobile phase (partition chromatography)
3. A solid polymeric stationary phase containing replaceable ions and ionic liquid mobile phase (ion exchange chromatography)
4. An inert gel, which acts as a molecular sieve, and a liquid mobile phase (gel chromatography)

2.3.3.1 Partition Chromatography

Separation in liquid–liquid, liquid–solid and gas–liquid chromatography is based on the equilibrium distribution of each sample component between two phases, a mobile phase and a stationary phase.

The relative distribution of the solute between the two phases is determined by the intermolecular interactions of the solute molecules with the molecules of the mobile phase and the stationary phase. When the interactions are greater in one phase than in the other, the concentration of solute species in that corresponding phase will be greater. In a fundamental sense, the kinds and magnitudes of different intermolecular forces are responsible for the sample/solvent interactions.

A combination of solvent partitions and conventional technique is to be tried for resolving the plant extract into various fractions.

Column Chromatography

Column chromatography is a separation process involving the uniform percolation of liquid solute through a column packed with finely divided material. The separation in the column is affected either by direct interaction between the solute components and surface or stationary phase or by adsorption of solute by stationary phase (Table 2.2). Silica gel is the most commonly used adsorbent for the resolution of mixture of natural products. Preliminary separation of the crude mixture is still carried out by this technique. Polarity and composition of solvents depend upon the nature of compounds to be eluted. Polarity is changed by observing the mobility of the compounds as follows:

Pet. ether or *n*-Hexane 100 %

Pet. ether : Benzene (95:5, v/v)

Pet. ether : Benzene (90:10, v/v)

Pet. ether : Benzene (85:15, v/v) and so on till eluted with methanol

Paper Chromatography

Paper partition chromatography was developed by Shukla (2002) as a technique for the separation of amino acids. Paper is used as the support or adsorbent, but partition probably plays a greater part than adsorption in the separation of the components of mixtures, as the cellulose fibres have a film of moisture round them even in the air-dry state. The technique is therefore closely allied to column partition chromatography, but whereas the latter is capable of dealing with a gram or more material, the former requires micrograms. It is therefore an extremely sensitive technique of enormous value in chemical and biological fields.

The amino acid profile of various lichen species has been carried out by adopting this technique (Table 2.3). The paper functions in the same way as a column but, since evaporation of the mobile phase would occur as development progressed, it must be enclosed in a chamber to prevent such loss. Chromatographic tanks are therefore used to avoid such loss.

Chemical reagents for the development of colour are numerous as they either react with a specific group in the molecule to be detected or they are nonspecific.

Both one- and two-dimensional ascending and descending paper chromatographic methods were applied using Whatman nos. 1 and 3 chromatographic filter papers.

One-Dimensional Paper Chromatography. One-dimensional paper chromatography, i.e. the development of chromatogram in only one direction, is quite satisfactory for the separation of many mixtures. A wide variety of solvents can be tested in order to get best separation. This technique is also useful for the identification of pure compounds by determining their R_f values and comparing the co-chromatogram.

Two-Dimensional Paper Chromatography. For the complex mixtures, where the separation remains incomplete by the above method, two-dimensional development, i.e. the development of chromatogram using another solvent at right angle to the first, can be carried out to obtain a better separation of the components. This technique is useful to achieve separation of the compounds with nearly equal R_f values.

Thin Layer Chromatography (TLC)

The wider choice of media in this technique offers separations by partition, adsorption and gel filtration. TLC is mainly used for separation and isolation purposes as well as checking the purity of the isolated compound(s). Basically it is considered for qualitative identification of organic

2.3 Isolation and Modern Spectroscopic Techniques Involved in Characterisation of Lichen Substances

Table 2.2 Adsorbent and solvents

	Adsorbent		Solvent
Weak	Sucrose		Petroleum ether
	Starch		Carbon tetrachloride
	Inulin		Cyclohexane
	Talc		Carbon disulphide
Medium	Sodium carbonate		Ether (ethanol free)
	Calcium carbonate		Acetone
	Calcium phosphate	Increasing polarity	Benzene
	Magnesium carbonate		Toluene
	Magnesium oxide		Esters
	Calcium hydroxide		Chloroform
Strong	Activated Magnesium silicate		Acetonitrile
	Activated alumina		Alcohol
	Activated charcoal	↓	Water
	Activated silica		Pyridine
			Organic acids
			Mixture of acids or base with ethanol or pyridine

natural products. TLC is often employed for preparative scale separation.

The lipophilic spots may be conveniently located by spraying with water, whereas coloured compounds form distinct bands. However, the recovery of components from the stationary phase is time consuming. To some extent, this is overcome by carrying out separation over a new and elegant centrifugally accelerated radial TLC instrument called chromatotron (Burnouf-Radosevich and Delfel 1986).

Although the term 'adsorbent' is frequently used but it must be remembered that adsorption may not always be the principle on which the separation of components of a mixture may be achieved. Thin layer separations may involve any one of the mechanism as enlisted in Table 2.4.

Standardised Two-Dimensional Thin Layer Chromatography. Two-dimensional thin layer chromatography is useful for microchemical studies on mixtures difficult to resolve by the standardised one-dimensional TLC method. The standardised 2D procedure also allows more reliable comparisons of chromatograms and the determination of R_f classes of components of complex mixtures (Culberson 1974; Culberson et al. 1981). The method has been effectively employed to resolve the orcinol-type depsides of two closely related species, *Parmelia loxodes* and *P. verruculifera* (Culberson and Johnson 1976; Culberson 1972).

Gas Chromatography (GC)

In GC, the various compounds in a sample mixture are separated in the gas phase between mobile and stationary phases. This technique can be applied almost to all types of compounds, which have reasonable volatility and are stable under the chromatographic conditions. GC in combination with MS has been successfully employed to analyse the volatile content of oak moss (*Evernia prunastri*) (Heide et al. 1975); it enabled the isolation of 61 components.

The application of GC has its limitations owing to the thermal lability and low volatility of most well-known ingredients (Shultz and Albroscheit 1989).

High Performance Liquid Chromatography

HPLC is more versatile technique than GC. This method was applied to identify and to quantify

Table 2.3 Amino acid profile of some Indian lichens separated by paper chromatography

S. no.	Lichen species	Alanine	Arginine	Aspartic acid	Glutamic acid	Glycine	Histidine	Glutamine	Isoleucine	Leucine	Lysine	Methionine	Phenylalanine	Proline	Serine	Threonine	Tryptophan	Tyrosine	Valine	Cystine	α-Alanine	Unidentified	References
1.	*Anaptychia diademata*	+	+	+	+	−	−	+	−	−	−	−	+	−	+	−	−	−	−	−	−	−	Badhe (1976)
2.	*A. podocarpa*	+	+	+	+	+	−	+	−	−	−	−	+	−	+	−	−	−	−	−	−	−	Badhe (1976)
3.	*Parmelia wallichiana*	−	−	−	+	+	+	−	+	+	−	−	+	+	+	+	−	+	−	+	+	−	Badhe and Patwardhan (1972)
4.	*Leptogium azureum*	−	−	−	+	−	+	−	+	+	−	−	−	−	−	+	−	−	−	+	+	−	Badhe and Patwardhan (1972)
5.	*Peltigera canina*	+	−	−	−	+	−	−	−	+	+	−	+	−	+	+	+	+	+	−	−	−	Subramanian and Ramakrishnan (1964)
6.	*Lobaria subisidiosa*	+	+	+	+	+	+	−	+	+	+	+	+	+	+	+	+	+	+	−	−	+	Ramakrishnan and Subramanian (1966)
7.	*Umbilicaria pustulata*	+	+	+	+	+	−	−	+	+	+	+	+	+	+	+	+	+	+	−	−	+	Ramakrishnan and Subramanian (1966)
8.	*Parmelia nepalensis*	+	+	+	+	+	−	−	+	+	+	+	+	−	+	+	+	+	+	−	−	+	Ramakrishnan and Subramanian (1966)
9.	*Ramalina sinensis*	+	+	+	+	+	−	−	+	+	−	+	−	−	+	+	+	+	−	−	−	+	Ramakrishnan and Subramanian (1966)
10.	*Dermatocarpon moulinsii*	+	+	+	+	+	−	−	+	+	+	+	+	+	+	+	+	+	+	−	−	+	Ramakrishnan and Subramanian (1966)

11. *Cladonia rangiferina*	+	+	+	+	−	+	+	+	+	−	−	+	Ramakrishnan and Subramanian (1965)
12. *Cladonia gracilis*	+	+	+	−	+	+	+	+	+	−	−	+	Ramakrishnan and Subramanian (1965)
13. *Lobaria isidora*	+	+	+	+	+	+	+	+	+	−	−	+	Ramakrishnan and Subramanian (1965)
14. *Roccella montagnei*	+	−	+	−	−	+	+	+	+	−	−	−	Ramakrishnan and Subramanian (1964)
15. *Parmelia tinctorum*	+	−	+	−	−	−	+	+	+	−	−	−	Ramakrishnan and Subramanian (1964)
16. *Usnea venosa*	−	−	+	−	−	−	+	−	+	−	−	−	Ramakrishnan and Subramanian (1964)
17. *Usnea flexilis*	+	−	+	−	−	−	+	+	+	−	−	−	Ramakrishnan and Subramanian (1964)
18. *Everniastrum nepalense*	−	−	−	−	−	−	+	+	+	−	−	−	Shukla (2002)
19. *Parmotrema nilgherrense*	+	−	−	−	+	−	+	−	−	−	−	−	Shukla (2002)
20. *Parmelia subtinctoria*	−	+	+	−	+	−	+	−	−	−	−	−	Shukla (2002)
21. *P. habibana*	−	−	−	+	+	−	+	−	+	−	−	−	Shukla (2002)
22. *P. reticulata*	−	−	+	−	+	−	−	−	−	−	+	−	Shukla (2002)
23. *P. chhirata*	−	−	+	−	+	−	−	−	−	−	−	−	Shukla (2002)

Table 2.4 Examples of materials for thin layer chromatography

Coating substance	Solvent system	Mechanism of separation
Silica or alumina (activated)	Chloroform: methanol (9:1)	Adsorption
Toluene: acetone (1:1)	Diethylaminoethyl cellulose 0.1 M aqueous NaCl	Ion exchange
Cellulose or silica	Butanol: acetic acid: water	Partition (unactivated)
Paraffin oil or silicone oil	Acetic acid: water (3:1)	Reversed-phase partition
Sephadex G-50	Aqueous buffer	Molecular exclusion

characteristic substances in commercially available oak moss products (Shultz and Albroscheit 1989). The advantages of the HPLC method are: rapid characterisation of both the typical lichen compounds and of the artefacts resulting from the extraction procedure, the possibility of quantifying characteristic oak moss substances in commercially available products for quality control purposes and possibilities of establishing a process control correlating the individual concentrations of the detected ingredients with a desired odour quality.

HPLC is used for nonvolatile lichen substances. HPLC technique, has proved to be almost universally applicable to chromatographic problems, was first applied to lichen substances in 1972 (Strack et al. 1979).

This technique provides a powerful complement to the established TLC methods. The bonded reverse phase columns are used here and all the aromatic lichen products are suitable for analysis with this method. Samples are dissolved in methanol and injected in to the appropriate portion column, through which an appropriate solvent or sequence of solvents is passed under high pressure. The substance separates and is detected using UV detector. The retention time (Rt or time of passage) and peak intensity are recorded by a chart recorder. HPLC is also used to measure either absolute or relative concentrations of lichen compounds, because the peak intensity (area under curve) is proportional to the concentration. Most workers use HPLC to detect lichen compounds combine this technique with TLC and/or mass spectrometry to verify the identification of the peaks.

There are several other methods used for identification of lichen compounds, but later discontinued because of their inefficiency.

Standardised TLC

It is a relatively simple and inexpensive technique which can be performed by anyone with access to basic laboratory facilities. Lichen substances are extracted in acetone and the extract is spotted on to glass or aluminium plates coated with silica gel. The plate is placed in a sealed tank so that the base of the plate is immersed in a shallow layer of a specific mixture of organic solvents. The different lichen substances present in the sample are separated from each other during the passage of solvent through the silica gel layer and are later made visible by spraying with sulphuric acid. The resulting spots are provisionally identified by their colour and relative position in comparison to the control sample.

High Performance Thin Layer Chromatography (HPTLC):

This is the modification of the standard TLC method that utilises TLC plates comprising a thin layer of smaller grained silica particles (average 5–6 μm compared with 10–12 μm for ordinary TLC plates). It is reported to be a more sensitive method, requiring shorter run times and less solvent. However, HPTLC is much more sensitive to humidity and not a preferred method.

Gas Chromatography and Lichen Mass Spectrometry (GCLMS)

This technique is well suited for xanthone, anthraquinones, dibenzofurans, terpenes and pulvinic acid derivatives that lack ester linkages. Most of the lichen substances, especially depsides and depsidones, contain thermally labile ester linkages which decompose during the volatilisation process of gas chromatography. This complicates the analysis and identification of lichen substances. Gas chromatography is always coupled with mass spectroscopy, and it is slightly modified for lichens

2.3 Isolation and Modern Spectroscopic Techniques Involved in Characterisation of Lichen Substances

Table 2.5 IR absorption bands of compounds isolated from lichen species (Shukla 2002)

S. no.	Compounds	IR stretchings –O–H st.		–COO st.				
1.	4-O-Methyl orsellinic acid	3,600	3,300	3,000		1,620		
2.	Protocetraric acid		3,300	1,750	1,660	1,640		
3.	Stictic acid	3,400		1,750	1,720	1,670	1,590	
4.	Salazinic acid	3,520	3,250	1,760	1,730	1,650	1,620	1,540
5.	β-Orcinol carboxylic acid	3,400				1,635		
6.	Orsellinic acid	3,530	3,440				1,605	
7.	4-O-Methyl-β-orcinol carboxylic acid	3,400		3,050		1,620		
8.	2,4-O-Dimethyl-β-orcinol carboxylic acid			3,050		1,690		

and called 'Lichen Mass Spectrometry'. Here the small pieces of lichen (<50 ng) are injected directly in to the inlet system of mass spectrometer and volatilised with high temperature (100–150 °C) under very low pressure. Only xanthones or some terpenoids can resist this temperature.

2.3.4 Spectroscopic Techniques

The pure compounds are identified by a combination of spectroscopic techniques in micro- and milligram quantities.

2.3.4.1 Infrared Spectroscopy

IR spectra of natural substances may be measured by an automatic recording IR spectrophotometer either in solution, as a mull with Nujol oil, or in the solid state, mixed with KBr. The IR spectra provide valuable information about the various functional groups and stereochemistry of some positions in various compounds. The range of measurement is from 4,000 to 600 cm^{-1}. The region in the IR spectrum above 1,200 cm^{-1} shows spectral bands or peaks due to the vibrations of individual bonds or functional groups in the molecule under examination. The region below 1,200 cm^{-1} shows bands due to the vibrations of whole molecule and, because of its complexity, is known as 'fingerprint' region. The fact that many functional groups can be identified by their characteristic vibration frequencies makes the IR spectrum the simplest and often the most reliable method of assigning a compound to its class. The IR stretching (Nishitoba et al. 1987) of some of the compounds isolated from lichen species are as in Table 2.5.

2.3.4.2 UV–Visible Spectroscopy

UV and visible spectroscopy is helpful in the determination of some functional groups (chromophores) and ethylenic bonds present in the molecules. For colourless compounds, measurements are made in the range 200–400 nm; for coloured compounds, the range is 400–700 nm. UV–Vis spectroscopy of various compounds dissolved in MeOH (or EtOH) and with the addition of the classical shift reagents is a method of choice in structure determination. The shape of these spectra and the accurate calculation of the maxima, shoulders and inflections provide very useful information for identification purposes. The advantage of this is the requirement of minute amounts of purified compound to run a complete UV spectrum.

2.3.4.3 Circular Dichroism

When two or more chromophores absorbing strongly around the same wavelength are closely located in space, the system gives rise to a circular dichroism (CD) curve. The sign of CD curve reflects the chiralities of the interacting transitions (Liu and Nakanishi 1981). The CD curve is the only means other than the X-ray method for determining the absolute configuration or conformation of molecules in a non-empirical

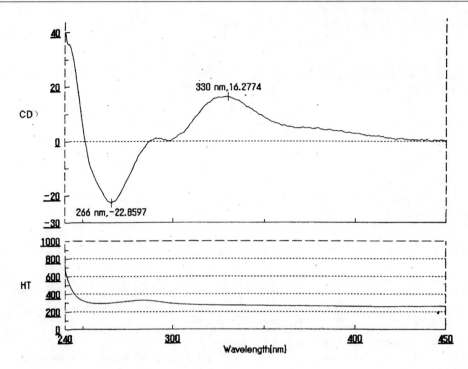

Fig. 2.3 CD curve of usnic acid affirms the optical rotation of the usnic acid obtained from the *Usnea* and *Ramalina* species was found to have positive rotation (From Shukla 2002)

manner. The CD spectroscopy has been applied to a variety of compounds (Hirada and Nakanishi 1981).

Circular dichroism method has been applied to establish the stereochemistry of usnic acid. Both the (+) and (−) forms of usnic acid can be affirmed by recording their CD curves (Fig. 2.3), and by the application of the above technique, the optical rotation of the usnic acid obtained from the *Usnea* and *Ramalina* species was found to have positive rotation.

2.3.4.4 Mass Spectroscopy

Mass spectrometry is one of the most important analytical tools currently available to chemists and biochemists that has made complex analytical problems more easier (McLafferty 1980). It provides information concerning molecular composition, molecular weight, structure of unknown compound and the concentration of a target analyte present in complex matrices. The various techniques of MS include electron ionisation (EI), chemical ionisation (CI), fast atom bombardment (FAB), field desorption (FD), laser desorption (LD), desorption and chemical ionisation (DCI), thermospray and electron spray ionisation.

Fast Atom Bombardment (FAB-MS)

The determination of the molecular weight of relatively large polar molecules has been facilitated enormously since the discovery of the FAB-MS (Barber et al. 1981). The sample of the 0.0321 µ mol level is dissolved in 1 µl of glycerol and placed on FAB probe and bombarded with a beam of 2–8 keV Ar/Xe beam (Dall and Morris 1983). When the sample is dissolved in glycerol and exposed to particle bombardment, the sputtered sites on the surface of the liquid matrix are continuously replenished with fresh sample by the process of diffusion. It requires a small amount of the sample (0.5 mg) in a solvent (glycerol). FAB can, therefore, provide extremely stable intense and long-lived ion current (lasting up to 30–45 min) for signal averaging for improved signal to noise ratio, high-resolution mass measurements and host of MS/MS and collisional activation experiments.

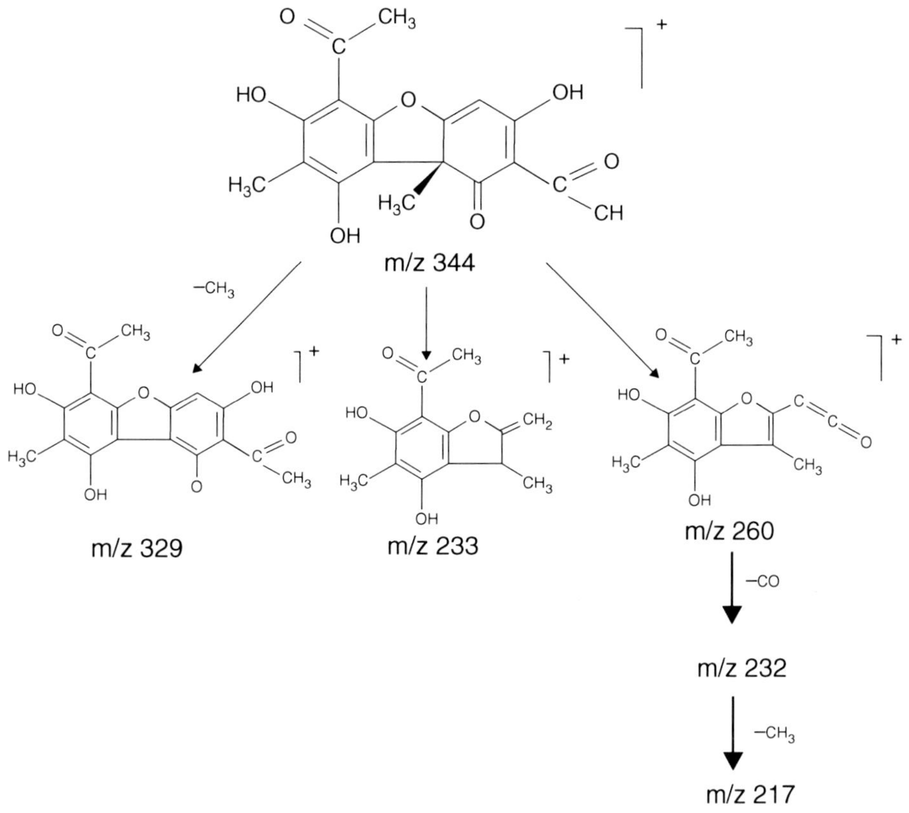

2.3 Isolation and Modern Spectroscopic Techniques Involved in Characterisation of Lichen Substances

Fig. 2.4 Mass fragmentation of usnic acid (From Shukla 2002)

Electron Ionisation (EI-MS)

Electron ionisation is the oldest and most common method of ionisation for samples, which are volatile below 300 °C under vacuum. The sample molecules in gaseous state are bombarded by a beam of energetic (10–70 eV, usually 70 eV) electrons produced from a heated tungsten or rhenium filament. The electron bombardment leads to the formation of molecular ion and fragment ions. The EI technique often requires modification of sample by methylation/acetylation or silylation to produce volatile derivatives. The technique has its limitations as in case of higher glycosides, it is difficult to get the molecular ion peaks and derivatisation leads to the formation of a number of products.

EI-MS technique has proved to the most reliable technique to establish the structure of the lichen substance obtained. The fragmentation of the depsides under electron impact was of interest in connection with the mass spectrometric structural elucidation of lichen substances (Huneck et al. 1968).

The mass spectral fragmentation of usnic acid follows the major fragmentation pathways as are illustrated in Fig. 2.4. If the base peaks of the compounds isolated are observed, then it seems to be formed by typical hydrogen transfer rearrangements and the basic structure is as in Fig. 2.5.

Since the sterol composition is found to be very complex in lichens, MS plays an important role in their qualitative identification (Wojciechowski et al. 1973). As it is observable that the mass fragmentation follows a typical path therefore by the peaks observed for the fragment ion and the base peak gives accurate information about the structure of the compound under investigation.

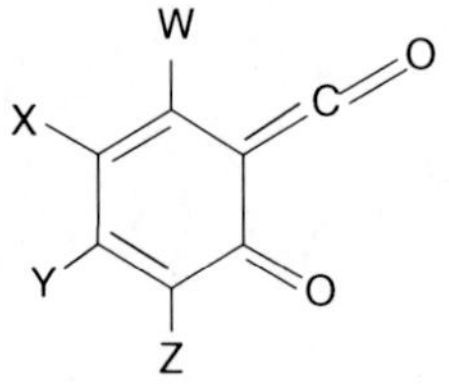

Fig. 2.5 Typical hydrogen rearrangement responsible for formation of base peak of compounds in lichens

Table 2.6 Nuclear properties of ^{1}H and ^{13}C

	^{1}H	^{13}C
Nuclear spin (I)	$^1/_2$	$^1/_2$
Resonance frequency at 235 T (MHz)	100	25.2
Natural abundance (%)	99.9	1.1
Sensitivity for an equal number of nuclei (%)	1.00	0.016
Shift range (ppm)	20	600

2.3.4.5 Nuclear Magnetic Resonance Spectroscopy (NMR)

NMR spectroscopy is the technique that permits the exploration of a molecule at the level of the individual atom and affords information concerning the environment of that atom. Thus NMR offers an excellent physical means of investigating molecular structure and molecular interactions.

The introduction of Fourier transformation (FT) technique in the pulsed NMR spectroscopy (Ernst and Anderson 1966) started a new era in this branch of spectroscopy.

Both the ^{1}H and ^{13}C NMR are important from the viewpoint of the organic chemist. The precise frequency from which energy is absorbed gives an indication of how an atom is bound to, or located spatially with respect to other atoms, but there are important differences between both the techniques (Table 2.6).

2.3.4.6 One-Dimensional (1D) NMR Spectroscopy

^{1}H-NMR

All the protons in organic molecules produce NMR signals at different field strength when definite radio frequency is applied. Thus, very conclusive results have been obtained to determine the structure and stereochemistry of organic compounds. NMR spectral (Table 2.7) analysis provided important information for the identification of secondary products of chemotaxonomic significance in the lichen genera *Umbilicaria* and *Lasallia* (Narui et al. 1998).

^{13}C-NMR

^{13}C-NMR spectrum with chemical shift range about 220 ppm (downfield to TMS) has proved to be distinctly advantageous in structure elucidation of natural products (Wehrli and Wirthlin 1976). For such studies, three types of ^{13}C-NMR spectra are generated.

In proton noise decoupled (pnd) spectrum, nonequivalent carbons resonate as separate signal lines and provide information about the number and nature of carbon atoms on the basis of their chemical shifts.

The signal frequency off-resonance decoupled (s-fold) spectrum gives the hydrogen substitution pattern where carbon signals are split according to the number of attached hydrogen atoms; the proton-coupled spectrum, which gives J_{CH} coupling values extending up to three bonds; and the ^{13}C-NMR of the depsides and tridepsides important for chemotaxonomic investigation and identification (Table 2.8).

Nuclear Overhauser Effect (NOE)

Unlike spin decoupling and spin tickling, which are based upon changes in energy levels, the Nuclear Overhauser Effect results from population changes in the energy levels which cause corresponding intensity changes in that spectra. The NOE experiment consists of saturating one signal and observing the intensities of other signals. The resonance bands due to protons spatially close to the group being saturated will show an increase in intensity. Such intensity changes are best measured by integration rather than peak height. The theory of the NOE is dependent upon relaxation mechanisms, and the relaxation of any nucleus is affected by all surrounding magnetic nuclei. Since the contribution of any nucleus to the relaxation of a second nucleus is dependent upon the mean square value of the magnetic field produced by the former, then the magnitude of the Nuclear Overhauser Effect will relate to the internuclear distance.

Table 2.7 ¹H NMR spectral data of depsides (chemical shifts (ppm) in DMSO$_6$ at 500 MHz) (Shukla 2002)

Proton no.	Methyl evernate	Gyrophoric acid	Evernic acid	Papulosic acid	Lasallic acid	Deliseic acid
3	6.35(d)	6.24(d)	6.35(d)	6.21(d)	6.25(d)	6.41(s)
	J = 2.44 Hz	J = 2.14 Hz	J = 2.14 Hz	J = 1.83 Hz	J = 1.82 Hz	
5	6.39(d)	6.24(d)	6.39(d)	6.27(d)	6.24(d)	
	J = 2.44 Hz	J = 2.14 Hz	J = 2.14 Hz	J = 1.83 Hz	J = 1.82 Hz	
8	2.37(s)	2.37(s)	2.37(s)	2.49(s)	2.38(s)	2.10(s)
3′	6.61(d)	6.70(d)	6.62(d)		6.70(d)	6.67(d)
	J = 1.83 Hz	J = 1.68 Hz	J = 2.14 Hz		J = 1.98 Hz	J = 0.92 Hz
5′	6.59(d)	6.67(d)	6.59(d)	6.33(s)	6.69 (d)	6.65(d)
	J = 1.83 Hz	J = 1.68 Hz	J = 2.14 Hz		J = 1.98 Hz	J = 0.92 Hz
8′	2.22(s)	2.38(s)	2.36(s)	2.45(s)	2.51(s)	2.35(s)
3″		6.65(d)				6.60(d)
		J = 1.98 Hz				J = 1.83 Hz
5″		6.62(d)			6.39(s)	6.58(d)
		J = 1.98 Hz				J = 1.83 Hz
8″		2.40(s)			2.46(s)	2.37(s)
2-OCH$_3$						
4-OCH$_3$	3.75(s)		3.75(s)			
2′-OCH$_3$						
7′-OCH$_3$	3.80(s)					
7″-OCH$_3$						
2-OH		10.29(s)	10.38(s)	10.26(br s)		
4-OH					10.30(s)	10.15(s)
5-OH						
2′-OH		10.42(s)		10.09(s)	10.45(s)	10.56(s)
2″-OH		9.96(br s)			9.96(br s)	10.22(br s)
5-OAc						2.26(s)

Bold terms indicate atom in question

Interactions giving rise to Nuclear Overhauser Effect may be intermolecular or intramolecular. In the dilute solutions in aprotic solvents, only the later will be observed. It is important from an experimental standpoint in all NOE experiments to degas samples. Oxygen, being paramagnetic, will affect and often dominate the relaxation process and its presence will invalidate results (Backers and Shaefer 1971).

Insensitive Nuclei Enhancement by Polarisation Transfer (INEPT)

In this technique the polarisation is transferred from a sensitive nuclei (e.g. ¹H) to insensitive nuclei (e.g. ¹³C), i.e. to say from a high magnetogyric ratio to low magnetogyric ratio. In the specific pulse sequence, the proton transitions are put into anti-phase, as in selective polarisation transfer, but using nonselective polarisation transfer in a manner independent of chemical shift.

The important elements of INEPT sequence are the modulation of transverse magnetisation of the sensitive nucleus (A) by scalar coupling to the insensitive nucleus (X) and the simultaneous application of two 180° X pulses in the 'A' and 'X' region.

The most important aspect of the INEPT method is the fact that it allows a much larger intensity increase for insensitive nuclei than the NOE. Furthermore, negative t-factors have no disadvantage in INEPT than in NOE. For example, between ¹⁵N and ¹H, NOE observed is negative, whereas INEPT experiments contribute to the population difference of the final signals of the ¹⁵N nucleus (Morris 1980).

Table 2.8 ^{13}C NMR spectral data of depsides (chemical shifts (ppm) in DMSO-d$_6$ at 125 MHz) (Shukla 2002)

Carbon no.	Methyl evernate	Gyrophoric acid	Evernic acid	Papulosic acid	Lasallic acid	Deliseic acid
1	110.68	108.23	110.54	105.49	108.19	110.11
2	159.13	160.08	159.25	162.52	160.15	155.14
3	98.98	100.49	98.98	100.47	100.52	101.55
4	162.09	161.08	162.14	162.00	161.12	156.29
5	108.04	109.85	108.10	110.87	109.90	130.28
6	139.60	140.21	139.72	142.33	140.30	130.57
7	166.72	167.07	166.68	166.47	167.40	167.13
8	20.83	21.22	20.91	22.88	21.26	13.58
1′	118.74	117.88	116.70	105.42	116.39	118.27
2′	156.14	156.30	158.75	156.36	157.78	156.37
3′	107.01	107.17	107.32	123.75	107.34	107.18
4′	151.64	152.18	152.21	153.20	152.52	152.20
5′	114.02	114.17	114.67	110.03	114.72	114.24
6′	137.82	137.95	139.47	139.09	139.75	138.16
7′	168.02	165.52	170.53	172.70	165.06	165.79
8′	19.42	19.26	20.91	22.50	20.37	19.40
1″		116.80			105.07	117.45
2″		158.77			156.35	158.89
3″		107.14			123.86	107.21
4″		152.08			153.64	152.14
5″		114.43			110.50	114.42
6″		139.56			139.46	139.67
7″		170.37			173.22	170.50
8″		20.84			23.17	20.98
2-OCH$_3$						
4-OCH$_3$	55.19		55.19			
2′-OCH$_3$						
7′-OCH$_3$	51.92					
5-OAc(C=O)						168.95
5-OAc (CH$_3$)						20.34

Bold terms indicate atom in question

Distortionless Enhancement by Polarisation Transfer (DEPT)

In practical applications of polarisation transfer experiments for resonance assignments, the DEPT sequence is usually preferred. The experiments bring about polarisation transfer in similar fashion to INEPT, but with the important difference that all the signals of the insensitive nucleus are in phase at start of acquisition. The DEPT experiments yield multiplets with uniform phase after three J/2 delays, with the application of ^{1}H-decoupling singlet signals for all types of ^{13}C-resonances. The pulse angle θ of the last A pulse can be optimised for individual groups in order to allow signal selection.

The DEPT experiment with flip angle 45° showed positive signals for all three multiplicities, with $\theta = 90°$ only methine signals should appear and $\theta = 135°$ methylene signals will appear negative, while methine and methyl signals remain positive. Indeed, it is possible in favourable cases to achieve a complete separation of the decoupled ^{13}C-NMR spectrum into CH, CH$_2$ and CH$_3$ sub-spectra by taking linear combinations of the DEPT or INEPT spectra with the three different values for θ (Morris 1986).

2.3.4.7 Two-Dimensional (2D) NMR Spectroscopy

Jenner first introduced the concept of 2D-FT NMR in 1971, which was then analysed in detail

(Aue et al. 1976). The power of 2D-NMR lies in its ability to resolve overlapping spectral lines, to enhance the sensitivity and to provide information not available by 1D methods. It is utilised for investigation of cross-relaxation and chemical exchange processes (Benn and Gunther 1983).

The frequency dimensions of 2D-NMR originate from the two time intervals t_1 and t_2 during which the nuclei can be subjected to two different sets of conditions. The amplitude of the signals detected during time t_2 is a function of what happened to the nuclei during the evolution period t_1, i.e. s(t_1 t_2). If over n experiments we increase each evolution period t_1 by a constant time increment Δt_1 varying from zero to several hundred milliseconds, a set of spectra is obtained with an amplitude of the resonances modulated with the frequencies that existed during the evolution period t_1. A Fourier transformation with respect to t_2 yields conventional 2D spectra, whose data points on the time-axis t_1 define the modulation frequency, which can be determined by a second Fourier transformation with respect to t_1.

In the 2D spectra a line detected in t_2 may have components lines in t_1; it therefore shows cross peaks to those lines. The peaks along the diagonal in this spectrum arise from magnetisation of components, which have same frequency during both t_1 and t_2, i.e. from the portion of magnetisation that was not transferred elsewhere during the second pulse. Thus, the diagonal gives the essence of the normal 1D spectrum, whereas the off-diagonal peaks (cross peaks) which are either a part of the same multiplet or a part of different multiplets that have a coupling. Since the magnetisation is transferred in both the direction between transitions, a cross-peak ν_1 and ν_2 have symmetrical partners at ν_1 ν_2.

2D ^{1}H-^{1}H COSY (Homonuclear Correlated Spectroscopy)

It is the homonuclear correlation through J coupling; the information obtained from the spectrum are the scalar coupling connectivity network pattern of the molecule concerned by the help of cross peaks. The rows of the spectra illuminated the nature of data, but it is clustered and confusing as the spectral complexity increases. Therefore, the clear representation of the spectrum can be obtained as contour plots. Removing several artefacts by proper phase, cycling and choosing suitable experimental parameters (Morris 1986) make the COSY spectrum an elegant approach for making connections through bonds.

2D ^{1}H-^{13}C COSY (Heteronuclear Correlated Spectroscopy)

Heteronuclear correlated spectroscopy is one of the most powerful 2D experiments, combining the excellent resolving power of decoupled 13C-NMR with ease for interpretation of proton chemical shifts. It offers a good chemical information and allows the resolution of single sites in all but the intractable spin systems. As in INEPT and DEPT, the usual experiment relies on transferring ^{1}H spin polarisation to ^{13}C through one-bond couplings. The fundamental concept is to use the evolution period t_1 for the precession motion of the ^{1}H spins and to measure the degree of precession with the ^{13}C channel. The transfer of information between ^{1}H and ^{13}C occurs in the mixing period, which is introduced between the evolution and the detection time. The coupled nuclei yield signals with the coordinates s(A), a(X) (Benn and Gunther 1983).

Heteronuclear Multiple Quantum Correlation (HMQC)

The protonated carbon atoms can be detected by HMQC experiments. In HMQC experiment, a 180° pulse refocuses chemical shift evolution and the Δ_1 delay is tuned for the heteronuclear coupling so that at the end of this period a coupled 'A' magnetisation is in anti-phase, while uncoupled magnetisation of homonuclear 'A' couplings is neglected which still points along the y-axis. The 90° X pulse transforms this magnetisation to z-direction. The delay Δ_2 is tuned equal to Δ_1. The HMQC experiment thus provides one-bond heteronuclear correlation between two nuclei, i.e. the attached nuclei (Gunther 1992).

Heteronuclear Multiple-Bond Correlation (HMBC)

HMBC is the long-range version of HMQC experiment and is the only way to establish connectivities between proton and non-protonated ^{13}C sites. In this experiment, the delays Δ_1 and Δ_2 are tuned to one-bond and long-range heteronuclear A_1X couplings, respectively. The first two 90° pulses separated by the delays $\Delta_1 = \frac{1}{2}\,^1J(A_1X)$ eliminate one-bond correlations. The second 90° X pulse then creates after the appreciably longer delay Δ_2 (60 ms) the desired multiple quantum coherence based on long-range couplings. This is the most effectively performed technique using gradient spectroscopy, which significantly improves the elimination of t_1-noise from the residual signals of molecules with non-magnetic X-nuclei (Gunther 1992).

2D Nuclear Overhauser Enhancement Spectroscopy (NOESY)

There is a component of magnetisation created by the COSY sequence, which is the part of the transverse magnetisation returned to the z-axis by the second pulse. In COSY this component generates no signals. At least two informative interactions are involved in which the frequency levelled as z-component. They are chemical exchange with other nuclei and Nuclear Overhauser Effect. Thus, an introduction of a delay (τ_m) and a $\pi/2$ pulse after the second $\pi/2$ pulse in the COSY pulse sequence makes the experiment NOESY pulse sequence $\pi/2$-t_1- $\pi/2$- τ_m- $\pi/2$-t_2 (Wider et al. 1984; Kumar et al. 1980). In this sequence, the NOE allows the changes in the z-magnetisation of one nucleus to lead to variations in the z-magnetisation of another, and so pairs of nuclei, which could show an NOE in a 1D-experiment, may show cross peaks in the 2D experiment. A suitable choice of τ_m delay allows maximum NOE build-up before the pulse samples the z-magnetisation.

Since NOE for small molecules is positive in nature, a phase-sensitive NOESY in which negative peaks are observed for positive NOE will offer stereo structure of the molecule in solution state, whereas for moderate size and bigger molecules, NOESY in magnitude mode is the method of choice (Wider et al. 1984).

Thus, the combination of spectroscopic techniques is helpful to elucidate structures of lichen compounds (Table 2.9).

Table 2.9 Five hundred mega hertz of ^{1}H NMR and One hundred and twenty-five of ^{13}C NMR data (in ppm relative to TMS) of usnic acid (Shukla 2002)

C/H	δ ^{13}C	DEPT	δ ^{1}H	HMBC connectivities $^2J_{CH}$	$^3J_{CH}$
C-1	198.05	–C–	–	–	–
C-2	105.22	–C–	–	–	–
C-3	191.15	–C–	–	–	–
C-4	98.32	–CH	6.20(s)	C-4a	C-1a, C-2
C-6	101.53	–C–	–	–	–
C-7	163.55	–C–	–	–	–
C-8	109.34	–C–	–	–	–
C-9	157.20	–C–	–	–	–
C-1a	59.07	–C–	–	–	C-14
C-4a	179.38	–C–	–	–	–
C-6a	155.20	–C–	–	–	–
C-9a	103.95	–C–	–	–	–
C-1′	201.77	–C–	–	–	–
C-2′	27.89	–CH$_3$	2.65(s)	–	C-2
C-1″	200.10	–C–	–	C-1′	–
C-2″	31.27	–CH$_3$	2.69(s)	C-1″	C-6
C-1a-CH3	32.34	–CH$_3$	1.70(s)	C-1a	C-9a, C-4a, C-1
C-8-CH3	7.50	–CH$_3$	2.10(s)	C-8	C-7, C-9
3-OH	–	–	18.80(s)	C-3	C-4, C-2, C-1′ [a]
7-OH	–	–	13.30(s)	C-7	C-6, C-8
9-OH	–	–	11.03(s)	C-9	C-8

[a] $^4J_{CH}$
Bold terms indicate atom in question

References

Ahmadjian V, Jacobs JB (1981) Relationship between fungus and alga in the lichen *Cladonia cristatella* Tuck. Nature 289:169–172

Alaimo MG, Dongarra G, Melati MR, Monna F, Varrico D (2000) Recognition of environmental trace metal contamination using pine needles as bioindicators. The urban area of Palermo (Italy). Environ Geol 39(8):914–924

Armaleo D (1995) Factors affecting depside and depsidone biosynthesis in a cultured lichen fungus. Cryptogam Bot 5:14–21

Armaleo D, Zhang Y, Cheung S (2008) Light might regulate divergently depside and depsidone accumulation in the lichen *Parmotrema hypotropum* by affecting thallus temperature and water potential. Mycologia 100:565–576

Asahina Y, Shibata S (1954) Chemistry of lichen substances. Japan Society for the Promotion of Science, Tokyo

Aue WP, Bartholdi E, Ernst RR (1976) Two-dimensional spectroscopy. Application to nuclear magnetic resonance. J Chem Phys 64:2229–2246

Backers GE, Shaefer T (1971) Applications of the intramolecular NOE in structural organic chemistry. Chem Rev 71:617

Badhe PD (1976) Free amino acids in two species of *Anaptychia* (lichens). The Bryologist 79:354–355

Badhe PD, Patwardhan PG (1972) Qualitative and quantitative determination of free amino acids in *Parmelia wallichiana* and *Leptogium azureum*. Bryologist 75:368–369

Barber M, Bordoli RS, Garner GV, Gordon DB, Sedgwick RD, Tetler LW, Tyler AN (1981) Fast-atom-bombardment mass spectra of enkephalins. Biochem J 197:401–404

Beart JE, Lilley TH, Haslam E (1985) Plant polyphenols secondary metabolism and chemical defence: some observations. Phytochemistry 24(1):33–38

Begora MD, Fahselt D (2001) Usnic acid and atranorin concentrations in lichens in relation to bands of UV irradiance. Bryologist 104:134–140

Benn R, Gunther H (1983) Modern Pulse Methods in High-Resolution NMR Spectroscopy. Angew Chem Int Ed Engl 22(5):350–380

Bennett JW, Ciegler A (1983) Secondary metabolism and differentiation in fungi. Marcel Dekker, Inc., New York, p 478

Bjerke JW, Dahl T (2002) Distribution patterns of usnic acid-producing lichens along local radiation gradients in West Greenland. Nova Hedwig 75:487–506

Bjerke JW, Lerfall K, Elvebakk A (2002) Effects of ultraviolet radiation and PAR on the content of usnic and divaricatic acids in two arctic-alpine lichens. Photochem Photobiol Sci 1:678–685

Bjerke JW, Joly D, Nilsen L, Brossard T (2004) Spatial trends in usnic acid concentrations of the lichen Flavocetraria nivalis along local climatic gradients in the Arctic (Kongsfjorden, Svalbard). Polar Biol 27:409–417

Bjerke JW, Elvebakk A, Dominiguez E, Dahlback A (2005) Seasonal trends in usnic acid concentrations of Arctic, alpine and Patagonian populations of the lichen Flavocetraria nivalis. Phytochemistry 66:337–344

Boustie J, Grube M (2005) Lichens – a promising source of bioactive secondary metabolites. Plant Genet Resour Characterization Util 3:273–287

Burnouf-Radosevich M, Delfel EL (1986) High-performance liquid chromatography of triterpene saponins. J Chromatogr 368:433–438

Cocchietto M, Skert N, Nimis PL, Sava G (2002) A review of usnic acid, an interesting natural compound. Naturwissenschaften 89:137–146

Coley PD, Bryant JP, Chapin FS (1985) Resource availability and plant antiherbivore defense. Science 230:895–899

Conti ME, Cecchetti G (2001) Biological monitoring: lichens as bioindicators of air pollution assessment – a review. Environ Pollut 114:471–492

Cooper-Driver G, Bhattacharya M (1998) Role of phenolics in plant evolution. Phytochemistry 49(5):1165–1174

Cuellar M, Quilhot W, Rubio C, Soto C, Espinoza L, Carrasco H (2008) Phenolics, depsides, triterpenes from Chilean lichen *Pseudocyphellaria nudata* (Zahlbr.) D.J. Galloway. J Chilean Chem Soc 53(3):1624–1625

Culberson CF (1969) Chemical and botanical guide to lichen products. University of North Carolina Press, Chapel Hill

Culberson CF (1970) Supplement to "chemical and botanical guide to lichen products". Bryologist 73:177–377

Culberson CF (1972) Improved conditions and new data for the identification of lichen products by a standardized thin-layer chromatographic method. J Chromatogr 72:113–125

Culberson CF (1974) Conditions for the use of Merck silica gel 60 F254 plates in the standardized thin-layer chromatographic technique for lichen products. J Chromatogr B 97:107–108

Culberson CF, Elix JA (1989) Lichen substances. In: Dey PM, Harbourne JB (eds) Methods in plant biochemistry, vol 1, Plant phenolics. Academic, London, pp 509–535

Culberson CF, Johnson A (1976) A standardized two-dimensional thin layer chromatographic method for lichen products. J Chromatogr 128:253–259

Culberson CF, Culberson WL, Arwood DA (1977) Physiography and fumarprotocetraric acid production in the *Cladonia chlorophaea* group in North Carolina. Bryologist 80:71–75

Culberson CF, Culberson WL, Johnson A (1981) A standardized TLC analysis of ß–orcinol depsidones. Bryologist 84:16–29

Culberson CF, Culberson WL, Johnson A (1993) Occurrence and histological distribution of usnic acid in the *Ramalina siliquosa* species complex. Bryologist 96:181–184

Dall A, Morris HR (1983) Glycoprotein structure determination by mass spectroscopy. Carbohydr Res 115:41

Dayan FE, Romagini JG (2001) Lichens as a potential source of pesticides. Pestic Outlook 12:229–232

Dayan FE, Romagini JG (2002) Structural diversity of lichen metabolites and their potential for use. In: Upadhyaya R (ed) Advances in microbial toxin research and its biotechnological exploration. Kluwer Academic/Plenum Publisher, New York, p 151

Deduke C, Timsina B, Piercey-Normore MD (2012) Effect of environmental change on secondary metabolite production in lichen-forming fungi. In: Young S (ed) International perspectives on global environmental change. InTech, ISBN: 978-953-307-815-1. Intech publisher, Europe. Available from: http://www.intechopen.com/books/international-perspectives-on-global-environmental-change/effect-of-environmental-change-on-secondary-metabolite-production-in-lichen-forming-fungi

Ernst RR, Anderson WA (1966) Applications of Fourier transform spectroscopy to magnetic resonance. Rev Sci Instrum 37:93–102

Fox EM, Howlett BJ (2008) Secondary metabolism: regulation and role in fungal biology. Curr Opin Microbiol 11:481–487

Frisvad JC, Anderson B, Thrane U (2008) The use of secondary metabolite profiling in chemotaxonomy of filamentous fungi. Mycol Res 112:231–240

Gauslaa Y (2005) Lichen palatability depends on investments in herbivore defence. Oecologia 143:94–105

Golojuch ST, Lawrey JD (1988) Quantitative variation in vulpinic and pinastric acids produced by *Tuckermannopsis pinastri* (lichen-forming Ascomycotina, Parmeliaceae). Am J Bot 75:1871–1875

Gunther H (1992) NMR spectroscopy, 2nd edn. Wiley, Chichester, U.K.

Hafellner J (1997) A world monograph of *Brigantiaea* (lichenized Ascomycotina, Lecanorales). Symb Bot Ups 32(1):35–74

Hager A, Brunauer G, Türk R, Stocker-Wörgötter E (2008) Production and bioactivity of common lichen metabolites as exemplified by Heterodea muelleri (Hampe) Nyl. J Chem Ecol 34:113–120

Hagerman AE, Robbins CT (1987) Implications of soluble tannin–protein complexes for tannin analysis and plant defense mechanisms. J Chem Ecol 13(5):1243–1259

Hamada N (1982) The effect of temperature on the content of the medullary depsidone salazinic acid in *Ramalina siliquosa* (lichens). Can J Bot 60:383–385

Hamada N (1991) Environmental factors affecting the content of usnic acid in the lichen mycobiont of *Ramalina siliquosa*. Bryologist 94:57–59

Haslam E (1989) Plant polyphenols. University Press Publishers, Cambridge

Hauck M (2008) Metal homeostasis in *Hypogymnia physodes* is controlled by lichen substances. Environ Pollut 153:304–308

Hauck M, Huneck S (2007a) Lichen substances affect metal adsorption in *Hypogymnia physodes*. J Chem Ecol 33:219–223

Hauck M, Huneck S (2007b) The putative role of fumarprotocetraric acid in the manganese tolerance of the lichen *Lecanora conizaeoides*. Lichenologist 39: 301–304

Hauck M, Jürgens SR (2008) Usnic acid controls the acidity tolerance of lichens. Environ Pollut 156:115–122

Hauck M, Jürgens SR, Brinkmann M, Herminghaus S (2008) Surface hydrophobicity causes SO_2 tolerance in lichens. Ann Bot 101:531–539

Hauck M, Jürgens SR, Huneck S, Leuschner C (2009a) High acidity tolerance in lichens with fumarprotocetraric, perlatolic or thamnolic acids is correlated with low pKa1 values of these lichen substances. Environ Pollut 157:2776–2780

Hauck M, Jürgens SR, Willenbruch K, Huneck S, Leuschner C (2009b) Dissociation and metal-binding characteristics of yellow lichen substances suggest a relationship with site preferences of lichens. Ann Bot 103:13–22

Hawksworth DL (1976) Lichen chemotaxonomy. In: Brown DH, Hawksworth DL, Bailey RH (eds) Lichenology: progress and problems. Academic, London, pp 139–184

Hawksworth DL, Rose F (1970) Qualitative scale for estimating sulfur dioxide air pollution in England and Wales using epiphytic lichens. Nature 227:145–148

Heide R, Provatoroff N, Traas PC, Valois PJ, Plasse N, Wobben HJ, Timmer R (1975) Qualitative analysis of the odoriferous fraction of oakmoss (*Evernia prunastri* (L.) Ach.). J Agric Food Chem 23(5):950–957

Hesbacher S, Froberg L, Baur A, Baur B, Proksch P (1996) Chemical variation within and between individuals of the lichenized Ascomycete Tephromela atra. Biochem Syst Ecol 8:603–609

Hirada N, Nakanishi K (1981) Circular dichroism spectroscopy. University Science, Mill Valley

Howell RK (1970) Influence of air pollution on quantities of caffeic acid isolated from leaves of *Phaseolus vulgaris*. Phytopathology 60(11):1626–1629

Huneck S (1999) The significance of lichens and their metabolites. Naturwissenschaften 86:559–570

Huneck S, Yoshimura I (1996) Identification of lichen substances. Springer, Berlin, pp 1–493

Huneck S, Djerassi C, Becher D, Barber M, Ardenne M, Steinfelder K, Tummler R (1968) Flechteninhaltsstoffe – XXXI: Massenspektrometrie und ihre anwendung auf strukturelle und streochemische probleme – CXXIII Massenspektrometrie von depsiden, depsidonen, depsonen, dibenzofuranen und diphenylbutadienen mit positiven und negativen ionen. Tetrahedron 24:2707

Hyvärinen M, Koopmann R, Hormi O, Tuomi J (2000) Phenols in reproductive and somatic structures of lichens: a case of optimal defence? Oikos 91:371–375

Ingólfdóttir K (2002) Usnic acid. Phytochemistry 61:729–736

Johansson S, Søchting U, Elix JA, Wardlaw JH (2005) Chemical variation in the lichen genus Letrouitia (Ascomycota, Letrouitiaceae). Mycol Prog 4:139–148

Karolewski P, Giertych MJ (1994) Influence of toxic metal ions on phenols in needles and roots and on root respiration of scots Pine seedlings. Acta Sociaetatis Botanicorum Poloniae 63(1):29–35

Kumar A, Wagner G, Ernst RR, Wüthrich K (1980) Studies of J-connectivities and selective 1H–1H Overhauser effects in H_2O solutions of biological macromolecules by two-dimensional NMR experiments. Biochem Biophys Res Commun 96:1156

LaGreca S (1999) A phylogenetic evaluation of the *Ramalina americana* chemotype complex (lichenized Ascomycota, Ramalinaceae) based on rDNA ITS sequence data. Bryologist 102:602–618

Lawrey JD (1977) Adaptive significance of O-methylated lichen depsides and depsidones. Lichenologist 9:137–142

Lawrey JD (1986) Biological role of lichen substances. Bryologist 89:111–122

Liao CRJ, Piercey-Normore MD, Sorenson JL, Gough KM (2010) In situ imaging of usnic acid in selected *Cladonia* spp. by vibrational spectroscopy. Analyst 135:3242–3248

Liu H, Nakanishi K (1981) A micromethod for determining the branching points in oligosaccharides based on circular dichroism. J Am Chem Soc 103(23): 7005–7006

Luo H, Yamamoto Y, Kim JA, Jung JS, Koh YJ, Hur J-S (2009) Lecanoric acid, a secondary lichen substance with antioxidant properties from *Umbilicaria antarctica* in maritime Antarctica (King George Island). Polar Biol 32:1033–1040

MacGillvray T, Helleur R (2001) Analysis of lichens under environmental stress using TMAH thermochemolysis-gas chromatography. J Anal Appl Pyrolosis 58–59:465–480

Macheix JJ (1996) Les composes phenoliques des vegetaux: quelles perspectivesala fin du XXeme siecle? Acta Bot Gallica 143(6):473–479

McEvoy M, Nybakken L, Solhaug KA, Gauslaa Y (2006) UV triggers the synthesis of the widely distributed secondary compound usnic acid. Mycol Prog 5:221–229

McEvoy M, Gauslaa Y, Solhaug KA (2007a) Changes in pools of depsidones and melanins, and their function, during growth and acclimation under contrasting natural light in the lichen *Lobaria pulmonaria*. New Phytol 175:271–282

McEvoy M, Solhaug KA, Gauslaa Y (2007b) Solar radiation screening in usnic acid containing cortices of the lichen *Nephroma arcticum*. Symbiosis 43:143–150

McLafferty FW (1980) Interpretation of mass spectra, 3rd edn. University Science, Mill Valley

Meehl GA, Stocker TF, Collins WD, Friedlingstein P, Gaye AT, Gregory JM, Kitoh A, Knutti R, Murphy JM, Noda A, Raper SCB, Watterson IG, Weaver AJ, Zhao ZC (2007) Global Climate Projections. In: Solomon S, Qin D, Manning M, Chen Z, Marquis M, Averyt KB, Tignor M, Miller HL (eds) Climate change 2007: the physical science basis. Contribution of Working Group I to the fourth assessment report of the Intergovernmental Panel on Climate Change. Cambridge University Press, Cambridge/New York

Monna F, Aiuppa A, Varrica D, Dongarra G (1999) Pb isotope composition in lichens and aerosols from eastern Sicily: insights into the regional impact of volcanoes on the environment. Environ Sci Technol 33:2517–2523

Moore D (1998) Fungal morphogenesis. Cambridge University Press, New York, 469

Morris GA (1980) Sensitivity enhancement in N-15-NMR polarization transfer using the INEPT pulse sequence. J Am Chem Soc 102:428

Morris GA (1986) Modern NMR techniques for structure elucidation. Magn Reson Chem 24:371

Mosbach K (1969) Biosynthesis of lichen substances, products of a symbiotic association. Angew Chem Int Ed 8:240–250

Muzika RM (1993) Terpenes and phenolics in response to nitrogen fertilization: a test of carbon/nutrient balance hypothesis. Chemoecology 4(1):3–7

Narui T, Sawada K, Takatsuki S, Okuyama T, Culberson CF, Culberson WL, Shibata S (1998) NMR assignments of depsides and tridepsides of lichen family Umbilicariaceae. Phytochemistry 48(5):815–822

Nishitoba Y, Nishimura H, Nishiyama J, Mizutani J (1987) Lichen acids, plant growth inhibitors from *Usnea longissima*. Phytochemistry 26:3181–3186

Nybakken L, Julkunen-Tiitto R (2006) UV-B induces usnic acid in reindeer lichens. Lichenologist 38:477–485

Nybakken L, Solhaug KA, Bilger W, Gauslaa Y (2004) The lichens *Xanthoria elegans* and *Cetraria islandica* maintain a high protection against UV-B radiation in Arctic habitats. Oecologia 140:211–216

Penuelas J, Estiarte M, Kimball BA, Idso SB, Pinter PJ, Wall GW, Garcia RL, Hansaker DJ, LaMorte RL, Hendrix DL (1996) Variety of responses of plant phenolic concentration to CO_2 enrichment. J Expt Bot 47(302):1463–1467

Pisani JM, Distel RA (1998) Inter and intraspecific variations in production of spines and phenols in Prosopis caldemia and Prosopis flexuosa. J Chem Ecol 24(1):23–36

Ramakrishnan S, Subramanian SS (1964) Amino acids of *Roccella montagnei* & *Parmelia tinctorum*. Indian J Chem 2:467

Ramakrishnan S, Subramanian SS (1965) Amino-Acid Composition *Cladonia rangiferina*, *Cladonia gracilis* and *Lobaria isidiosa*. Curr Sci 34:345–347

Ramakrishnan S, Subramanian SS (1966) Amino-acids of *Lobaria subisidiosa*, *Umbilicaria pustulata*, *Parmelia nepalensis* and *Ramlina sinensis*. Curr Sci 5:124–125

Reverberi M, Ricelli A, Zjalic S, Fabbri AA, Fanelli C (2010) Natural functions of mycotoxins and control of their biosynthesis in fungi. Appl Microbiol Biotechnol 87:899–911

Rhodes MJC (1994) Physical role for secondary metabolites in plants: some progress, many outstanding problems. Plant Mol Biol 24:1–20

Seaward MRD (1993) Lichens and sulphur dioxide air pollution: field studies. Environ Rev 1:73–91

Seo C, Choi Y, Ahn JS, Yim JH, Lee HK, Oh H (2009) PTP1B inhibitory effects of tridepside and related metabolites isolated from the Antarctic lichen *Umbilicaria antarctica*. J Enzyme Inhib Med Chem 24(5):1133–1137

Shirtcliffe NJ, Pyatt FB, Newton MI, McHale G (2006) A lichen protected by a super-hydrophobic and breathable structure. J Plant Physiol 163:1193–1197

Shukla V (2012) Physiological response and mechanism of metal tolerance in lichens of Garhwal Himalayas. Final technical report. Scientific and Engineering Research Council, Department of Science and Technology, New Delhi. Project No. SR/FT/LS-028/2008

Shukla V (2002) Chemical study of macrolichens of Garhwal Himalayas. PhD thesis, Garhwal University Srinagar, Garhwal

Shukla V, Negi S, Rawat MSM, Pant G, Nagatsu A (2004) Chemical Study of *Ramalina africana* (Ramaliniaceae) from Garhwal Himalayas. Biochem Syst Ecol 32:449–453

Shukla V, Joshi GP, Rawat MSM (2010) Lichens as potential natural source of bioactive compounds. Phytochem Rev 9:303–314. doi:10.1007/s11101-010-9189-6

Shultz H, Albroscheit G (1989) Characterization of oak moss products used in perfumery by high-performance liquid chromatography. J Chromatogr 466:301–306

Solhaug KA, Gauslaa Y (1996) Parietin, a photoprotective secondary product of the lichen *Xanthoria parietina*. Oecologia 108:412–418

Solhaug KA, Gauslaa Y, Nybakken L, Bilger W (2003) UV-induction of sun-screening pigments in lichens. New Phytol 158:91–100

Spaink HP (1998) Flavonoids as regulators of plant development. New insights from studies of plant-rhizobia interactions. In: Romeo JT et al (eds) Phytochemical signals and plant-microbe interactions. Plenum Press Publishers, New York, pp 167–177

Stocker-Wörgötter E (2001) Experimental studies of the lichen symbiosis: DNA- analyses, differentiation and secondary chemistry of selected mycobionts, artificial resynthesis of two- and tripartite symbioses. Symbiosis 30:207–227

Strack D, Feige GB, Kroll R (1979) Screening of aromatic secondary lichen substances by high performance liquid chromatography. Z Naturforsch 34c:695–698

Subramanian SS, Ramakrishnan S (1964) Amino acids of *Peltigera canina*. Curr Sci 33:522

Swanson A, Fahselt D (1997) Effects of ultraviolet on polyphenolics of *Umbilicaria americana*. Can J Bot 75:284–289

Verpoorte R, Choi YH, Kim HK (2009) NMR-based metabolomics at work in phytochemistry. Phytochem Rev 6:3–14. doi:10.1007/s11101-006-9031-3

Waterman PG, Mole S (1994) Analysis of phenolic plant metabolites. Blackwell Scientific Publishers, New York

Wehrli FW, Wirthlin T (1976) Interpretation of carbon-13 NMR spectra. Heydon, London

White CS (1994) Monoterpenes: their effects on ecosystem nutrient cycling. J Chem Ecol 20(6): 1381–1406

Wider G, Macura S, Kumar A, Ernst RR, Wuthrich K (1984) Homonuclear two-dimensional 1H NMR of proteins. Experimental procedures. J Magn Reson 56:207–234

Wojciechowski ZA, Goad LJ, Goodwin TW (1973) Sterols of lichen *Pseudevernia furfuracea*. Phytochemistry 12:1433

Selection of Biomonitoring Species 3

Lichens lack significant cuticle or epidermis and are devoid of a well-developed root system, therefore they absorb nutrients directly from the atmosphere. Along with nutrients, pollutants are also absorbed and/or adsorbed on the lichen thalli without having any visible signs of injury to the thallus. Lichens show differential sensitivity towards wide range of pollutants. Certain species are inherently more sensitive, while some species shows tolerance to high levels of pollutants. These characteristics make certain lichen species suitable for being utilised as an indicator species (based on their sensitivity and tolerance). These features of lichens, combined with their extraordinary capability to grow in a large geographical area, rank them among an ideal and reliable bioindicators of air pollution. Periodically monitoring lichen community and physiological changes in the lichen species may be effectively utilised to monitor air quality of an area. This chapter summarises the criteria for selection of biomonitoring species and how characters of host plant influences lichen diversity. Details about different lichen species utilised for biomonitoring have been discussed.

3.1 Introduction

Gilbert in 1969 recommended the use of cryptogamic epiphytes as bioindicators in the First European Congress on the Influence of Air Pollution on Plants and Animals, and European countries were the pioneers in implementing the recommendations. Lichens and mosses are reliable indicators of terrestrial air quality which due to lack of significant cuticle or epidermis make them well suited, bioindicators and biomonitors. Due of the lack of a well-developed root system, lichens absorb both nutrients and pollutants directly from the atmosphere. Higher plants (tracheophytes) are also employed in biomonitoring studies, but as they also accumulate pollutants from soil, interpretation of data requires more understanding of contributing factors. The ability of lichen and bryophytes to sequester heavy metals yet remaining unharmed makes them good biomonitors (Garty 2001).

Lichens have been used worldwide as air pollution monitors because they show sensitivity towards relatively low levels of sulphur, nitrogen and fluorine-containing pollutants (especially oxides of sulphur and nitrogen, fluoride gas and fertiliser used). Phytotoxic gases adversely affect many sensitive species, altering lichen community composition, growth rates, reproduction, physiology and morphological appearance. Whereas toxitolerant lichens concentrate a variety of pollutants in their tissues without being harmed.

Lichen species are often classified on the basis of the requirements of substrates, pH and ambient nutrient status (Seaward and Coppins 2004). Therefore, studying lichen communities can illuminate the surrounding environmental change (Table 3.1). Indeed epiphytic lichens have been recognised as indicators of air pollutions since the 1800s (Nash 2008). Lichens are useful bioindicators, especially where technical instruments

Table 3.1 Succession of lichen communities in disturbed and undisturbed environment and changing pH of substrate

	Age of substrate	Industrial pollution pH < 3.5	Intensive agriculture pH 6+	Ecological continuity, unpolluted air pH 5–6
1	1–10	Lecanoriod	Lecanoriod	Graphidiod
2	5–20	Pseudeverniod	Xanthoriod	Usneoid
3	15–50	Lecanoriod	Xanthoriod	Parmeliod
4	100+	Caliciod	Dimorphic (Cladoniod)	Lobariod

Modified from Wolseley et al. (2003)

are not economically feasible (Seaward 2008; Guidotti et al. 2009). Moreover, a correlation between air pollution and lung cancer in NE Italy by studying lichen biodiversity (Cislaghi and Nimis 1997) suggests the potential use of lichens to monitor human health.

Lichen biomonitoring is an integral part of managing well-documented air quality data in wide areas of Europe and North America carried out by different lichenologists and air quality regulatory agencies that include Nash and Gries (2002), Nimis et al. (2002), Garty (2001), Hyvärinen et al. (1993) and Richardson (1992).

Nash (2008) points out that the sensitivity to different pollutants varies among different species. Moreover, Insarova et al. (1992) state that the response to different pollutants and substrates may vary even within a species.

As lichen species are differentially sensitive to pollutants, the presence of these pollutants in the environment produces changes at community level. Survival strategy of epiphytic lichens growing close to a copper smelter in the Middle Urals was investigated, and it was found that in *Hypogymnia physodes*, an asexually reproducing strong competitor, and *Tuckermanopsis sepincola*, a sexually reproducing weak competitor, *T. sepincol*, benefit from the Cu sensitivity of *H. physodes*. As under Cu stress, *H. physodes* produced only esorediate or low sorediate thalli and sharply decreased in abundance, while abundance of *T. sepincola* on birch trunks increased (Mikhailova 2007).

Specific lichen communities on rock and soil occurring in heavy metal-polluted areas worldwide, mainly related to mining of metals, have been investigated by several authors (e.g. Nash 1989; Purvis and Halls 1996; Bačkor et al. 2004; Banásová et al. 2006). Some lichens associated with heavy metal-rich substrates are common species that tolerate metals and occur in both polluted and unpolluted areas. Most lichens requiring metal-rich substrates are crustose lichens belonging to the genera *Acarospora, Aspicilia, Lecanora, Lecidea, Porpidia, Rhizocarpon* or *Tremolecia* (Nash 1989; Purvis and Halls 1996; Bačkor et al. 2004; Bačkor and Loppi 2009).

The most commonly used lichen biomonitoring methods are community analysis, lichen tissue analysis and transplant studies. Lichens are long-lived and have wide distribution, so they can be monitored in field conditions permitting comparing air quality (spatial and temporal evaluation) of wide geographical area that too on low cost. Toxitolerant/resistant lichens have extensive geographical ranges, allowing study of pollution gradients over large areas.

Human activities related to agricultural and livestock uses cause the impoverishment of lichen communities, including the local disappearance of the most demanding species (Loppi and de Dominicis 1996; Hedenås and Ericson 2004; Bergamini et al. 2005; Nascimbene et al. 2007). The higher occurrence of Xanthorion species in agricultural and livestock stands compared with unmanaged forest may result from two factors: the high irradiance in more open woodland (Fuertes et al. 1996; Hedenås and Ericson 2004) and the high deposition of nutritious dust (Hedenås and Ericson 2004; Motiejûnaite and Faútynowick 2005).

Lichens grow on a wide range of substrates, both natural and man-made, and obtain their required nutrients and water directly from the atmosphere. This uptake of nutrients from the atmosphere means lichens are good indicators of environmental disturbance as they bioaccumulate airborne pollutants. These features of lichens, combined with their extraordinary capability to grow in a large geographical range and to accumulate mineral elements far above their needs, rank them among the best bioindicators of air pollution. Epiphytic lichens (lichens growing on trees or plants) are often best suited to the study of air pollution effects on lichen communities, lichen growth or physiology, and to the study of pollutant loading and distribution. Being located above the ground, epiphytic lichens usually receive greater exposure to air pollutants and do not have access to soil nutrient pools (Purvis et al. 2008), and as lichens usually grows on dead bark, therefore chances of contamination by nutrient cycling is also minimal (Loppi et al. 1998; Shukla et al. 2013). As lichens depend on deposition, water seeping over substrate surfaces, atmospheric gases and other comparatively dilute sources for their nutrition, the tissue content of epiphytic lichens largely reflects atmospheric sources of nutrients and contaminants (Purvis et al. 2008). Lichens on soils and rock substrates are more likely to be influenced by elements and chemicals from these substrates, but otherwise share morphological and physiological characteristics of epiphytes. Under certain conditions, lichen floristic and community analyses can be used in conjunction with measured levels of ambient or depositional pollutants accumulated by lichens to detect effects of changing air quality on vegetation. This information can demonstrate the extent of damage caused by air pollutants. Contaminants cause undesirable changes in species composition or presence/absence of lichen species. It is important that any alternative factors (e.g. drought, grazing, habitat alteration) for changes observed in species condition or composition (in addition to air pollution) should be considered while using lichens floristic and community data in an air pollution assessment.

Lichen species can be divided into three categories according to bark type and nutrient needs: nitrophytic species which prefer to grow on deciduous tree bark enriched with dust or nutrients, acidophytic species which prefer acidic barks and neutrophytic species which are indifferent species. This categorisation is widely employed as an ecological tool and is also a standard feature of assessing lichen floras (Gombert et al. 2004).

Lichens exhibit differing levels of sensitivity to pollution. In general, air pollution sensitivity increases among growth forms in the following series: crustose (flat, tightly adhered, crust-like lichens) < foliose (leafy lichens) < fruticose (shrubby lichens), though there are exceptions to this gradation. The kind and the level of pollution to be monitored in a research area should be taken into consideration in selection of the species of lichens that will be used as a biomonitor (Riga-Karandinos and Karandinos 1998). Foliose lichens are better accumulators in comparison with fruticose ones (St. Clair et al. 2002a, b).

3.2 Criteria for Selection of Biomonitoring Species

Not all biological organisms are suitable for use as biomonitoring tools. The general considerations which must be kept in mind in selecting biomonitoring organisms include (Wolterbeek et al. 2003; Conti and Cecchetti 2001):

- The organism must be capable of accumulating pollutants in measurable amounts, i.e. accumulate the pollutants without, however, being killed by the levels with which it comes into contact.
- Have a wide geographical distribution: be abundant, sedentary or of a scare mobility, as well as being representative of the collection area.
- It should be available throughout the years or for the whole period of study, with relative easy for collection.
- The organism should show a differential uptake/accumulation which is related to exposure, thus allowing either (1) relative pollution levels to be determined or (2) the establishment of a more quantitative relationship to deposition rate or air concentrations.
- For assessing air borne contamination, the organism must not be subject to substantial uptake or ingestion of metals from other sources.
- Cost of collection and analysis should be acceptable.

In order to determine the contaminants accumulated in the biomonitors, organisms may be selected on basis of their accumulative and time-integrative behaviour (Wolterbeek and Bode 1995). In most of the studies, biomonitoring species for assessment of air pollution are often selected on basis of criteria such as specificity, which implies that accumulation is considered to occur from the atmosphere only and provides a well-defined representation of a sampling site. Wolterbeek et al. (1996) asserted that selection should be on the basis of the differences between local and survey variances, and the almost implicit criterion for selection is the common occurrence of biomonitors. Earlier air pollution was indexed by geographical variances in biodiversity and species richness, but the recent studies are aimed at clarification of the impact of variable levels of atmospheric pollution based on the response of the biomonitors (such as photosynthesis, respiration, transpiration, element accumulation) (Paoli and Loppi 2008; Garty 1993). These studies are based on ecophysiological responses and relating impact to response in terms of elemental accumulation; knowledge is gathered on the dose–response relationships for the biomonitor of interest (Wolterbeek 2002).

Of all biological species used in biomonitoring, lichens have the most common occurrence. Lichens usually have considerable longevity, which led to their use as long-term integrators of atmospheric depositions. Unlike higher plants, they have certain characteristic, which meet several requirements of the ideal biological monitor. They are perennial, slow growing, maintain a uniform morphology in time and highly dependent on atmosphere for nutrients because root system is absent, and these plants do not shed parts as readily as vascular plants. Other characteristic of lichens is rapid uptake and accumulation of cations. The lack of waxy cuticle and stomata allows many contaminants to be absorbed over the whole lichen surface without exhibiting damage there by permitting monitoring over wide areas. In addition, the lichen surface, structure and roughness facilitate the interception and retention of particles. The absorption of metals in lichens involves following mechanisms (Nieboer and Richardson 1981):
- Intercellular absorption through an exchange process
- Intracellular accumulation
- Entrapment of particles that contain metals

Heavy metal content in lichen thallus tends to alternate over time in phases of accumulation and subsequent release. The metal absorption in lichens is influenced by geographical variations (altitude), temporal variations (seasonal changes), acid precipitation, soil dust and local pollution sources (commercial, industrial, vehicular, mining areas).

3.3 Biomonitoring Species (World and India)

Epiphytic lichen is being widely employed in different ways as air quality bioindicator and/or biomonitor, starting with phytosociological approach pioneered by De Sloover and Le Blanc (1968) to recent geostatistical modelling in source apportionment of pollutants (Augusto et al. 2009). Biomonitoring with lichens at regional scale employing various lichens species has been well studied in Europe and Northern America (Loppi and Frati 2006; Thormann 2006; Pinho et al. 2004). *Hypogymnia physodes*, *Parmelia sulcata*, *Evernia prunastri* and *Pseudevernia furfuracea* are established bioindicator species (Table 3.2) (Herzig et al. 1989; Arb et al. 1990; Loppi et al. 2004; Bari et al. 2001). It is recognised that a wide range of other substances like phytotoxic gases, alkaline dust (pesticides, fertilisers), heavy metals, polycyclic aromatic hydrocarbons (PAHs) and radionuclides, chlorinated hydrocarbon and unintentional man-made chemicals (PCDFs/PCDDs and PCBs) may also be detected and monitored using lichens (Garty 2001).

Several gaseous air quality monitoring methods utilising lichens are available (Conti and Cecchetti 2001). The lichen monitoring can be qualitative or quantitative and employ single indicator species or community changes. The choice of method depends on the pur-

Table 3.2 Variation in the concentration (µg g^{-1} dry weight) of metallic content in lichen species utilised worldwide for biomonitoring

S. no.	Lichen species	Country	Al	Cd	Cr	Cu	Fe	Hg	Mn	Ni	Pb	Zn	As	References
1	*Parmelia caperata* (N)	Italy	110–277	0.35–0.65	0.1–1.7	4.4–8.1	119–246	0.10–0.22	30.4–128	0.4–2.3	3.8–14.9	38.6–66.8	ND	Loppi and Pirintos (2003)
2	*Parmelia sulcata* (N)	Italy	2,384	0.191	3.6	9.1	1,800	ND	38.2	ND	15.9	65.7	ND	Loppi et al. 1998
3	*Flavoparmelia caperata* (N)	Serbia	ND	0.1310–0.2558	1.755–3.714	6.14–15.37	476.6–493.2	ND	13.67–14.12	1.27–1.59	9.8–22.3	17.6–20.7	0.0031–0.0037	Mitrović et al. (2012)
			ND	0.1546–0.1942	1.4572–1.4879	5.73–6.51	374.5–399.2	ND	12.4–15.62	1.09–1.301	7.3–10.1	17.0–24.3	0.0027–0.0038	
4	*Hypogymnia physodes* (T)	Southern Poland	ND	0.54–7.70	0.22–75.85	3.7–10.8	350–1,202	ND	ND	1.08–9.18	9.3–123.7	55–583	ND	Bialoska and Dayan (2005)
5.	*Ramalina lacera* (T)	Israel	1,228–1,587	ND	4.4–10.1	5.1–9.9	1,033–1,403	ND	21.8–29.5	4.0–8.9	7.3–11.2	34.6–43.1	ND	Garty et al. (2003)
6	*Pseudevernia furfuracea* (T)	Turkey	ND	ND	ND	ND	ND	ND	2.47	0.28	0.60	0.33	ND	Yildiz et al. (2008)

N native, *T* transplant

Table 3.3 Epiphytes responding to increasing influence of 'traffic' or 'agriculture' with decreasing or increasing frequencies

Frequency decreases with increasing vehicular activity	Frequency increases with increasing vehicular activity	Frequency decreases with increasing agricultural activity	Frequency increases with increasing agricultural activity
Lecanora expallens	*Phaeophyscia nigricans*	*Evernia prunastri*	*Amandinea punctata*
Lepraria incana	*Phaeophyscia orbicularis*	*Hypogymnia physodes*	*Caloplaca holocarpa*
Parmelia saxatilis	*Phaeophyscia hispidula*	*Lecanora conizaeoides*	*Lecidella elaeochroma*
Physcia tenella	*Xanthoria parietina*	*Lepraria incana*	*Physconia grisea*
	Pyxine cocoes	*Orthotrichum diaphanum*	*Xanthoria parietina*
		Melanelia subaurifera	
		Parmelia sulcata	
		Phaeophyscia orbicularis	
		Physcia caesia	

Modified from after Wolseley and James (2000), includes some prominent Indian lichen species

pose of the survey, the size of the study area, resources available and the desired detail of the output. Thus, lichens have a long history of use as monitoring of environmental pollution. Several authors have advocated the use of biological monitoring to assess and understand the status of trends within natural ecosystem (Jeran et al. 2002; Godinho et al. 2008).

Although the majority of lichens are sensitive to pollutants in the environment, some are able to survive on substrates rich in heavy metals (Purvis and Halls 1996). Additionally, differential sensitivity to pollutants has led to production of pollution scales based on delimitation of zones characterised by specific lichen flora (Hawksworth and Rose 1970). Numerous investigations of the interaction of lichens with its ambient environment reveal that lichens may be assigned to three categories in terms of their responses to air pollution (Garty et al. 2003): sensitive species, with varying degrees of sensitivity to the detrimental effects of pollutants, but ultimately succumbing to air pollution; tolerant species, resistant to pollution, belonging to the native community and remaining intact in their native habitat; and replacement species, making their appearance after destruction of the major part of the native lichen community as a result of pollution.

In the presence of pollution, each species of lichens responds in a different biological stress degree and consequently shows a different biomonitoring capacity (Table 3.3). Therefore, lichens provide different options as biomonitors of the air pollution, depending on the selected species. The biomonitoring capability of lichens has been studied in depth for various contaminants (Blasco et al. 2011).

As certain lichen species are inherently more sensitive to airborne contaminants, air quality can be effectively monitored by occasionally re-evaluating lichen community and/or assessing physiological parameters. Pollution related changes can then be documented by comparing follow-up data to original baseline data (St. Clair 1989).

Lichens can be collected from substrates of determinable age such as twigs, or mean tissue concentrations of selected species can be compared over time. However, caution should be used in such correlations. Garty's (2001) review of a dozen studies of age-related differences in lichen thalli (vegetative bodies) revealed that differences are not always significant, nor always size related, and vary with growth rate, target element and lichen species.

In particular, the epiphytic species *Hypogymnia physodes* has been shown to be one of the most suitable lichen monitors (in situ or transplanted) of atmospheric metal concentrations in central Europe (Jeran et al. 1995; Vestergaard et al. 1986; Herzig et al. 1989). *Hypogymnia physodes* is relatively commonly occurring lichen for which baseline levels of heavy metals have been established using data from the species

Table 3.4 Baseline values for some common metallic pollutants in *Hypogymnia physodes*

S. no.	Elements (mg g^{-1} dry weight)	Background average	Enriched average
1	Pb	19.5	126.9
2	Zn	73	427
3	Cd	0.56	2.56
4	Cr	2.11	43.50
5	Ni	1.72	12.28
6	Cu	6.0	28.6
7	Fe	621	3081
8	S	738	1695

After Bennett (2000)

collected worldwide (Table 3.4) (Bennett 2000). In northwest Britain, *Lobaria pulmonaria* is being utilised with its limitation being present on trees with bark pH >5 (Farmer et al. 1991).

In Italy epiphytic foliose lichen species, belonging to the genera *Xanthoria* and *Parmelia*, were mostly employed as bioaccumulators (Nimis et al. 2000). The use of a single species in the same survey is recommended by several authors (Bargagli 1998) to minimise the data variability (Minganti et al. 2003).

Trace metal pollution in natural ecosystems in Serbia have been monitored utilising the lichen species *Flavoparmelia caperata* because of its wide distribution and proven sentinel functions (Mitrović et al. 2012).

L. pulmonaria prefers light and exposed areas as long as its water demand is met (Barkman 1958; Renhorn et al. 1997; Hazell and Gustafsson 1999). Similarly, some species of *Collema* and *Leptogium* even survive on remnant trees and acclimatise to the new exposure conditions after clear cut (Hazell and Gustafsson 1999; Hedenås and Hedström 2007).

The presence of cyanolichens is closely related to microclimate, as they prefer sites with high humidity and moisture availability (Barkman 1958). Cyanolichen richness was also larger in unmanaged forests with high tree cover and diameter. Mikhailova et al. (2005), studying the influence of environmental factors on the local-scale distribution of cyanolichens in Russia, concluded that most of the cyanolichens studied preferred habitats with larger trees.

Humidity, temperature and light conditions inside forests can be altered by forest clearing and logging (Franklin and Forman 1987; Murcia 1995; Moen and Jonsson 2003; Belinchón et al. 2009), causing the systematic reduction and local extinction of some of the most outstanding representatives of sensitive communities as in the case of Thelotremataceae lichens.

In order to compare the bioaccumulation capacity of four established bioindicator species (*Hypogymnia physodes, Parmelia sulcata, Pseudevernia furfuracea and Usnea hirta*) with the help of transplant study carried out in urban site of North Italy revealed that *H. physodes, P. furfuracea and U. hirta* have a similar accumulation capacity, while *P. sulcata* has lower capacity Bergamaschi et al. (2007).

Tolerance of a lichen species to excessive concentrations of metals is based on different biochemical mechanisms of detoxification, among which the dominant role is played by reactions of metal ion complexation with organic acids contained in lichen thalli and extracellular decomposition of the complexes on the thallus surface or on the walls of mycobiont hyphae (Mikhailova and Sharunova 2008).

Nitrophytic species need both a relatively high bark pH and at least some additional nitrogen as well. On acid bark, the pH rather than nitrogen is a limiting factor for their occurrence. On basic bark, nitrogen can be limiting, but this is hardly the case in urban or rural circumstances, where nitrogen is usually widely available while acidophytic species require an acid substrate, but many of them are sensitive to increased levels of nitrogen as well (van Herk et al. 2002).

In India till date biomonitoring studies which have been conducted have employed *Pyxine cocoes* (Swartz) Nyl. in active biomonitoring studies. These studies have provided the air pollution level of various major cities of India which included Kolkata (capital city of West Bengal), Bengaluru (capital city of Karnataka), Pune (IT hub) and Lucknow (capital city of Uttar Pradesh). These studies have showed that Kolkata is highly polluted metropolitan in India. Crustose lichen *Rinodina sophodes* has been employed for air quality monitoring in the industrial city of Kanpur (Uttar Pradesh).

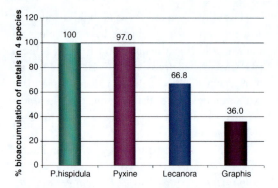

Fig. 3.1 Comparative bioaccumulation potential of four lichen species for metals

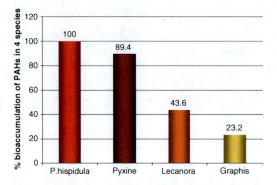

Fig. 3.2 Comparative bioaccumulation potential of four lichen species for PAHs

In the temperate regions of Garhwal Himalayas, lichen *Phaeophyscia hispidula* is a common lichen of urban areas and one of the most widely utilised lichen species in Garhwal Himalayas, along with other species such as *Pyxine cocoes* and *Pyxine subcinerea*. Nitrophilous lichen *P. hispidula* presents a large contact surface for atmospheric pollutants and is able to accumulate high amounts of heavy metals in polluted areas. However, it did not show visible signs of injury, which can correlate with increased tolerance of this species to atmospheric pollution (Shukla and Upreti 2007).

Morphology plays a predominant role in nutrient accumulation based on the available surface area for adsorption and/or absorption (Blasco et al. 2011). Among the four lichen species (*Phaeophyscia hispidula*, *Pyxine*, *Lecanora* and *Graphis*), the bioaccumulation potential of metals (Fig. 3.1) and PAHs (Fig. 3.2) was found significantly different among the growth forms.

It was found that *P. hispidula* had maximum accumulation of metals as well as PAHs followed by *Pyxine*, while *Graphis* accumulated 64 % less than *P. hispidula*. Findings establish the use of *Pyxine* and *Phaeophyscia hispidula* as bioindicator species in India (Bajpai et al. unpublished).

It has been observed that ultrastructural feature also affects the adsorption of pollutants associated with particulate matter $PM_{2.5}$ and PM_{10} (Fig. 3.3).

In a developing country like India, biomonitoring is in its preliminary stage and needs more elaborate and extended programmes (like nationwide projects) to utilise biomonitoring techniques for pollution monitoring in vast different geographical conditions where employing instrument is not feasible task.

3.3.1 Are Lichens Suitable Phytoremediator?

The lichen species accumulates huge amount of metals, metalloid, pesticides and radionucleotides in their thallus, but they cannot be considered as phytoremediator (bioremediator). According to Zabludowska et al. (2009), phytomediators should have the shortest life cycle, larger biomass and capability to hyperaccumulate toxic contaminants with no harm to self. The lichens are also capable as hyperaccumulating of various toxicants and remaining unharmed, but they fail to produce larger biomass due to its slow growth rate. Due to this reason, it is universally accepted that though lichens are excellent bioindicator and bioaccumulator, but are not suitable phytoremediator. Although it is still difficult to point out the exact sources of metals that are accumulated by lichens, but their distribution helps to elucidate their origin. Exploratory analysis revealed that the accumulation of toxic metals in lichens may be used in determining the air quality of the city and can be used for future biomonitoring studies. However, for economical and practical reasons, biomonitoring is absolutely necessary to establish and maintain region-wide monitoring systems and for retrospective studies.

3.4 Host Plants for Lichen Colonisation

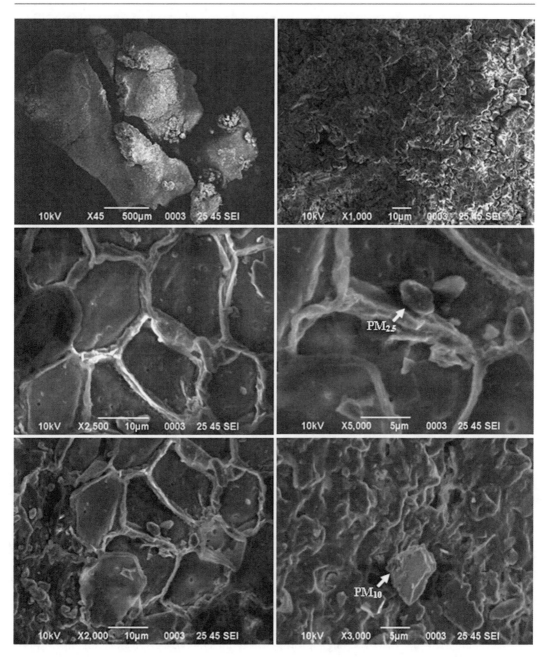

Fig. 3.3 SEM image showing ultrastructural features of lichen species, *Pyxine subcinerea* (hexagonal compartments), which readily adsorb and retain PM of diameter <5 µm in the compartment while PM of diameter >10 µm are not adsorbed on the surface (Shukla 2012)

3.4 Host Plants for Lichen Colonisation

Small-scale distribution of epiphytic lichens is controlled by the spatial and temporal variation of several biotic and physicochemical factors including microclimate, atmosphere and substrate chemistry as well as substrate structure (Hauck and Spribille 2005). Lichens grow on any substratum that provides a convenient foothold to them. This may be tree trunk, rotting wood (*corticolous*), soil (*terricolous*), humus (*humicolous*), rocks and stones and bricks (*saxicolous*), lime plaster (*calcicolous*), moss (*musicolous*), leaves (*foliicolous*), iron pillars, glass panes or insects.

The nature of the substratum is an extremely important factor determining the growth of lichens. Few lichen species are able to grow successfully on a wide range of substrata. Substrate-indifferent species tend to be widespread, while substrate-specific species tend to be restricted in distribution. Thus, lichen communities growing on trees will be different from those growing on rock or on soil (Hauck et al. 2007). Communities growing on rock will be further different based on whether particular rock is acidic (*siliceous*) or basic (*calcareous*).

Bark with rich nutrient (eutrophicated bark) are favoured by some species of *Caloplaca*, *Phyllospora*, *Phaeophyscia* and *Physcia*, while *Arthonia*, *Bacidia*, *Graphis* and *Lecanora* colonise on non-eutrophicated bark. Foliicolous lichens require perennial leaves for their growth.

Texture of the bark smoothness, hardness, relative stability and surface features restricts the lichen diversity on the substratum. Ease of colonisation is a prominent factor dependent on the surface texture. Lichen diaspores get trapped and develop on rough surfaces more easily than on smooth surfaces. The exfoliating nature of the bark does not support lichen colonisation as in *Pinus roxburghii*, while smooth textured and fissured bark supports highest lichen diversity as in the case of *Mangifera indica* (Sheik et al. 2009).

The lichen community occurring smooth twigs are quite different from that on rough and older twigs from the same tree as in the case of *Graphioid* lichens which prefer smooth bark. In North America, the leaves of the tree *Thuja plicata* are colonised by twig-dwelling species which has been attributed to scurfy and rough nature of the leaf surface (Brodo 1973).

Availability of moisture is contributing factor on distribution of lichen on the tree trunk as substrate moisture acts in hydration of the thalli, which influences rate of photosynthesis and respiration in lichens (Harris 1971). While considering moisture relations of bark, two sources are there: firstly, moisture originating due to metabolic activity of the phorophytes and, secondly, from external environment like rain, snow fog, dew, etc. Of the two externally derived, moisture content has been found to be important source for lichens.

Barks with different densities, porosities, texture and internal structure differ in their water holding capacities. Among *oak* and *pines*, oak has been found to retain more moisture and liberate at regular intervals for a longer period of time. Bark water capacities of *Quercus* species and *Pinus albicaulis* have been found to be a factor resulting in difference in the lichen diversity. Moisture capacities of bark differ with exposure and age of tree (Brodo 1973).

A study on the different diameter of twigs of *Quercus semecarpifolia* and *Q. leucotrichophora* exhibits that pH of twig varies with diameter of the twig and so the resulting lichen diversity also varies. The diversity of twig lichens exhibits a decreasing trend with increasing diameter as well as pH. The younger twigs with 1–2 cm diameter are dominated by crustose lichens, and as the diameter increases, crustose lichens are replaced by foliose and fruticose lichens (Kholia et al. 2011). Rock and soil also absorb, bind and release water at different rates. Lichens least prefer colonisation on sand as it has least water holding capacity. Sand stone being porous has moderate water holding capacities and is much preferred by lichens for colonisation.

Availability of nutrients has great significance in lichen distribution (Hauck et al. 2001). Barkman (1958) concluded that it is important to characterise the chemical composition of water in direct contact with the lichen and bark on which it colonises than the substrate itself as many minerals in the substrate are either insoluble or not available to the plant.

It has been observed that the stem flow (water flowing from the crown of the tree down to the trunk after rain) is rich with nutrients especially potassium, calcium, sodium, organic matter, soluble carbohydrate and polyphenols, which is an major source of nutrients to the epiphytic lichens (Hauck et al. 2002; Hauck and Runge 2002).

Table 3.5 Influence of bark chemistry (based on the ash content) on epiphytic vegetation

S. no.	Type of bark	Ash content (%)	Vegetation	Epiphytic lichen vegetation
1	Eutrophic	[5–] 8–12	*Acer, Sambucus Mangifera* and *Prunus*	Physcoid
2	Mesotrophic	[2–] 3–5	*Quercus, Fagus* and *Fraxinus* species	Arthoniod and Parmeliod
3	Oligotrophic	0.4–2.7	*Betula, Picea* and *Abies* species	Caliciod

After Brodo (1973)

A close correlation is known to exist between total electrolyte concentration (determined by ash content of the bark and epiphytic vegetation) (Table 3.5). In rock as substratum, acidic and basic nature of the rock determines lichen diversity (see Sect. 1.3). Serpentine rocks are rich in magnesium, silicon and iron with low soluble potassium and phosphate with nickel and chromium which does not support colonisation of plant communities (Rune 1953).

Although lichens absorb and accumulate nutrients, the nutrient profile does not accurately reflect the composition in the substratum, and it has been concluded that probably of the total nutrient content in lichens, the contribution of the substratum is least.

Distribution of lichen species and communities is strongly correlated with substrate acidity of which hydrogen ion concentration (pH) is an important determining factor. Acidity and alkalinity affects the bioavailability of the element. Surface temperature of the substrate affects lichen distribution. Under experimental conditions, wet lichens perish between 35 and 45 °C although some could tolerate temperature over 70 °C. The indirect effect of substrate heating, by increasing water loss and respiration rate, may be more pronounced. In the case of rock substrates, it provides protection mechanism for regulation of thallus temperature. Rock, being an effective heat sink, loses warmth accumulated during the day at night and that it helps to keep lichen cool during the heat of the afternoon by its slow heat absorption, while in cold climates the ability of substrate to absorb and hold heat might be a factor in substrate selection by lichens. In Antarctica, Rudolph (1963) recorded surface temperatures of 90 °F (32 °C) on rocks covered with vegetation. Barkman also suggested that probably lichens which are anheliophytic (against shade) and thermophobous (disliking heat) are restricted on north face of the tree trunk.

Lichen–bark interface is dependent on the extent penetration of lichen appendages into the tree bark which is fairly dependent on the type of the substratum and morphology of the lichen species. Fruticose lichens penetrate deep into the substratum up to the cambium, while foliose lichens penetrate less but firm attachment with the help of rhizinae.

Substrate specificity is defined as the extent to which a lichen is restricted to a narrowly defined substrate type, and it primarily depends upon the plants requirements and tolerances. If the lichen species requires more nutrient, it will grow on substrate containing large amount of nutrients like calcicolous lichens which prefer calcium-rich substrate, limestone (Ahamadjian and Hale 1973).

References

Ahamadjian V, Hale ME (eds) (1973) The lichens. Academic Press, London

Arb CV, Mueller C, Ammann K, Brunold C (1990) Lichen Physiology and air pollution II. Statistical analysis of the correlation between SO2, NO2, NO and O3 and chlorophyll content, net photosynthesis, sulphate uptake and protein synthesis of *Parmelia sulcata* Taylor. New Phytol 115:431–437

Augusto S, Máguas C, Matos J, Pereira MJ, Soares A, Branquinho C (2009) Spatial modeling of PAHs in lichens for fingerprinting of multisource atmospheric pollution. Environ Sci Technol 43:7762–7769

Bačkor M, Loppi S (2009) Interactions of lichens with heavy metals. Biol Plant 53(2):214–222

Bačkor M, Fahselt D, Wu CT (2004) Free proline content is positively correlated with copper tolerance of the lichen photobiont *Trebouxia* erici (Chlorophyta). Plant Sci 167:151–157

Banásová V, Horák O, Čiamporová M, Nadubinská M, Lichtscheidl I (2006) The vegetation of metalliferous and nonmetalliferous grasslands in two former mine regions in Central Slovakia. Biologia 61:433–439

Bargagli R (1998) Trace elements in terrestrial plants, an ecophysiological approach to biomonitoring and biorecovery. Springer, Berlin/Heidelberg, 324 pp

Bari A, Rosso A, Minciardi MR, Troiani F, Piervittori R (2001) Analysis of heavy metals in atmospheric particulates in relation to their bioaccumulation in explanted *Pseudevernia furfuracea* thalli. Environ Monit Assess 69:205–220

Barkman JJ (1958) Phytosociology and ecology of cryptogamic epiphytes. Van Gorcum, Assen

Belinchón R, Martínez I, Otálora MAG, Aragón G, Dimas J, Escudero A (2009) Fragment quality and matrix affect epiphytic performance in a Mediterranean forest landscape. Am J Bot 96:1974–1982

Bennett JP (2000) Statistical baseline values for chemical elements in the lichen *Hypogymnia physodes*. In: Agrawal SB, Agrawal M (eds) Environmental pollution and plant responses. Lewis Publishers, Boca Raton, pp 343–353

Bergamaschi L, Rizzio E, Giaveri G, Loppi S, Gallorini M (2007) Comparison between the accumulation capacity of four lichen species transplanted to a urban site. Environ Pollut 148:468–476

Bergamini A, Scheidegger C, Carvalho P, Davey S, Dietrich M, Dubs F, Farkas E, Groner U, Kärkkäinen K, Keller C, Lökös L, Lommi S, Máguas C, Mitchell R, Rico VJ, Aragón G, Truscott AM, Wolseley PA, Watt A (2005) Performance of macrolichens and lichen genera as indicators of lichen species richness and composition. Conserv Biol 19:1051–1062

Bialoska D, Dayan FE (2005) Chemistry of the lichen *Hypogymnia physodes* transplanted to an industrial region. J Chem Ecol 31(12):2975–2991. doi:10.1007/s10886-005-8408-x

Blasco M, Domeño C, López P, Nerín C (2011) Behaviour of different lichen species as biomonitors of air pollution by PAHs in natural ecosystems. J Environ Monit 13:2588–2596

Brodo IM (1973) Substrate ecology. In: Ahamadjian V, Hale ME (eds) The lichens. Academic Press, London

Cislaghi C, Nimis PL (1997) Lichens, air pollution and lung cancer. Nature 387:463–464

Conti ME, Cecchetti G (2001) Biological monitoring: lichens as Bioindicator of air pollution assessment – a review. Environ Pollut 114:471–492

Farmer AM, Bates JW, Bell JNB (1991) Seasonal variations in acidic pollutant inputs and their effects on the chemistry of stemflow, bark and epiphyte tissues in three oak woodlands in N.W. Britain. New Phytol 118:441–451

Franklin JF, Forman RF (1987) Creating landscape patterns by forest cutting: ecological consequences and principles. Landsc Ecol 1:5–18

Fuertes E, Burgaz AR, Escudero A (1996) Preclimax epiphyte communities of bryophytes and lichens in Mediterranean forests from the Central Plateau (Spain). Vegetatio 123:139–151

Garty J (1993) Lichens as biomonitors for heavy metal pollution. In: Market B (ed) Plant as biomonitors. VCH Publication, Weinheim, pp 265–294

Garty J (2001) Biomonitoring atmospheric heavy metals with lichens: theory and application. Crit Rev Plant Sci 20(4):309–371

Garty J, Tomer S, Levin T, Lehra H (2003) Lichens as biomonitors around a coal-fired power station in Israel. Environ Res 91:186–198

Godinho RM, Wolterbeek HT, Verburg T, Freitas MC (2008) Bioaccumulation behaviour of lichen *Flavoparmelia caperata* in relation to total deposition at a polluted location in Portugal. Environ Pollut 151:318–325

Gombert S, Asta J, Seaward MRD (2004) Assessment of lichen diversity by index of atmospheric purity (IAP), index of human impact (IHI) and other environmental factors in an urban area (Grenoble, southeast France). Sci Total Environ 324:83–199

Guidotti M, Stella D, Dominici C, Blasi G, Owczarek M, Vitali M, Protano C (2009) Monitoring of traffic-related pollution in a province of central Italy with transplanted lichen *Pseudevernia furfuracea*. Bull Environ Contam Toxicol 83:852–858

Harris GP (1971) The ecology of corticolous lichens. I. The zonation on oak and birch in south Devon. J Ecol 59:431–439

Hauck M, Runge M (2002) Stemflow chemistry and epiphytic lichen diversity in dieback-affected spruce forest of the Harz Mountains, Germany. Flora 197:250–261

Hauck M, Spribille T (2005) The significance of precipitation and substrate chemistry for epiphytic lichen diversity in spruce-fir forests of the Salish Mountains, northwestern Montana. Flora 200:547–562

Hauck M, Jung R, Runge M (2001) Relevance of element content of bark for the distribution of epiphytic lichens in a montane spruce forest affected by forest dieback. Environ Pollut 112:221–227

Hauck M, Hesse V, Runge M (2002) The significance of stemflow chemistry for epiphytic lichen diversity in a dieback-affected spruce forest on Mt. Brocken, northern Germany. Lichenologist 34:415–427

Hauck M, Dulamsuren C, Mühlenberg M (2007) Lichen diversity on steppe slopes in the northern Mongolian mountain taiga and its dependence on microclimate. Flora 202:530–546

Hawksworth DL, Rose F (1970) Qualitative scale for estimating sulphur dioxide air pollution in England and Wales using epiphytic lichens. Nature 227:145–148

Hazell P, Gustafsson L (1999) Retention of trees at final harvest – evaluation of a conservation technique using epiphytic bryophyte and lichen transplants. Biol Conserv 90:133–142

Hedenås H, Ericson L (2004) Aspen lichens in agricultural and forest landscapes: the importance of habitat quality. Ecography 27:521–531

Hedenås H, Hedström P (2007) Conservation of epiphytic lichens: significance of remnant aspen (*Populus tremula*) trees in clear-cuts. Biol Conserv 135:388–395

Herzig R, Libendörfer UM, Ammann K, Cuechheva M, Landolt W (1989) Passive biomonitoring with lichens as a part of integrated biological measuring system for monitoring air pollution in Switzerland. Int J Environ Anal Chem 35:43–57

Hyvärinen M, Soppela K, Halonen P, Kauppi M (1993) A review of fumigation experiments on lichens. Aquilo Serie Botanica 32:21–31

Insarova ID, Insarov GE, Bråkenhielm S, Hultengren S, Martinsson PO, Semenov SM (1992) Lichen sensitivity and air pollution (Swedish Environmental Protection Agency report 4007). Environmental Impact Assessment Department, Uppsala

Jeran Z, Byrne AR, Batic F (1995) Transplanted epiphytic lichens as biomonitors of air-contamination by natural radionuclides around the Zirovski vrh uranium mine, Slovenia. Lichenologist 27(5):375–385

Jeran Z, Jacimovic R, Batic F, Mavsar R (2002) Lichens as integrating air pollution monitors. Environ Pollut 120:107–113

Kholia H, Mishra GK, Upreti DK, Tiwari L (2011) Distribution of lichens on fallen twigs of *Quercus leucotrichophora* and *Quercus semecarpifolia* in and around Nainital city, Uttarakhand, India. Geophytology 41(1–2):61–73

Loppi S, de Dominicis V (1996) Effects of agriculture on epiphytic lichen vegetation in Central Italy. Israel. J Plant Sci 44:297–307

Loppi S, Frati L (2006) Lichen diversity and lichen transplants as monitors of air pollution in a rural area of central Italy. Environ Monit Assess 114:361–375

Loppi S, Pirintsos SA (2003) Epiphytic lichens as sentinels for heavy metal pollution at forest ecosystems (central Italy). Environ Pollut 121:327–332

Loppi S, Pirintsos SA, Dominics V (1998) Soil contribution to the elemental composition of epiphytic lichens (Tuscany, Central Italy). Environ Monit Assess 58:121–131

Loppi S, Kotzabasis K, Loppi S (2004) Polyamine production in lichens under metal pollution stress. J Atmos Chem 49:303–315

Mikhailova IN (2007) Populations of epiphytic lichens under stress conditions: survival strategies. Lichenologist 39:83–89

Mikhailova IN, Sharunova IP (2008) Dynamics of heavy metal accumulation in thalli of the epiphytic lichen *Hypogymnia physodes*. Russ J Ecol 39:346–352

Mikhailova I, Trubina M, Vorobeichik E, Scheidegger C (2005) Influence of environmental factors on the local-scale distribution of cyanobacterial lichens: case study in the North Urals, Russia. Folia Cryptogam Estonica 41:45–54

Minganti V, Capelli R, Dravai G, De Pellegrini R, Brunialti G, Giordani P, Modenesi P (2003) Biomonitoring of trace metals by different species of Lichens (*Parmelia*) in North-West Italy. J Atmos Chem 45:219–229

Mitrović T, Stamenković S, Cvetković V, Nikolić M, Bašosić R, Mutić J, Andelković A, Bojić A (2012) Epiphytic lichen *Flavoparmelia caperata* as a sentinel for trace metal pollution. J Serbian Chem Soc 77(9):1301–1310

Moen J, Jonsson BG (2003) Edge effects on Liverworts and lichens in forests patches in a mosaic of Boreal Forest and Wetland. Conserv Biol 17:380–388

Motiejûnaite J, Faútynowick W (2005) Effect of land use on lichen diversity in the transboundary region of Lithuania and north-eastern Poland. Ekologija 3:34–43

Murcia C (1995) Edge effects in fragmented forests: implications for conservation. Trends Ecol Evol 10:58–62

Nimis PL, Lazzarin A, Lazzarin G, Skert N (2000) Biomonitoring of trace elements with lichens in Veneto (NE Italy). Sci Total Environ 255:97–111

Nascimbene J, Marini L, Nimis PL (2007) Influence of forest management on epiphytic lichens in a temperate beech forest of northern Italy. For Ecol Manage 247:43–47

Nash TH (1989) Metal tolerance in lichens. In: Shaw AJ (ed) Heavy metal tolerance in plants: evolutionary aspects. CRC Press, Boca Raton, pp 119–131

Nash TH III (2008) Lichen biology, 2nd edn. Cambridge University Press, New York

Nash TH III, Gries C (2002) Lichens as bioindicators of sulfur dioxide. Symbiosis 33:1–21

Nieboer E, Richardson DHS (1981) Lichens as monitors of atmospheric deposition. In: Eisenreich SJ (ed) Atmospheric inputs of pollutants to natural waters. An Arbor Science Publishing, Michigan, pp 339–388

Nimis PL, Scheidegger C, Wolseley PA (eds) (2002) Monitoring with lichens – monitoring lichens, Nato Science Series IV: Earth and Environmental Sciences 7. Kluwer, Dordrecht

Paoli L, Loppi S (2008) A biological method to monitor early effect of the air pollution. Environ Pollut 155:383–388. doi:10.1016/j.envpol.2007.11.004

Pinho P, Augusto S, Branquinho C, Bio A, Pereira MJ, Soares A, Catarino F (2004) Mapping lichen diversity as a first step for air quality assessment. J Atmos Chem 49:377–389

Purvis OW, Halls C (1996) A review of lichens in metal-enriched environments. Lichenologist 28:571–601

Purvis OW, Dubbin W, Chimonides PDJ, Jones GC, Read H (2008) The multielement content of the lichen Parmelia sulcata, soil, and oak bark in relation to acidification and climate. Sci Total Environ 390:558–568

Renhorn KE, Esseen PA, Palmqvist K, Sundberg B (1997) Growth and vitality of epiphytic lichens.1. Responses to microclimate along a forest edge-interior gradient. Oecologia 109:1–9

Richardson DHS (1992) Pollution monitoring with lichens. Richmond, Slough

Riga-Karandinos NA, Karandinos GM (1998) Assessment of air pollution from a lignite power plant in the plain of megalopolis (Greece) using as a biomonitors three species of lichens; impact on some biochemical parameters of lichens. Sci Total Environ 215:167–183

Rudolph ED (1963) Vegetation of Hallett station area, Victoria Land, Antarctica. Ecology 44:585–586

Rune O (1953) Plant life on serpentines and related rocks in the north of Sweden. Acta Phytogeogr Suec 31:1–139

Seaward MRD (2008) Environmental role of lichens. In: Nash TH III (ed) Lichen biology, 2nd edn. Cambridge University Press, New York, pp 274–298

Seaward MRD, Coppins BJ (2004) Lichens and hypertrophication. In: Döbbeler P et al (eds). Bibliotheca Lichenologica 88:561–572

Sheik MA, Raina AK, Upreti DK (2009) Lichen flora of Surinsar-Mansar wildlife sanctuary, J & K. J Appl Nat Sci 1(1):79–81

Shukla V (2012) Physiological response and mechanism of metal tolerance in lichens of Garhwal Himalayas. Final technical report. Scientific and Engineering Research Council, Department of Science and Technology, New Delhi. Project No. SR/FT/LS-028/2008

Shukla V, Upreti DK (2007) Physiological response of the lichen *Phaeophyscia hispidula* (Ach.) essl. To the urban environment of Pauri and Srinagar (Garhwal), Himalayas. Environ Pollut 150:295–299. doi:10.1016/j.envpol.2007.02.010

Shukla V, Patel DK, Upreti DK, Yunus M, Prasad S (2013) A comparison of heavy metals in lichen (*Pyxine subcinerea*), mango bark and soil. Int J Environ Sci Technol 10:37–46. doi:10.1007/s13762-012-0075-1

St. Clair LL (1989) Report concerning establishment of a lichen biomonitoring program and base line for the Jarbige Wilderness Area, Humboldt National Forest, Naveda. U.S. Forest Service technical report

St. Clair BS, St. Clair LL, Mangelson FN, Weber JD (2002a) Influence of growth form on the accumulation of airborne copper by lichens. Atmos Environ 36:5637–5644

St. Clair BS, St. Clair LL, Weber JD, Mangelson FN, Eggett LD (2002b) Element accumulation patterns in foliose and fruticose lichens from rock and bark substrates in Arizona. Bryologist 105:415–421

Thormann MN (2006) Lichens as indicators of forest health in Canada. Forestry Chron 82(3):335–343

van Herk CM, Aptroot A, van Dobbin HF (2002) Longterm monitoring in the Netherlands suggests that lichen respond to global warming. Lichenologist 34:141–154

Vestergaard N, Stephansen U, Rasmussen L, Pilegaard K (1986) Airborne heavy metal pollution in the environment of a Danish steel plant. Water Air Soil Pollut 27:363–377

Wolseley P, James P (2000) Factors affecting changes in species of *Lobaria* in sites across Britain 1986–1998. Forest Snow Landsc Res 75(3):319–338

Wolseley P, James P, Sutton MA, Theobald MR (2003) Using lichen communities to assess changes in sites of known ammonia Concentrations (2003) In: Lichens in a changing pollution environment. workshop at Nettlecombe, Somerset, 24–27 February 2003 organised by the British Lichen Society and English Nature

Wolterbeek B (2002) Biomonitoring of trace element air pollution: principles, possibilities and perspectives. Environ Pollut 120:11–21

Wolterbeek HT, Bode P (1995) Strategies in sampling and sample handling in the context of large-scale plant biomonitoring surveys of trace element air pollution. Sci Total Environ 176:33–43

Wolterbeek HT, Bode P, Verburg TG (1996) Assessing the quality of biomonitoring via signal-to-noise ratio analysis. Sci Total Environ 180:107–116

Wolterbeek HT, Garty J, Reis MA, Freitas MC (2003) Biomonitors in use: lichens and metal air pollution. In: Markert BA, Breure AM, Zechmeister HG (eds) Bioindicators and biomonitors. Elsevier, Oxford, pp 377–419

Yildiz A, Aksoy A, Tug GN, Islek C, Demirezen D (2008) Biomonitoring of heavy metals by *Pseudevernia furfuracea* (L.) Zopf in Ankara (Turkey). J Atmos Chem 60:71–81. doi:10.1007/s10874-008-9109-y

Zabludowska E, Kowalsa J, Jedynak L, Wajas S, Sklodowska A, Antosiewicz DM (2009) Search for a plant for phytoremediation-what can we learn from field and hydrophonic studies. Chemosphere 77:301–307

Lichen Diversity in Different Lichenogeographical Regions of India

4

India is a megadiversity region having rich lichen diversity of 2,300 species belonging to 305 genera and 74 families, collected from different regions of the country. Owing to its vast geographical area and varied climate, different phytogeographical regions of India exhibit variation in the diversity of lichens. The detailed lichen diversity in different lichenogeographical region provides the distribution and occurrence of different lichen species including some common and toxitolerant species of the region.

One can retrieve preliminary information about the air quality based on the lichen community structure and distribution of bioindicator species as lichen communities/indicator species provide valuable information about the natural-/anthropogenic-induced changes in the microclimate and land-use changes due to human activity. For identification of species, a key to genera and species provides concise information to identify the lichen species based on their morphological and anatomical characters and chemicals present. Key provided in the chapter will help beginners to identify some common lichen species based on the distribution in different climatic zones of India. The section also provides comprehensive information about the bioindicator communities and bioindicator species.

4.1 Introduction

The lichenology in India started with the account of lichens from Peninsular India. The initiation of Taxonomic studies on Indian lichens dates back to the period of Linnaeus in the eighteenth century, who mentioned a single lichen species, now under the genus *Roccella* (*R. montagnei*) in the publication *Species Plantarum*. It was in 1810 and 1814 the father of lichenology, Erik Acharius, who described four species of lichens from India. During the nineteenth and the twentieth centuries, the lichens from the Indian subcontinent were collected largely by the European botanist. The lichens collected and identified by the European lichenologists are maintained in the different European herbaria. Quaraishi in 1928 was probably the first Indian worker who recorded 35 species of lichens occurring near Mussoorie in Western Garhwal Himalayas. Later Prof. S.R. Kashyap initiated the collection of lichens from different parts of India determined by A.L. Smith, and the data was published by Chopra in 1934. Dr. D.D. Awasthi started working on Indian lichens in a systematic way. During the period 1958–1990, Dr. Awasthi carried exhaustive study on lichens collected by different parts of the country, and the consolidated information is available in the form of keys for the identification of macrolichens and microlichens which described 1,850 species, 234 genera and 80 families (Upreti 1998, 2001).

Recently, Singh and Sinha (2010) published annotated checklists of Indian lichens which describe 2,300 species, collected from different regions of the country, belonging to 305 genera and 74 families (Fig. 4.1). Out of these species reported, about 520 species (22.5 %) are endemic (Singh and Sinha 2010). Family Parmeliaceae shows maximum diversity with 345 species (Fig. 4.2) followed by Graphidaceae with 279 species. Similarly, at generic level *Graphis* shows maximum diversity with 111 species followed by 90 species of *Pyrenula* distributed in tropical regions of the country (Fig. 4.3). The major centres of diversity and occurrence are the Eastern Himalayas including the Northeastern states, Western Ghats, Western Himalayas and Andaman and Nicobar Islands. Lichens grow up to 5,000 m altitude in the Himalayas and show its luxuriance and diversity in tropical to temperate areas (Fig. 4.4).

4.2 Distribution and Diversity of Widespread and Rare Lichen Species in Different Lichenogeographical Regions Along with Its Pollution Sensitivity

The climate of India is extremely varied, from cold snow clad mountain peaks in the north to some of the hottest places as Thar Desert and from zero rainfall regions of the west to the wettest areas in northeast India.

Awasthi catalogued the lichens enumerated by different European and Indian workers who studied in Indian subcontinent including Pakistan, Sri Lanka and Nepal. The knowledge of the lichen flora of India is not complete yet quite satisfactory. In the nineteenth century, most of the areas in the Himalayas and Western Ghats of India were explored exhaustively for their lichen wealth (Awasthi 1988, 1991, 2000, 2007). The areas that are more or less extensively explored lichenologically in India are Jammu and Kashmir, Himachal Pradesh, Manipur, Nagaland, Sikkim, Darjeeling districts, Kumaun and Garhwal Himalayas, Palni and Nilgiri hills in Western Ghats and Andaman Islands. The cursory collections from the state of Assam,

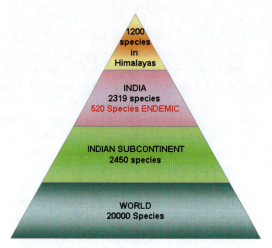

Fig. 4.1 Contribution of Indian lichen to the rest of the world

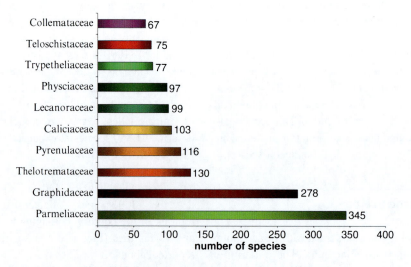

Fig. 4.2 Dominant lichen families in India, Parmeliaceae being the dominant family

4.2 Distribution and Diversity of Widespread...

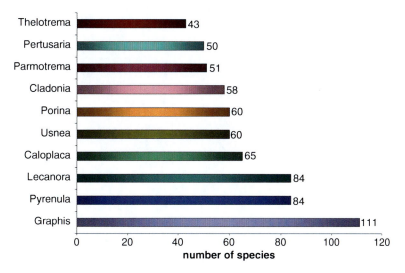

Fig. 4.3 Dominant lichen genera in India, Graphis being the predominant genera in the country

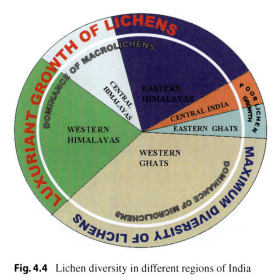

Fig. 4.4 Lichen diversity in different regions of India

Meghalaya, Arunachal Pradesh, Orissa and Rajasthan were also undertaken in the past. There are a number of isolated localities from where casual reports of few species of lichens have been made: Bihar, Punjab, Haryana, Madhya Pradesh, Andhra Pradesh and Tamil Nadu.

Owing to its vast geographical area and varied climate, different phytogeographical regions of India (Fig. 4.5 and Table 4.1) exhibit variation in the diversity of lichens as follows:

4.2.1 Western Himalayas

It occupies the extreme north-western margins of India and influences the climate of the entire region. It sustains remarkable assemblages of flora (Upreti 1998).

Jammu and *Kashmir.* The climate in the state varies from tropical to alpine; the annual precipitation ranges from 107 to 650 mm, with an average of 600 mm snowfall during winter. The forest spreads over 22,236 km^2, which accounts for 20 % of the total geographical area of the state. Over 19,236 km^2 is under coniferous softwood (pine) and 946 km^2 under non-coniferous softwood. The varied altitude and climate provide a variation in the vegetation of the state. The forest vegetation of maple, horse chestnuts, silver fir and Rhododendron provides diverse substrates for colonisation of a number of lichen taxa (Sheik et al. 2006a). According to Sheik et al. (2006b), Parmeliaceae and Physciaceae are the dominant lichen families in the state, while *Cladonia, Lecanora, Xanthoria, Caloplaca, Flavoparmelia, Phaeophyscia, Anaptychia, Dermatocarpon, Xanthoparmelia, Heterodermia, Peltigera, Parmelina, Chrysothrix, Parmelia* and *Physconia* are the dominant genera.

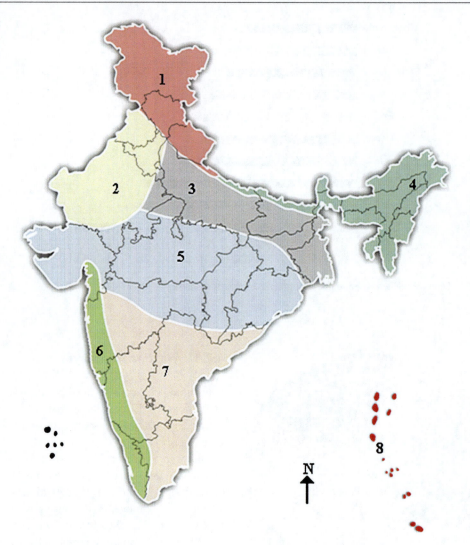

Fig. 4.5 Different lichenogeographical regions of India. *1* Western Himalayas, *2* Western Dry Region, *3* Gangetic Plain, *4* Eastern Himalayas, *5* Central India, *6* Western Ghats, *7* Eastern Ghats and Deccan Plateau and *8* Andaman and Nicobar Islands

Among the different trees, the cultivated trees of chinar, *Salix*, *Morus alba* and *Pyrus malus* bear luxuriant growth of *Xanthoria parietina*. Apart from trees the rocks and boulders in higher altitudes, particularly in cold desert areas, exhibit luxuriant growth of *Xanthoria elegans*. Both species of *Xanthoria* (*X. elegans* and *X. parietina*) are easily recognised by their yellow-orange-coloured thalli.

In lower altitudes between 1,000 and 1,500 m, *Hyperphyscia adglutinata* (Florke) Mayrh. & Poelt grows abundantly on *Acacia modesta* and *Melia azedarach*, *Phaeophyscia orbicularis* (Necker) Moberg, grows on *Enterolobium saman*, and *Pyxine subcinerea* colonises on *Mangifera indica* and *Melia azedarach* (Sheik et al. 2006a).

Ladakh. The lichen flora of Rumabak catchment of Hemis National Park in the Zanskar ranges of the Trans Himalayan mountains has been explored by Negi and Upreti (2000). The study revealed the occurrence of 21 species belonging

4.2 Distribution and Diversity of Widespread...

Table 4.1 Summary of distribution of lichens and available substrate for lichen colonisation in different climatic zones of India

	Climatic regions	Vegetation	Available substrates for lichen colonisation	Abundant lichen species
1	Alpine lichens	Usually devoid of trees	Rivulets	*Lecidea, Lecanora, Acarospora, Rhizocarpon, Caloplaca*
			Humus	*Cladonia, Stereocaulon, Thamnolia*
			Moraine	*Cladonia, Stereocaulon, Thamnolia*
2	Temperate lichens	Abundant trees	Tree (corticolous lichens)	Luxuriant growth of corticolous lichens *Parmelia, Ramalina, Alectoria, Usnea*
			Soil (terricolous lichens)	*Leptogium, Collema, Peltigera, Loeria, Sticta*
			Dry exposed rocks (saxicolous lichen)	*Diploschistes, Lecanora, Lecidea, Pertusaria, Rhizocarpon, Umbilicaria, Dermatocarpon.*
3	Moist tropical evergreen forest	Abundant trees	Thinned-out forest	Corticolous: Pyrenolichens and Graphidaceous lichens
			Deep shady dense forest	Moisture-loving lichens: *Leptogium* and *Collema*
			Fringes of forest and area near stream	Foliicolous lichen
4	Shola Forest of South India	Luxuriant growth of foliose and fruticose lichen on trees equally and sufficiently exposed to rain sunlight and water		
5	Coastal region	Palm trees	Smooth bark	Graphidaceous, *Physcia* and *Pyxine*
6	Mid-eastern and Peninsular plateau	Moderate number of corticolous lichen		
7	Desert or arid regions and Indo-Gangetic plain	Scarce growth of lichen	Rock (saxicolous lichens)	*Caloplaca*
8	Andaman and Nicobar Islands		Original forest	Good growth of foliose and fruticose lichens
			Thinned-out forest	Graphidaceous and Pyrenocarpous lichen

to 12 genera and 9 families. Being a cold desert region, big trees exhibit their absence in the area, and most of the lichens grow on rocks and soil.

Species of lichen genera *Acarospora, Lecanora* and *Rhizoplaca* exhibit their dominance in the area. The most abundant species is *Xanthoria elegans* on rocks, while on soil *Physcia dilatata* is the most abundant. Poor lichen diversity has been attributed to extreme cold condition and lack of rainfall because of its location of the catchment area falling in the leeward side of the mountain.

Himachal Pradesh. Among the Himalayan states, the state of Himachal Pradesh is one of the biodiversity-rich area in Western Himalayas. The state is represented by the occurrence of 503 species belonging to 107 genera and 44 families, of which *Lecanora, Pertusaria, Heterodermia, Cladonia, Caloplaca* and *Parmotrema* are the most dominant genera. Among the different protected areas of the state, the Great Himalayan National Park (GHNP) exhibits the occurrence of 192 species belonging to 65 genera and 31 families (Upreti and Nayaka 2000). Yadav (2005) explored the lichen diversity of 12 districts of Himachal Pradesh. Results revealed maximum diversity in and around Kullu district exhibited by 276 species belonging to 64 genera and 42 families species followed by Shimla, Solan Sirmaur and Chamba district represented by 203, 102, 125 and 102 species, respectively. Districts located in the foothills, Bilaspur, Hamirpur and Una, due to anthropogenic pressure exhibit lower lichen diversity of 59, 29 and 6, respectively.

Uttarakhand. The state of Uttarakhand is comprised of Kumaun and Garhwal regions. The lichen flora of the Kumaun region is well worked out which is represented by the occurrence of 630 species belonging to 134 genera and 44 families (Mishra 2012), which is approximately 27 % of the total 2,303 species known from Indian lichen flora by Singh and Sinha (2010). Parmeliaceae, Physciaceae, Lecanoraceae and Collomataceae are the dominant lichen families of the region. Among the different genera *Lecanora* is represented by 44 species followed by *Cladonia* (32 sp.), *Caloplaca* (29 sp.) and *Heterodermia* (27 sp.). The total lichen diversity shows different habitat preferences as 428 species are corticolous, 191 species are saxicolous, while 118 species are terricolous. Parmeliaceae and Lecanoraceae are the common families of the region, while *Parmotrema, Heterodermia, Lecanora* and *Caloplaca* are the most common genera of the area. *Parmotrema tinctorum, Heterodermia diademata, Heterodermia incana* and *Phaeophyscia hispidula* are the most abundant species. Among the different districts of Kumaun, Pithoragarh exhibited the maximum diversity of lichens represented by 391 species (Mishra et al. 2010). The probable reason for luxuriant growth of corticolous lichen lies on the fact that the region has rich floral diversity enabling lichen colonisation on different phorophytes which includes *Syzygium cumini, Shorea robusta, Quercus leucotrichophora* and *Quercus floribunda.*

Quercus trees in the temperate Himalayas form major vegetation and are the excellent host for colonisation of a large number of lichen taxa together with epiphytic ferns and orchids. *Quercus semecarpifolia* and *Q. dilatata* have dominance of *Usnea* and *Ramalina* species.

Both *Quercus* and *Pinus* shows dominance of Parmeliod lichens (Kumar and Upreti 2008; Upreti and Chatterjee 1999).

Q. leucotrichophora grows abundantly in lower altitude between 1,500 and 2,000 m sometimes pure or mixed with *Rhododendron* and coniferous trees (*Cedrus deodara* and *Cupressus torulosa*), while *Q semecarpifolia* dominates in the higher altitudes between 2,500 and 3,000 m. Lichen diversity on the fallen twigs also provides valuable information about the environmental condition. The diameter of the fallen twig has a prominent influence on lichen colonisation (Kholia et al. 2011). Young smooth barks with pH range 6.5–8 provide optimum condition for colonisation of crustose and foliose lichens, while mature barks having acidic pH of 6 restricts growth of some specific lichens. Other than diameter of the twig, incident light also influences lichen diversity. According to Kantvillas and Jarman (2004), lichen diversity of tree twig in open and exposed areas has more lichen diversity in comparison to the areas with dense tree canopy.

Pindari and Milam valley areas have been extensively investigated by Joshi (2009). Out of the 394 species belonging to 94 genera and 41 families, Pindari glacier exhibited occurrence of 286 species belonging to 78 genera and 36 families, while Milam glacier region is represented by 234 species belonging to 72 genera and 33 families. Dominant lichen genera are *Anthracothecium, Caloplaca, Lecanora, Pertusaria* and *Pyrenula* (Joshi et al. 2011).

Badrinath area situated in the Garhwal region has scanty growth of trees. As the area is devoid of big trees, the vegetation is dominated by herbs, scrubs and bushes. The lichens in the area are widespread on the barren sedimentary rocks, boulders and on the soil. Saxicolous crustose lichen taxa exhibit their dominance with 57 % in the area. *Xanthoria elegans* and *Lecanora muralis* are prominent species of the area (Shukla and Upreti 2007a, b, c).

Nanda Devi Biosphere Reserve spread over in both the regions of the state has unique flora and sustains diverse plant groups including lichens at different altitudes ranging from 1,000 to 1,817 m

(Upreti and Negi 1995). The blue pine forest (*Pinus wallichiana*) occupies the localities between altitudes of 2,100 and 2,400 m. The fruticose lichen species *Usnea longissima* grows profusely together with foliose species *Parmelia* and *Cetraria*. The species of *Peltigera* and *Lobaria* grows luxuriantly on soil.

Mixed forest of *Cupressus–Betula–Taxus* between altitudes of 2,500 and 3,000 m provides suitable substratum for the growth of fruticose lichen; *Usnea longissima* together with many foliose lichen species of the genera *Parmelia*, *Cetraria*, *Cetrelia*, *Evernia*, *Heterodermia* and *Ramalina* species grows on diverse trees.

The Betula–Rhododendron forest (3,000–3,700 m) provides suitable substrates for lichen species of the genera *Cladonia*, *Sticta* and *Lobaria*. *Cladonia* prefers deadwood bark and humus-rich soil. *Lobaria* sp., prefers wet, shaded, woody habitat and *Stereocaulon* prefers rocks and soils.

In alpine meadows the shrubs such as *Cotoneaster*, *Juniperus* and *Rhododendron* provide suitable substrate to most of the lichen taxa. Species of the lichen genera *Heterodermia*, *Parmelia*, *Ochrolechia* and *Pertusaria* prefer to grow on shrubs, while between the altitude of 4,000 and 4,500 m, the meadows rich in moraine support growth of saxicolous lichens species of *Lecanora*, *Nephroma*, *Squamarina*, *Verrucaria* and *Xanthoria* (Upreti and Negi 1995).

Lichen plays an important role in nutrient cycling through litter fall. It has been observed that among different phorophytes, *Quercus semecarpifolia* has the maximum lichen biomass followed by *Acer oblongum* and *Pinus wallichiana* (Kumar and Upreti 2008). Parmelioid lichen communities being bigger in shape and size has maximum lichen litter fall biomass followed by *Usnea* type (Rawat et al. 2011).

4.2.2 Western Dry Region

The states of Punjab, Haryana and Rajasthan in the western dry region have not been much explored for their lichen flora. The Mt. Abu region in the Sirohi district of the state of Rajasthan has been explored exhaustively by Awasthi in the year 1985, and the lichen taxa belonging to lichen genera *Caloplaca* and members of the lichen family Lichinaceae dominate the area.

4.2.3 Gangetic Plain

This region comprises of the largest provinces of the Indian republic, i.e. Uttar Pradesh, Bihar, Jharkhand and West Bengal. Being situated in the plains of the holy river Ganga, the area has fertile land which is being used for agriculture since ancient times. The region has poor forest cover due to dense population and urbanisation in large areas.

This region has scarce growth of lichens due to severe inhibitory factors such as insufficient rainfall, high temperature, long dry seasons in summers and dense human population and vast expansion of cultivated land in Indo-Gangetic plain.

It is interesting to note that cultivated trees of *Mangifera indica*, *Anacardium* sp., *Citrus* sp. and *Artocarpus* sp. growing in orchards and avenue trees along the roadside are much preferred by lichen species of genera *Dirinaria*, *Rinodina* and *Lecanora*.

Certain pollution-tolerant, lime-loving lichen species such as *Endocarpon*, *Phylliscum* and *Peltula* species having in-built tolerance against atmospheric pollution can also withstand extreme desiccations (Upreti 1998) and grow luxuriantly on walls and plasters of building and monuments in urban areas.

Uttar Pradesh. An inventory of lichens in the state of Uttar Pradesh compiled based on the lichen diversity of 15 districts of Uttar Pradesh revealed occurrence of 90 species belonging to 24 families and 33 genera (Nayaka and Upreti 2011). Out of the 15 districts, Bahraich district had maximum diversity of 45 species followed by Lucknow district with 34 species. The state has dominance of crustose lichens, while foliose lichens belonging to genera *Collema*, *Dirinaria*, *Hyperphyscia*, *Parmotrema*, *Phaeophyscia*, *Physcia* and *Pyxine* are common. The state equally has large number of squamulose lichen belonging to the genera *Endocarpon*, *Phylliscum*, *Heppia* and *Peltula*. It is interesting to note that the state exhibits complete absence of fruticose lichens.

West Bengal. The available lichen record from Kolkata dates back to 1865. Kurz described 53 species (19 new species) from Kolkata. In the 1960s, Awasthi revisited the area and found that there was significant loss of lichen flora (Upreti et al. 2005a, b).

Jagadeesh Ram (2006) carried out extensive exploration on lichens growing in Indian part of Sundarbans Biosphere Reserve (Jagadeesh Ram and Sinha 2002). The overall collection yielded a total of 165 species, of which 9 were new to science and 24 were new records for India, along with 28 endemic lichen species, which is a clear evidence for lichen uniqueness in mangrove forests. The members of Arthonioid, Pyrenocarpous group, family Roccellaceae and to some extent Graphidaceae, which prefer smooth bark trees (mangrove), are the most common inhabitants of mangrove forests. Among the foliose lichens, species of *Dirinaria* and *Pyxine* are the most common in the mangrove forest together with *Roccella montagnei* Bél fruticose lichen abundant in the mangrove forests. Occasionally, a few species of *Ramalina* also occur in mangroves. Among the other lichens, some of the common representatives of mangrove forest are *Bacidia*, *Buellia*, *Lecanora*, *Physcia* and *Rinodina*. The species of *Parmelioid* and *Physcioid* (except for *Dirinaria* and *Pyxine*) lichens are rare in mangroves.

4.2.4 Eastern Himalayas

The Eastern Himalayan region of India (northeast India, Sikkim and Darjeeling) exhibits the maximum diversity of most of the plant groups including lichens (Singh 1999). The region is endowed with a rich lichen flora both in luxuriance and species diversity which shows closer affinity with the Sino-Japanese and Southeast Asian countries. The eastern Himalayan region is represented by the occurrence of 843 species; the crustose lichens dominates with 457 species followed by 243 foliose and 143 fruticose taxa. Parmeliaceae with 115 species is the largest family followed by Graphidaceae and Physciaceae (Rout et al. 2010). Shrubs and small trees of *Rhododendron*, *Contoneaster*, *Rosa* and *Juniperus* provide convenient substrata for the corticolous species of the lichen genera, *Cetraria*, *Heterodermia*, *Parmelia*, *Ramalina*, *Usnea*, *Buellia*, *Lecanora*, *Ochrolechia* and *Pertusaria*.

The alpine region exhibits luxuriant growth of saxicolous or terricolous lichens, generally growing intermixed or on decaying mosses and forming a felt-like growth. The species of *Cladonia*, *Stereocaulon*, *Thamnolia* and *Umbilicaria* either grow on rocks or on soil over rocks or meadows. Boulders and stones in small rivulets coming from the glaciers bear plenty of crustose and squamulose lichen species of the genera *Acarospora*, *Rhizocarpon*, *Caloplaca*, *Lecidea* and *Rinodina* (Upreti 1998).

In the Eastern Himalayas especially in the state of Sikkim, *Alnus*, *Michelia*, *Cryptomeria* and *Cupressus* are common phorophytes which show luxuriant growth of corticolous lichens. *Alnus nepalensis* is inhabited by mostly Parmeliod, Graphidaceous and Pyrenocarpous lichens. *Michelia* trees having hardwood with thin smooth surface are preferred by Graphidaceous and Pyrenocarpous lichens. *Cryptomeria* having dry hard bark is preferred by Lecanoroid and other crustose lichens. The bark of *Cupressus* is dry and peels off at maturity; therefore, it rarely allows few lichens to colonise (Bajpai et al. 2011).

Arunachal Pradesh. The floristic account of lichens of few districts of Arunachal Pradesh is available (Dubey et al. 2007). Out of the 843 species known from the Northeast states, Arunachal Pradesh is represented by 112 species (Rout et al. 2004). The Sessa Orchid Sanctuary located in West Kameng district of the state is represented by occurrence of 42 species with dominance of corticolous taxa growing on diverse tree of *Taxus baccata*, *Canarium strictum*, *Ailanthus grandis*, *Ammora wallichi*, *Setreospermum chelonoides*, *Pinus roxburghii*, *Citrus sinensis*, *Albizia* sp. and *Caesalpinia* sp. Among corticolous lichens, Parmeliaceae is the dominant family with 20 species followed by Physciaceae.

The lichen flora of Along town in West Siang district revealed the occurrence of 130 lichen taxa belonging to 55 genera and 26 families. The members of lichen family Physciaceae exhibit their dominance in the area. Graphidaceae and Pyrenulaceae are the dominant families of the area. *Pyxine cocoes* is the widespread species in the area found growing on most of the available substratum. Among phorophytes *Castanopsis indica* having smooth bark exhibits maximum occurrence of lichens followed by *Artocarpus* and *Terminalia myriocarpa* trees (Dubey 2009). *Assam.* In tropical rainforest of Assam, the *Areca catechu* (betel nut) trees are an excellent host for lichen colonisation. Singha (2012) reported an occurrence of 71 lichen species belonging to 32 genera and 15 families on betel nut trees of the area. It was observed that crustose lichen dominates the lichen flora on *Areca*. Graphidaceae followed by Calicaaceae is the dominant family of the area. *Graphis* is the dominant genera followed by *Pyxine, Trypethelium, Arthonia* and *Anthracothecium*. *Dirinaria aegialata* is the most luxuriantly growing species in the region, and it is used for transplant studies in the area.

In another study carried out in a reserve forest in southern Assam, Rout et al. (2010) reported dominance of Graphidaceae and Pyrenulaceae. *Pyrenula, Sarcographa* and *Graphis* accounted for approximately 30 % of the total lichen diversity.

4.2.5 Central India

The central region covers Madhya Pradesh, Chattisgarh, Orissa, Gujrat and portion of Jharkhand. The central states are well known for rich cultural heritage and have many scared grooves and historical monuments of India. Especially Gujrat and Madhya Pradesh being hinterland of sea-route trade in the medieval time have a glorious history, which has been preserved till today as magnificent monuments built of stones. The major construction material of monuments is sandstone and lime plaster which favour good growth of calcicolous lichens (lime loving) represented by species of *Endocarpon* and *Phylliscum* together with members of the family Lichiniaceae (Bajpai et al. 2008).

Madhya Pradesh and Chattisgarh. The forest of Madhya Pradesh and Chattisgarh fall under tropical dry deciduous forest type. Both states have 23.27 % of the total tribal population of India (Goswami 1995). In tropical regions of India, *Shorea robusta* (Serga/sal) trees are the excellent host for lichen growth. In younger (regenerated) sal, the soft and smooth trunks are preferred by *Dirinaria* and *Pertusaria* species, while dense moist places along stream trees show occurrence of Graphidaceous member together with species of Parmelioid species (Upreti 1996; Mohabe 2011). The exposed rock surfaces of Bhimbetka (a rock shelter with paintings) are a suitable abode for the lichen genus *Caloplaca* and other crustose lichen genus (Joshi and Upreti 2007).

Upreti et al. (2005a) enumerated 179 species from Madhya Pradesh and 6 species from Chattisgarh. *Mallotus philippensis* and *Shorea robusta* provide suitable conditions for the growth of lichens. Crustose lichens are common in the area, while Physciaceae is the dominant family of the area with 49 species.

Madhya Pradesh is endowed with rich and diverse forest resources. The Vindhyan region comprised of Katni, Satna, Panna, Umaria, Rewa and Singrauli districts of the state of MP have been extensively surveyed for their lichen wealth, and Mohabe et al. (2010) enumerated occurrence of 44, 40, 37, 34, 30 and 19 species from Katni, Satna, Panna, Umaria, Rewa and Singrauli districts, respectively. *Mangifera indica, Shorea robusta* and *Diospyros melanoxylon* are the common trees of the region, which showed the dominance of members of the family Caliciaceae and Lecanoraceae. *Caloplaca, Lecanora, Pertusaria* and *Peltula* are the dominant genera of the region, while *Pertisaria leioplaca, Rinodina sophodes, Lecanora tropica* and *Peltula euploca* are some of the abundant lichen species.

Mangifera indica is the common phorophyte with the dominance of Physciaceae members. *Pyxine cocoes, Phaeophyscia hispidula, Hyperphyscia adglutinata* and *Rinodina sophodes* are the dominant epiphytic lichens, while *Caloplaca tropica* and *Peltula euploca* are the dominant rock-inhabiting species in the area (Ingle et al. 2012).

Bhopal, capital of Madhya Pradesh, has a distribution of 45 species belonging to 22 genera and 13 families. The lichen family Teloschistaceae with species under single genus *Caloplaca* dominates the area, while Physciaceae and Caliciaceae are also abundant. Along with *Caloplaca* genus *Hyperphyscia*, *Peltula* and *Pertusaria* are widely distributed in the area. Despite heavy anthropogenic activity in and around the city centre, presence of *Hyperphyscia adglutinata* and *Pyxine petricola* exhibits their toxitolerant nature.

Nayaka et al. (2010) described five lichens belonging to the family Roccellaceae from coastal regions of Gujrat. All the species, *Cresponea flava* (Vain.) Egea and Torrente, *Dirina paradoxa* subsp. *africana* (Fée) Tehler, *Enterographa pallidella* (Nyl.) Renger, *Opegrapha arabica* (Müll. Arg.) Vain. and *O. varians* (Müll. Arg.) Vain., are new records for Indian lichen flora.

Achanakmar–Amarkantak area, an interstate biosphere reserve falling in Madhya Pradesh as well as Chattisgarh, has been designated as the 14th Biosphere Reserve by the Government of India in the year 2005. Satya and Upreti (2011) studied the floristic account of lichens from the Biosphere and enumerated species and found maximum coverage of physciod lichen in the area, which indicates human interference in the area.

The Amarkantak Biosphere Reserve and Achanakmar Wildlife Sanctuary show the dominance of crustose lichens. *Collema ryssoleum* (Tuck.) A. Schneider and *Pyxine cocoes* (Swartz) Nyl. were the only foliose lichens in the biosphere. A folicolous lichen is also reported from the biosphere, *Fellhanera semecarpi* (Vainio) Vezda, leaf-inhabiting lichen collected from the area (Nayaka et al. 2007). Two common trees of the biosphere *Mallotus philippensis* (Lam.) Muell Arg. and *Shorea robusta* Gaertn. F are less abundant in the core area of the sanctuary, while *Terminalia cuneata* Roth which grows abundantly and has thick rough and dry peeling bark shows poor lichen diversity.

4.2.6 Western Ghats

Western Ghats, an ecologically sensitive zone declared by the Ministry of Environment and Forests, Government of India, New Delhi, in the year 2009, sustains wide variety of flora and fauna (Bajpai et al. 2013). Western Ghats stretches from Tapti valley in the north to Kanyakumari in the south and covers six provinces of the country; forest vegetation is dense tropical moist broad-leaved trees. A total of 1,155 taxa belonging to 1,136 species, 19 intraspecific taxa belonging to 193 genera and 54 families has been reported so far from the area. Nilgiri Biosphere Reserve has the highest number of species (745). The region is dominated by crustose lichens followed by corticolous lichens. Graphidaceous, Pyrenocarpous and Parmelioid lichen community dominates the lichen flora of the region. *Graphis*, *Pyrenula*, *Parmotrema* and *Usnea* are abundant genus (Nayaka and Upreti 2012).

Goa. Goa is a part of Western Ghats and Cotigao Wildlife Sanctuary is the oldest protected area of the state. About 20 % of the land has been declared protected area which includes four wildlife sanctuaries of the state (Bondla, Cotigao, Bhagwan Mahavir, Salim Ali Bird Sanctuary) (Nayaka et al. 2004).

The Bondla and Bhagwan Mahavir Wildlife Sanctuaries are represented by the occurrence of and recorded 7 and 18 species, respectively (Nayaka et al. 2004), with the dominance of Pyrenocarpous and Graphidaceous lichens.

Cotigao Wildlife Sanctuary also exhibits dominance of Pyrenocarpous and Graphidaceous lichens represented by 14 and 9 species, respectively (Pathak et al. 2004). Among the Pyrenocarpous lichens, *Porina* and a common foliose lichen *Pyxine cocoes* grow abundantly on cashew nut and mango trees.

4.2.7 Eastern Ghats and Deccan Plateau

Eastern Ghats and Deccan plateau encompass Eastern Maharashtra, Karnataka and Tamil Nadu, and southern tips of Chattisgarh and Orissa are less explored for their lichen wealth. Coastal area of Eastern Ghats is a broken chain of hills that extends from Orissa to Tamil Nadu, and because of such topography, the hilly terrain and surrounding planes are densely populated (Shyam et al. 2011).

Maharashtra. Bajpai and Upreti (2011) revealed the occurrence of 65 lichen species belonging to 29 genera and 20 families from Koyna and Mahabaleshwar area from Satara district. Graphidaceae is the dominant family in the area followed by Ramalinaceae, Lecanoraceae and Collemataceae. *Lecanora, Graphis, Pertusaria, Bacidia* and *Caloplaca* show their dominance in the area. *Cladonia scabriuscula, Caloplaca amarkantakana* and *Lepraria lobificans* also grow abundantly on rocks in the area.

Palm trees in coastal region provide a suitable habitat for Graphidaceous and Pyrenocarpous lichens along with some members of *Physcia, Dirinaria* and *Pyxine*. A fruticose lichen *Roccella montagnei* is very luxuriantly growing in the eastern coast of India on all available substrates in the coastal region (Upreti 1998).

Tamil Nadu. Mohan and Hariharan (1999) studied the distribution pattern of lichens in Pichavaram mangroves in Tamil Nadu and reported ten species *Buellia* sp., *B. montana* Magnusson, *Dirinaria confluens* (Fr.) D.D. Awasthi, *D. consimilis* (Stirton) D.D. Awasthi, *Graphis* sp., *G. scripta* (L.) Ach., *Lecanora* sp., *Pyrenula* sp1, *Pyrenula* sp2 and *Roccella montagnei* Bél. The study showed the dominance of crustose lichens in the forest. The sites exposed to heavy anthropogenic disturbances favoured luxuriant growth of lichens due to increased availability of light and moisture from sea breeze.

Lichens present in the mangroves are termed as manglicolous lichens. The population of lichens on mangroves is low when compared to the lichens of terrestrial ecosystems, as their growth is arrested by the high level of salinity and moisture.

Lichen flora in Pichavaram and Muthupet mangroves in southeast coast of India is represented by the occurrence of 21 species belonging to 14 genera and 10 families found growing on *Avicennia*, *Rhizophora* and *Excoecaria* vegetation (Logesh et al. 2012).

The Kollihills of Tamil Nadu have 48 species, 23 genera belonging to 12 lichen families, of which *Heterodermia, Parmotrema* and *Pertusaria* dominate the lichen flora (Shyam et al. 2011).

4.2.8 Andaman and Nicobar Islands

Corticolous genera belonging to Graphidaceous and Pyrenocarpous groups are the major lichens of this region. Mostly the original forest cover in Andaman has been thinned out (Upreti 1998).

4.3 Key for Identification of Toxitolerant and Common Lichens in Different Lichenogeographical Regions

Keys consist of sequential pairs of parallel, but opposing, statements that can be compared against its growth form collected in different regions. To identify a lichen species, begin with the first statement or 'lead' in the key and select the statement (i.e. 1a or 1b) that most accurately describes the specimen in hand. Next lead is selected based on the characteristic features matching with the genus in question and process is repeated till appropriate genus is identified and then same process is followed for the species keys. The end point in the keying process is reached when the selected lead yields a species name. If it does not, then the process must be repeated to determine where a wrong turn was taken. It may prove helpful to jot down the identification sequence so as to retrace it more quickly. Many species are keyed out at more than one location: where a specimen seems well described by both leads of a pair, it can usually be looked for under both leads (Goward et al. 1994).

Key to Genera

1	Thallus crustose, leprose	2
1(a)	Thallus squamulose, foliose or fruticose	21
2	Thallus leprose	3
2(a)	Thallus crustose	4

3	Apothecia absent, K+, C−, Pd+ orange	**Lepraria**
3(a)	Asci arrested bitunicate 8 spored, spore colourless	**Chrysothrix**
4	Ascocarps mazedium	**Calicium**
4(a)	Ascocarp otherwise	5
5	Ascocarp perithecia	6
5(a)	Ascocarp otherwise	10
6	Perithecia in stroma	**Trypethelium**
6(a)	Perithecia not in stroma, single	7
7	Ascospore simple, mostly growing on rocks or hard soil	**Verrucaria**
7(a)	Ascospore septate, hyaline or brownish	8
8	Asci 6–8 spored, spore colourless	**Porina**
8(a)		9
9	Spores 3-septate	**Pyrenula**
9(a)	Spore multicelled muriform, primary transverse septa darker	**Anthraco-thecium**
10	Ascocarps apothecia, regular	11
10(a)	Ascocarps modified, or otherwise irregular	17
11	Apothecia immersed in fertile verrucae	**Pertusaria**
11(a)	Apothecia not immersed in fertile verruca	12
12	Apothecia with thalline margin	13
12(a)	Apothecia lacking thalline margin	15
13	Ascocarps brown 1-septate	**Rinodina**
13(a)	Ascospores colourless	14
14	Asci multispored	**Acarospora**
14(a)	Asci up to 8 spored	**Ochrolechia**
15	Ascospore simple, hyaline	**Lecidea**
15(a)	Ascospore transversely septate or muriform	16
16	Ascospores transversely septate colourless	**Bacidia**
16(a)	Ascospores muriform, smoky brown or brown	**Rhizocarpon**
17	Ascocarps modified into lirellae	18
17(a)	Ascocarps irregular, not modified into lirellae	**Arthonia**
18	Lirellae in stroma	**Sarcographa**
18(a)	Lirellae not in stroma	19
19	Asci clavate to subcylindrical up to 8 spored	**Graphis**
19(a)	Asci otherwise	20
20	Asci bitunicate globular-elongate 4–8 spored	**Opegrapha**
20(a)	Asci thick walled, 8 spored, spore colourless	**Enterographa**
21.	Thallus squamulose	22
21(a)	Thallus otherwise	27
22	Thallus with cyanobacterial photobiont	23
22(a)	Thallus with green algal photobiont	24
23	Asci multispored	**Peltula**
23(a)	Asci 8 spored	**Phylliscum**
24	Ascocarps perithecia	**Endocarpon**
24(a)	Ascocarps apothecia	25
25	Thalline margin with alga not up to the top; with parietin as chemical	**Caloplaca**
25(a)	Thalline margin with alga up to the top	26
26	Asci with non-amyloid to faintly amyloid tholus	**Squamarina**
26(a)	Asci with distinct amyloid tholus, 8-spored ascospore colourless usnic acid present in upper cortex (K−)	**R hizoplaca**

4.3 Key for Identification of Toxitolerant and Common Lichens...

27	Thallus foliose	28
27(a)	Thallus fruticose or dimorphic	52
28	Thallus and apothecia yellow	29
28(a)	Thallus otherwise, grey, white grey, bluish grey, olive green	30
29	Upper surface orange, K+ purple	**Xanthoria**
29(a)	Upper surface greenish yellow, K−	**Candelaria**
30	Thallus with cyanobacterial photobiont (*Nostoc*)	31
30(a)	Thallus with photobiont green or blue green	32
31	Thallus corticated on one or both the surfaces	**Leptogium**
31(a)	Thallus ecorticated or with pseudocortex	**Collema**
32	Thallus umbilicate	33
32(a)	Thallus not umbilicate	34
33	Ascocarps perithecia immersed in upper surface, spore simple ellipsoid colourless 8 per ascus	**Dermato-carpon**
33(a)	Ascocarps apothecia, apothecial disc even or variously fissured or with central protruding button	**Umbilicaria**
34	Thallus pseudocyphellate corticated on both sides	35
34(a)	Thallus not pseudocyphellate, corticated on one side or both	39
35	Thallus pseudocyphellate on upper and lower side	**Cetrelia**
35(a)	Pseudocyphaellae either on upper or on lower side	36
36	Pseudocyphaellae on upper side	37
36(a)	Pseudocyphaellae on lower side	38
37	Pseudocyphaellae linear, effigurate or punctuate, upper side often white	**Parmelia**
37(a)	Pseudocyphaellae punctiform to suborbicular	**Pyxine**
38	Apothecia marginal ± nephromoid, photobiont *Nostoc* or *Trebouxia*	**Nephroma**
38(a)	Apothecia otherwise	39
39	Thallus corticated only on upperside	40
39(a)	Thallus corticated on both side	42
40	Cortex composed of longitudinally oriented thick-walled conglutinate hyphae	41
40(a)	Cortex paraplectenchymatous, thallus large with veins on lower side, photobiont *Nostoc* or *Trebouxia*	**Peltigera**
41	Cortex K+ yellow, atranorin present	**Heterodermia**
41(a)	Cortex K−, atranorin absent	**Anaptychia**
42	Ascospores brown 2-septate	43
42(a)	Ascospore colourless, simple or transversely septate	48
43	Hypothecium brown to pale brown	44
43(a)	Hypothecium colourless to pale yellow	45
44	Rhizinae absent on lower side epithecium K−	**Dirinaria**
44(a)	Rhizinae present, epithecium K+ purple	**Pyxine**
45	Thallus ± adglutinate to substratum pycnoconidia more than 10 µm long rhizinae absent	**Hyperphyscia**
45(a)	Thallus adnate, rhizinae present	46
46	Upperside purinose Ascospores *Physconia*-type	**Physconia**
46(a)	Upperside epurinose or purinose, ascospore *Physcia* or *Puechysporaria*-type	47
47	Upper cortex K+ yellow, atranorin present	**Physcia**
47(a)	Upper cortex K−, atranorin absent	**Phaeophyscia**
48	Ascospores transversely septate, photobiont *Nostoc* or *Trebouxia*	**Lobaria**
48(a)	Ascospores simple	49

49	Thallus yellow green, upper cortex K−, usnic acid present.	50
49(a)	Thallus grey to dark grey, upper cortex K+ yellow	51
50	Thallus lobes round to subrotund	**Flavopar-melia**
50(a)	Thallus lobes elongate	**Xanthopar-melia**
51	Lobes large lower side with wide marginal zone lacking rhizinae	**Parmotrema**
51(a)	Lobes narrower, rhizinae up to the margin	**Parmelina**
52	Thallus dimorphic	53
52(a)	Thallus uniform not dimorphic	54
53	Thallus centrally hollow	**Cladonia**
53(a)	Thallus solid.	**Stereocaulon**
54	Thallus podetia like, vermiform, milky white	**Thamnolia**
54(a)	Thallus not vermiform, cylindrical	55
55	Thallus flat to strap shaped	56
55(a)	Thallus cylindrical	**Usnea**
56	Thallus with cyanobacterial photobiont, *Chroococcus*	**Lichinella**
56(a)	Thallus with green photobiont.	57
57	Thallus with chondroid tissue	**Ramalina**
57(a)	Thallus without chondroid tissue.	58
58	Cortex double layered outer paraplectenchymatous, inner prosoplectenchymatous.	**Cetraria**
58(a)	Cortex otherwise	59
59	Cortex of palisade hyphae at right angle to longitudinal axis	**Roccella**
59(a)	Cortex composed of erect or horizontally oriented hyphae.	**Evernia**

Key to Species

1	Thallus crustose, leprose	2
1(a)	Thallus otherwise.	12
2	Thallus leprose, powdery, sorediate.	**Lepraria lobificans**
2(a)	Thallus crustose.	3
3	Apothecia with thalline margin	4
3(a)	Apothecia lacking thalline margin.	8
4	Asci multispored, ascospore small	**Acarospora gwynii**
4(a)	Asci 8 spored, ascospore bigger	5
5	Ascospore brown, 1-septate.	6
5(a)	Ascospore simple.	7
6	Apothecial disc brown to dark brown	**Rinodina sophodes**
6(a)	Apothecial disc olive green	**Rinodina olivaceobrunnea**
7	Lobes 2–4 mm long greenish yellow to yellow brown areolate or lobate at centre . . . . **Lecanora muralis**	
7(a)	Younger apothecia immersed in thallus becoming sessile apothecial disc pale orange to yellowish brown	**Lecanora leprosa**
8	Thallus and apothecia or thallus yellow orange.	9
8(a)	Thallus and apothecia not yellow	10
9	Thallus isidiate, spore polariocular	**Caloplaca bassiae**
9(a)	Thallus not isidiate, spore muriform	**Rhizocarpon geographicum**
10	Ascospores muriform, brown	**Rhizocarpon flavum**
10(a)	Ascospore transversely 1-septate, brown.	11

11	Apothecia not pruinose medulla I+ blue or I–.	**Buellia aethalea**
11(a)	Apothecia pruinose, medulla I–.	**Buellia meiosperma**
12	Thallus squamulose	13
12(a)	Thallus foliose or fruticose	18
13	Thallus and apothecia or thallus yellow orange.	14
13(a)	Thallus and apothecia or thallus not yellow orange	15
14	Thallus saxicolous, suborbicular lobes radiating, densely crowded apothecia on central part	**Xanthoria elegans**
14(a)	Thallus up to 10 cm in diameter, lobes 0.5–2.0 (–3.5) mm wide, concave.	**Xanthoria parietina**
15	Thallus with cyanobacterial photobiont.	16
15(a)	Thallus with green photobiont.	17
16	Thallus peltate, sorediate, ascus multispored	**Peltula euploca**
16(a)	Thallus not peltate, esorediate, ascus 8–16 spored	**Phylliscum indicum**
17	Thallus with apothecia, K+ yellow orange	**Lobothallia praeradiosa**
17(a)	Thallus with perithecia, K–.	**Endocarpon subrosettum**
18	Thallus foliose	19
18(a)	Thallus fruticose or dimorphic.	35
19	Thallus umbilicate, rhizinate on lower side	**Dermatocarpon vellereum**
19(a)	Thallus not umbilicate	20
20	Thallus yellow orange	**Candelaria concolor**
20(a)	Thallus not yellow orange	21
21	Ascospores brown, 1-septate transversely	22
21(a)	Ascospore simple, hyaline.	32
22	Lower side lacking rhizinae, thallus tightly adpressed to substratum	23
22(a)	Lower side with rhizinae	26
23	Thallus sorediate	24
23(a)	Thallus lacking soredia	**Dirinaria confluens**
24	Soralia on apex of isidioid dactyls.	**Dirinaria aegialata**
24(a)	Soralia on lamina of thallus, capitate.	25
25	Thallus with divaricatic acid	**Dirinaria applanata**
25(a)	Thallus lacking divaricatic acid	**Dirinaria consimilis**
26	Epithecium K+ purple	27
26(a)	Epithecium K–.	29
27	Thallus sorediate	28
27(a)	Thallus lacking soredia, UV+ yellow	**Pyxine petricola**
28	Soralia laminal white, upper surface with plaques of pruina and pseudocyphellate	**Pyxine cocoes**
28(a)	Soralia marginal yellow, non-pseudocyphellate	**Pyxine subcinerea**
29	Thallus K–, atranorin absent	30
29(a)	Thallus K+, atranorin present	31
30	Thallus larger up to 10 cm across, lobes 2–3 (–5) mm wide	**Phaeophyscia hispidula**
30(a)	Thallus smaller, up to 3.0 cm, across lobes 1.0 mm wide	**Phaeophyscia orbicularis**
31	Thallus suborbicular, lobes	**Heterodermia diademata**
31(a)	Thallus rosulate.	**Heterodermia incana**
32	Thallus lower side rhizinate up to the margin	34
32(a)	Rhizinae restricted only in the central part of the lower surface	33

33	Thallus C−, KC, P−	**Parmotrema praesorediosum**
33(a)	Thallus C+ red, KC+ red	**Parmotrema austrosinesis**
34	Rhizinae dichotomously branched	**Remototrachyna awasthii**
34(a)	Rhizinae simple	**Parmelinella wallichiana**
35	Thallus fruticose, cylindrical	36
35(a)	Thallus dimorphic	**Cladonia praetermissa**
36	Thallus filaments long, evanescent	**Usnea longissima**
36(a)	Thallus filaments short and persistent	**Usnea antarctica**

4.4 Bioindicator Lichen Species

Many lichens are habitat specific and thus a diversity of lichens at a site indicates habitat heterogeneity. Having the close relationship of lichens with other organisms, and their contribution to biodiversity, lichens provide an ideal group to monitor for changes in diversity in ecosystems. Lichens are being utilised in three ways to monitor changes in the air quality on the ecosystems which include elemental analysis, mapping of lichen diversity and transplant studies (Hale 1983). The most common approach involves floristic survey followed by quantification of pollutants accumulated in the lichen thalli (St. Clair 1989). The kind and the level of pollution to be monitored in a research area should also be taken into consideration in selection of the species of lichens that will be used as a biomonitors as accumulation of specific pollutant is largely dependent on the morphology and physiological requirements of the specific species (Riga-Karandinos and Karandinos 1998). Foliose lichens are better accumulators in comparison with fruticose ones (St. Clair et al. 2002a, b). Numerous investigations of the interaction of air pollution and lichens performed within the last three decades reveal that lichens may be assigned to three categories in terms of their responses to air Pollution (Garty et al. 2003): (1) sensitive species, with varying degrees of sensitivity to the detrimental effects of pollutants, but ultimately succumbing to air pollution; (2) tolerant species, resistant to pollution, belonging to the native community and remaining intact in their native habitat; and (3) replacement species, making their appearance after destruction of the major part of the native lichen community as a result of natural and/or man-made changes. Therefore, occurrence of specific lichen may be used in ecosystem monitoring as an indicator of alteration in the natural condition due to contaminants released by nature or due to various anthropogenic activities.

In India, lichen diversity varies due to the vast geographical area. Different regimes of humidity, temperature and rainfall support specific lichens to thrive. Therefore, there are different lichen species (based on their distribution and sensitivity) which may be utilised as an indicator species, and their indicator value may further be employed in air quality studies of an area of interest. The following are the predominant species in different climatic zones of the country based on data available on biomonitoring studies and lichen diversity studies carried out till now.

1. *Dirinaria consimilis* (Stirton) D.D. Awasthi (Fig. 4.6a)

    Description: A foliose grey-white, corticolous lichen commonly growing on trees as a circular tightly adpressed thallus 8 (−13) cm across 1–2 mm-wide lobes flabellate, plicate, upper side capitate soralia with granular soredia. Apothecia to 1.5 mm in diameter and ascospores 14–23 × 6–8 µm 1-septate brown ascospores.

    Chemistry: Medulla K−, C−, P−. Sekikaic acid present

    Habitat: Mostly corticolous, rarely saxicolous

    Distribution: Tropical areas

    Indicator Value: Tolerant; potential biodeteriorating species occurring in monuments of central India (Bajpai et al. 2008)

2. *Dirinaria applanata* (Fée) D.D. Awasthi (Fig. 4.8b)

    Description: A foliose grey-white, corticolous lichen commonly growing on trees as a circular tightly adpressed thallus plicate-rugose 1-septate brown ascospores.

4.4 Bioindicator Lichen Species

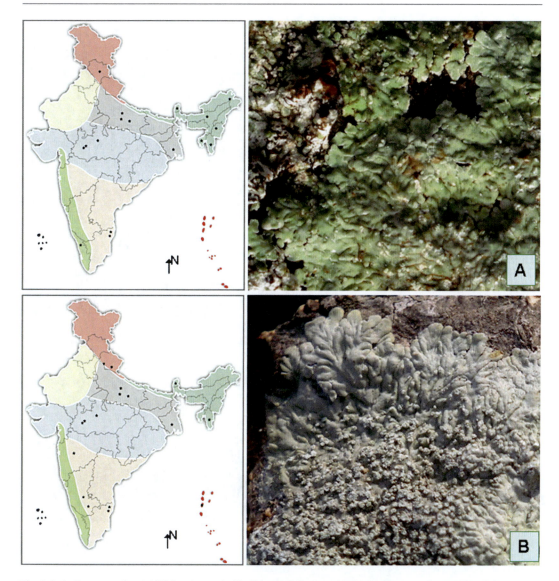

Fig. 4.6 Indicator species: (**a**) *Dirinaria consimilis* (Stirton) D.D. Awasthi; (**b**) *Dirinaria applanata* (Fée) D.D. Awasthi

Chemistry: Divaricatic acid, atranorin and triterpenoid present
Habitat: Mostly corticolous, rarely saxicolous
Distribution: Tropical areas
Indicator Value: Tolerant; potential biodeteriorating species occurring in monuments of central India (Bajpai et al. 2008)

3. *Remototrachyna awasthii* (Hale & Patw.) Divakar & A. Crespo (formally *Hypotrachyna awasthii* Hale & Patwardhan) (Fig. 4.7a)
Description: A foliose grey-white corticolous lichen commonly growing on trees, twigs and branches. More or less free-lobed thallus adnate, to 10 cm across with 5–10 mm-wide apically rotund lobes. Upper side grey; isidiate; isidia simple to branched, often black tipped; lower side black with simple and dichotomously branched rhizinae; medulla white. Apothecia to 2.5 mm in diameter; ascospores 10 × 8 µm, simple colourless.
Chemistry: Medulla K+ yellow turning red, C−, P+ red. Norstictic and salazinic acids present
Habitat: Mostly corticolous

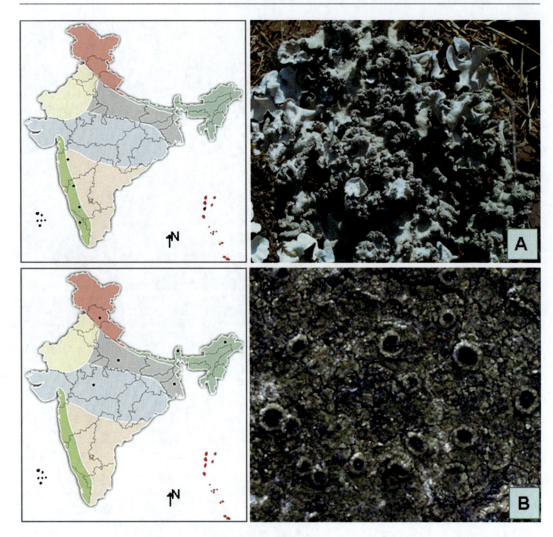

Fig. 4.7 Indicator species: (**a**) *Remototrachyna awasthii* (Hale & Patw.) Divakar & A. Crespo; (**b**) *Rinodina sophodes* (Ach.) Massal

Comments: The species has a wide distribution in South and Southeast Asia and found frequently growing on twigs and trunks of trees and shrubs.

Distribution: Tropical areas

Indicator Value: Tolerant (Bajpai et al. 2013)

4. *Rinodina sophodes* (Ach.) Massal. (Fig. 4.7b)

Description: A crustose grey to dark brown verrucose-areolate thalloid lichen with dark prothallus. Photobiont a green alga (Trebouxia). The apothecia are sunken to sessile, lecanorine or lecideine, 0.5–1 mm diameter, disc dark to black, hypothecium colourless, rarely brownish. Paraphyses simple, capitate and with end cells thickened forming brown to reddish-brown or blue-green epithecium; asci *Lecanora* type, 8 spored, spores 13–16 × 7–8 μm, brown and 1-septate.

Chemistry: Cortex I+ blue

Habitat: Crustose, saxicolous, terricolous, muscicolous or corticolous

Distribution: Tropical areas

Indicator Value: Toxitolerant; abundant at sites influenced by heavy vehicular and industrial pollution. Physiological estimation and PAHs studies have been carried out (Satya and Upreti 2009; Satya et al. 2012)

4.4 Bioindicator Lichen Species

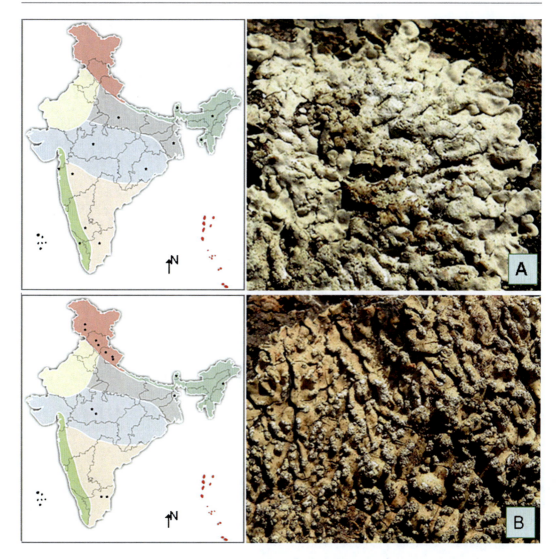

Fig. 4.8 Indicator species: (**a**) *Pyxine cocoes* (Sw.) Nyl.; (**b**) *Pyxine subcinerea* Stirton

5. *Pyxine cocoes* (Sw.) Nyl. (Fig. 4.8a)
   Description: A yellowish-grey foliose lobate thallus of 6 cm across; maculae laminal and marginal turning into pseudocyphellae and then into soralia; medulla stramineous. Apothecia to 1 mm in diameter; thalline margin soon blackened and excluded; internal stipe brown, K+ red violet; ascospore (12) 16–20×6–8(–10) μm.
   Chemistry: Upper cortex UV+ yellow; medulla K−, C−, P−. Lichexanthone and triterpenes present
   Habitat: Corticolous rarely saxicolous
   Distribution: A common lichen from subtropical to temperate regions
   Indicator Value: A potential bioindicator species employed for passive as well as transplant studies in India and other countries of Southeast Asia (Saxena et al. 2007; Wolseley et al. 1994)

6. *Pyxine subcinerea* Stirton (Fig. 4.8b)
   Description: A foliose lobate pseudocyphellate thallus of 7 cm across with 1–2 mm-wide lobes; upper side greyish; margins intermittently pseudocyphellate; pseudocyphallae developing into soralia and

spreading on to lamina; soredia white to stramineous; medulla yellow. Apothecia 1–2 mm in diameter, exciple pseudothalline in young stages, later black; internal stipe not well differentiated; ascospores 12–20 (22) × 6–8 µm, brown, 1-septate.

Chemistry: Upper cortex UV+ yellow; medulla K–, C–, P–.Lichexanthone in cortex and triterpenes in medulla

Habitat: Corticolous rarely saxicolous

Distribution: A common lichen of subtropical and temperate areas found growing on trees and sometimes on rocks

Indicator Value: Toxitolerant (Shukla and Upreti 2008; Shukla et al., unpublished, 2013)

7. *Phaeophyscia hispidula* (Ach.) Essl. (Fig. 4.9a)

Description: A foliose grey-brown lobate thalloid lichen with rounded apices grey brown on upper side; laminal, capitate soralia, often extending up to margin; soredia rarely becoming granular; black lower side rhizinae long, black, projecting beyond lobes; medulla white. Apothecia to 3 mm in diameter, coronate; ascospores, 18–27 (–30) × 9–12 (–15) µm, brown, 1-septate.

Chemistry: No lichen substance in colour test and TLC

Habitat: Corticolous, terricolous and saxicolous

Distribution: A common foliose lichen found growing on tree trunk on rocks, man-made artefacts, soil in tropical, subtropical, temperate and alpine areas

Indicator Value: Toxitolerant suitable bioindicator species for PAHs and heavy metal accumulation (Shukla and Upreti 2007a, b, c, 2009; Shukla et al. 2010, 2011, 2012a, b; Rani et al. 2011)

8. *Phaeophyscia orbicularis* (Neck.) Moberg (Fig. 4.9b)

Description: Thallus corticolous or saxicolous, orbicular, to 3 cm across; lobes 1 mm wide; upper usually grey brown, soralia laminal to marginal, capitate; soredia white to yellowish; lower side black, rhizinae projecting beyond lobes; medulla white. Apothecia to 1.5 mm in diameter, coronate; ascospores *Physcia*-type, 18–26 × 7–11 µm

Chemistry: Rarely Skyrin present

Habitat: Corticolous, terricolous and saxicolous

Distribution: A common foliose lichen found growing on tree trunk on rocks, man-made artefacts, soil in tropical, subtropical, temperate and alpine areas

Indicator Value: Toxitolerant suitable bioindicator species for various pollutants especially PAHs accumulation (Shukla et al. 2010)

9. *Parmotrema praesorediosum* (Nyl.) Hale (Fig. 4.10a)

Description: A foliose lobate lichen with 10 cm across adnate thallus; lobes 5–8 mm wide, eciliate; upper side grey to darker, emaculate, soralia usually marginal, linear or crescent-shaped; soredia granular; lower side centrally black, narrow marginal zone lighter tan, nude; medulla white. Apothecia to 4 mm in diameter, imperforate; ascospores 15–21 × 7–10 µm, simple colourless.

Chemistry: Medulla K–, C–, KC–, P–. Protopraesorediosic, praesorediosic and fatty acid present

Habitat: Corticolous or saxicolous

Distribution: A common foliose lichen found growing on trees, rocks in subtropical and temperate areas

Indicator Value: Intermediate (Shukla 2012)

10. *Parmelinella wallichiana* (Taylor) Elix & Hale (Fig. 4.10b)

Description: A foliose lobate lichen with 6–20 cm across thallus adnate to loosely adnate. Lobes subdichotomously to irregularly branched, sinuate, (2) 3–10 mm wide (rarely to 15 mm wide), apices rotund to subrotund; margin entire to crenate, eciliate or ciliate. Cilia sparse short, especially in the axils; cilia about 0.5 mm long. Upper surface mineral grey. Isidia cylindrical, simple to branched, brown tipped, 0.1–1.0 mm long. Medulla white. Lower surface black, with naked or papillate,

4.4 Bioindicator Lichen Species

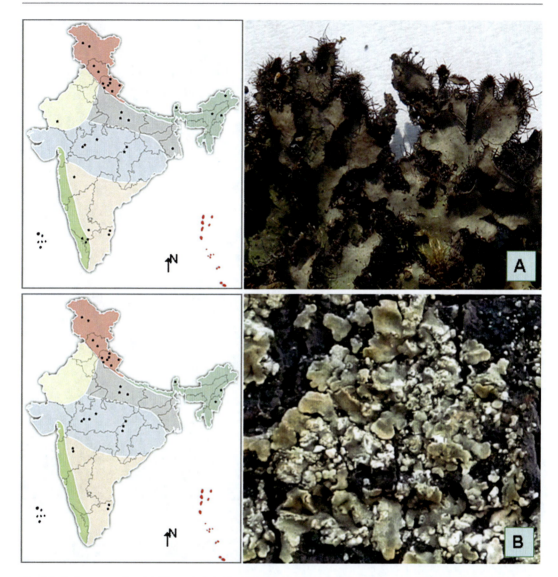

Fig. 4.9 Indicator species: (**a**) *Phaeophyscia hispidula* (Ach.) Essl.; (**b**) *Phaeophyscia orbicularis* (Neck.) Moberg

dark brown, 2–5 mm wide zone near lobe apices, sparsely rhizinate. Rhizinae mostly in the centre of the thallus, simple. Apothecia adnate, constricted at base, 2–12 mm in diameter; disc dark brown, imperorate; amphithecium smooth to faintly rugulose. Spores 8–20×4–13 μm, simple, colourless, oval ellipsoid.

Chemistry: Cortex K+ yellow; medulla K+ yellow turning red, C−, KC−, P+ orange red. TLC: Atranorin, Salazinic acid and consalazinic acid.

Habitat: Rock, boulders and soil

Distribution: A common species found growing on rocks, boulders, trees and soil in temperate and subtropical areas

Indicator Value: Intermediate (Dubey 2009; Shukla and Upreti 2011)

11. *Candelaria concolor* (Dicks.) Arnold (Fig. 4.11a)

Description: A small lobed foliose lichen with bright yellow thallus, forming small rosettes (less than 1 cm across) of overlapping lobes; lobes only 0.1–0.5 mm

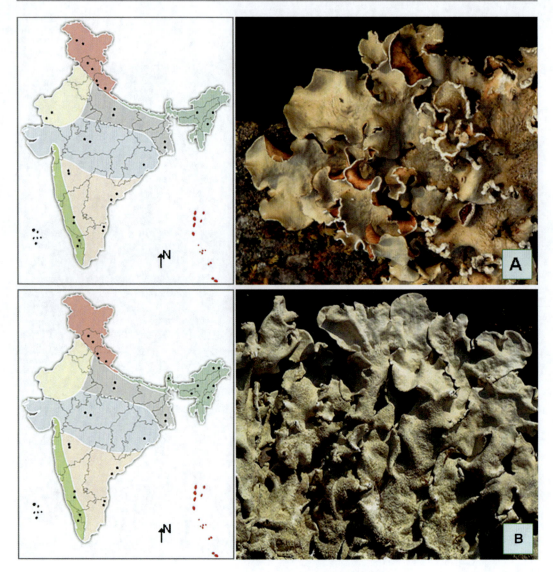

Fig. 4.10 Indicator species: (**a**) *Parmotrema praesorediosum* (Nyl.) Hale; (**b**) *Parmelinella wallichiana* (Taylor) Elix & Hale

across with lacy margins edged with granules and granular soredia, rarely almost entirely dissolving into granular soredia. Apothecia usually absent.

Chemistry: Cortex PD−, K+ pink or K−, KC−, and C−. Apothecial discs often K+ pink (yellow pigments are calycin and other compounds related to pulvinic acid). Medulla, no reactions

Habitat: Extremely common and widespread, especially on nutrient-rich substrates, often forms luxuriant colonies along rain tracks on tree trunks or on twigs of trees in temperate areas

Distribution: A common species found growing on trees and rocks in temperate areas

Indicator Value: Tolerant (Shukla and Upreti 2011)

12. *Heterodermia diademata* (Tayl.) D.D. Awasthi (Fig. 4.11b)

Description: A foliose laciniate smaller thalloid than Parmelioid lichens 15 cm across, branched thallus; lobes linear, to 2.5 mm wide, rarely secondary lobules in central

4.4 Bioindicator Lichen Species

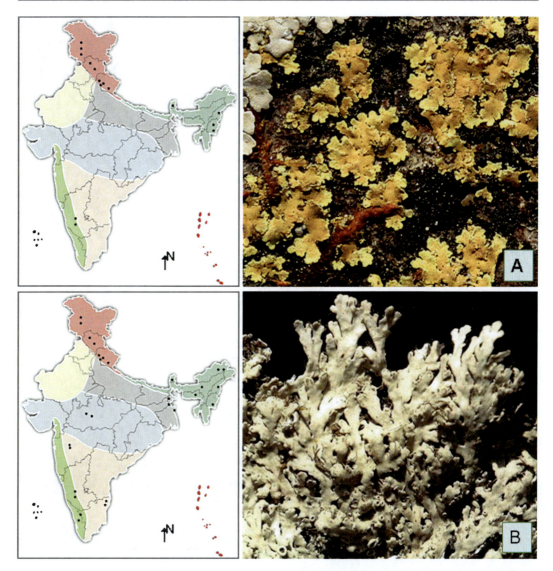

Fig. 4.11 Indicator species: (**a**) *Candelaria concolor* (Dicks.) Arnold; (**b**) *Heterodermia diademata* (Tayl.) D.D. Awasthi

part, corticated on both sides; upper side grey to grey white, lacking isidia and soredia; lower side pale brown with concolourous, sparse rhizinae. Apothecia to 7 mm in diameter, cortex of receptacles I– (negative); ascospores (16–) 22–32 (–40) × 10–18 μm, brown, 1-septate lacking sporoblatidia.

Chemistry: Medulla K+ yellow, C–, P+ pale yellow or P–. Zeorin present

Habitat: Corticolous, terricolous and saxicolous

Distribution: One of the most omnicolous lichen species found growing on bark of trees, shrubs, bushes, rocks, soil over rock and other man-made substrates in the temperate

Indicator Value: Tolerant (Bajpai and Upreti 2011; Shukla and Upreti 2011)

13. *Lecanora muralis* (Schreb.) Rabenh. Em. Poelt (Fig. 4.12a)

Description: A saxicolous squamulose lichen with rosettiform, to 3 cm across; often confluent areolate thalli with apothecia;

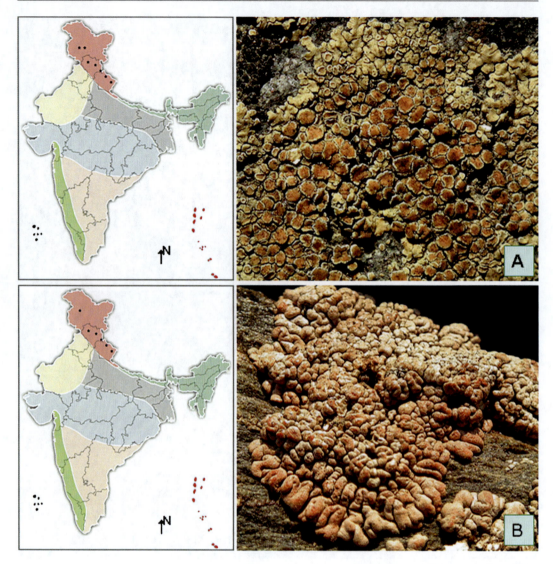

Fig. 4.12 Indicator species: (**a**) *Lecanora muralis* (Schreb.) Rabenh. Em. Poelt; (**b**) *Lobothallia praeradiosa* (Nyl.) Hafellner

marginal lobes to 4 mm long, 1.5 mm wide, bluish-grey-margined pruinose. Upper side yellowish green to grey. Apothecia dense, to 1.5 mm in diameter; disc yellowish to reddish brown, epruinose; ascospores 9–14×4–7 µm, simple and colourless.

Chemistry: Thallus KC+ yellowish. Usnic acid leucotylin and zeorin

Habitat: Saxicolous

Distribution: A common saxicolous lichen growing on exposed rocks both in moist and dry habitats, forming extensive circular patches on substrates in alpine and higher temperate regions

Indicator Value: Tolerant (Shukla and Upreti 2007a, b, c; Shukla 2007)

14. *Lobothallia praeradiosa* (Nyl.) Hafellner (Fig. 4.12b)

Description: A squamulose lichen with radiating laciniate brownish to reddish-grey thallus, convex but not rounded and not hollow in cross section. Apothecia not seen.

Chemistry: K+ yellow, C–, KC–, P+ yellow orange. Norstictic acid

4.4 Bioindicator Lichen Species

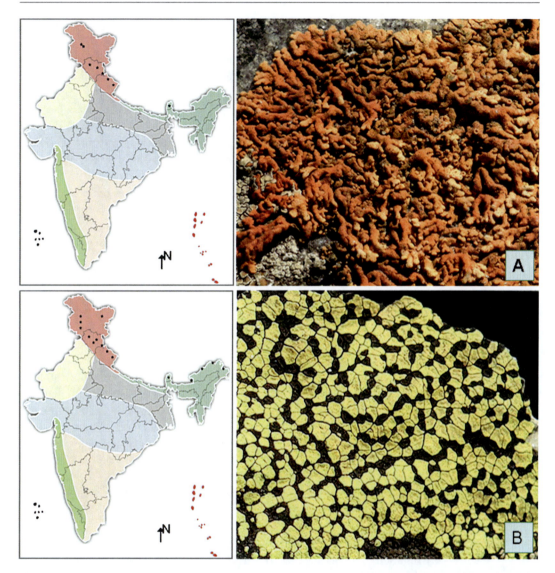

Fig. 4.13 Indicator species: (**a**) *Xanthoria elegans* (Link) Th. Fr.; (**b**) *Rhizocarpon geographicum* (L.) DC. in. Lam & DC.

Habitat: Saxicolous

Distribution: The species grow frequently on rocks or on soil over rocks near glacier and exposed areas in alpine and higher temperate areas.

Indicator value: Intermediate; PAHs (Shukla et al. 2010)

15. *Xanthoria elegans* (Link) Th. Fr. (Fig. 4.13a)

Description: A saxicolous squamulose lichen with suborbicular, to 3 (–7) cm across lobate thallus; lobes radiating compact, 0.25–01 mm wide, convex, nodulose with densely crowded apothecia in central part; upper side orange red to reddish brown, lacking isidia and soredia; lower side grey; medulla white, ± hollow. Apothecia 1 mm in diameter; ascospores 12–16 (–18)×6–8 (–10) μm with 4–5 μm-thick transverse septum, colourless.

Chemistry: K+ red purple C–, KC+, P–. Parietin (major) with minor amounts of emodin, teloschistin, fallacinal, fallacinol, xanthorin, parietinic acid, erythroglaucin, confluentic acid, notoxanthin, lutein, cryptoxanthin and â-carotene.

Habitat: Saxicolous

Distribution: The most common lichen found growing on exposed boulders and rocks in both moist and cold desert alpine and higher temperate areas

Indicator value: Tolerant (Shukla 2007)

16. *Rhizocarpon geographicum* (L.) DC. in. Lam & DC. (Fig. 4.13b)

    Description: A crustose lichen with areolate, verrucose or uniform, white, yellow or brown thallus, upper surface with corticiform gelatinised layer. Photobiont a protococcoid green alga. Apothecia black, circular or angular, usually between areoles, lecideine, hymenium colourless, reddish or greenish. Hypothecium dark. Paraphysis branched, net-like anastomosing, conglutinate, capitate, tips darkened.

    Chemistry: Medulla I+ blue, K–, P+ yellow, epithecium K+ reddish, or greenish

    Habitat: Saxicolous

    Distribution: The most common lichen found growing on exposed boulders and rocks mostly near the summit of glaciers, stones of rivulets in alpine areas and higher temperate regions

    Indicator Value: Tolerant; to reconstruct paleoclimate (lichenometry) (Joshi et al. 2012)

17. *Cladonia praetermissa*. A.W. Archer (Fig. 4.14a)

    Description: A dimorphic lichen with twofold thallus, squamules of primary thallus medium to large sized, crenate, persistent. Podetia 5 (–10) mm tall, 0.5 mm thick at base, simple, subulate, always escyphose; rarely brown hymenial discs on tips. Podetial surface corticated, squamulose at base; ecorticated and sorediate at apices.

    Chemistry: Podetia K+ weakly yellow or K–, KC–, P+ red. Atranorin, fumarprotocetraric acid, protocetraric acid, and rarely psoromic acid

    Habitat: Terricolous

    Distribution: A rarely occurring lichen in the temperate regions

    Indicator Value: Intermediate; undisturbed soil ecosystem (Bajpai et al. 2008)

18. *Dermatocarpon vellereum* Zschacke (Fig. 4.14b)

    Description: An umbilicate foliose lichen with usually monophyllus, to 12 cm across, umbilicate, 200–450 µm thick in marginal area, 600–1,000 µm thick in central part rather thick, leathery thallus; upper side light brownish red, white to dark pruinose; lower side black, with dense, thick, stumpy, coralloid rhizomorphs. Perithecia pale red; ascospores ellipsoid, 9–12×(5–) 6–9 µm, simple, colourless.

    Chemistry: No substances present

    Habitat: Saxicolous, rarely corticolous

    Distribution: The species is common in temperate regions found growing on boulders, dripped with water

    Indicator Value: Tolerant (Shukla et al. 2010, 2012a, b)

19. *Lepraria lobificans* Nyl. (Fig. 4.15a)

    Description: A powdery (leprose) pale blue-grey thallus forming fairly coherent but fragile membrane, with poorly delimited granules at centre; distinct granules mainly marginal, up to 100 µm diameter, a few marginal granules to 300 µm. Young thalli with a ± delimited margin and sometimes with very small and indistinct lobes. Underside pale; no dark hyphae seen, apothecia absent.

    Chemistry: K+, C–, Pd+ orange. Atranorin, zeorin; stictic, ±constictic, cryptostictic, connorstictic, constictic acids and an unidentified fatty acid

    Habitat: Muscicolous, saxicolous or corticolous

    Distribution: One of the most common lichen found growing on varied substrates covering large areas with granular powdery crust in tropical, subtropical up to lower temperate areas

    Indicator Value: Tolerant (Bajpai et al. 2009)

20. *Usnea longissima* Ach. (Fig. 4.15b)

    Description: A fruticose lichen with pendulous, filamentose branched to 60 cm long, pale yellow, greyish green to light brownish; 0.5–1 mm in diameter thallus; lateral branchlets dense, perpendicular, 2–5 cm long; surface of filamentose branches usually decorticated, rarely

4.4 Bioindicator Lichen Species

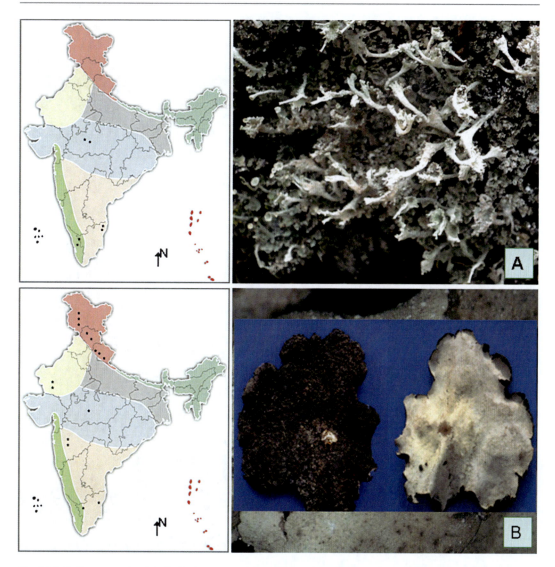

Fig. 4.14 Indicator species: (**a**) *Cladonia praetermissa*. A. W. Archer; (**b**) *Dermatocarpon vellereum* Zschacke

pulverulent to powdery, cortex of lateral branchlets persistent, cracked near base, with soredia or isidia; central axis solid. Apothecia rare to 5 mm in diameter, margin ciliate; ascospores 8×6 μm. Medulla with different lichen substances resulting 7 strains.

Chemistry: Strain (1) usnic acid; strain (2) usnic, barbatic, barbatolic; strain (3) usnic, barbatic, barbatolic and squamatic acids; strain (4) usnic and diffractic acids; strain (5) usnic, evernic acids and an unknown; strain (6) usnic acid, fumarprotocetraric acid and a purple unknown substance; strain (7) usnic, squamatic acids and an unknown substance

Chemistry: Central axis I+ blue, Usnic acid

Habitat: Corticolous

Distribution: A common fruticose lichen in upper temperate and alpine regions found growing on mostly coniferous and oak tree canopy and on small bushes and shrubs in alpine region

Indicator value: Sensitive (Rawat 2009)

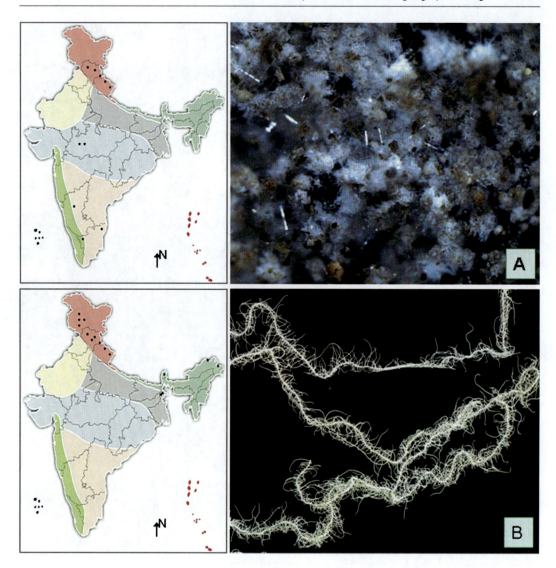

Fig. 4.15 Indicator species: (**a**) *Lepraria lobificans* Nyl. (**b**) *Usnea longissema* Ach.

4.5 Bioindicator Lichen Communities

Environmental changes produce varying responses in lichen symbionts, including variations in diversity, morphology, physiology, genetics and ability to accumulate pollutants. Lichens tend to be long-lived and highly habitat specific organisms; they tolerate extremes of heat and cold environments and grow on all types of substrata and habitats. Lichen communities are sensitive to landscape structure and land-use context and to forest management (Will-Wolf et al. 2002). The widespread distribution and ability to withstand extreme climatic condition are characteristic features which make lichens ideal monitors and can be used to estimate species diversity and habitat potential at all times of the year. Lichens differ substantially from higher plants because of their poikilohydric nature and unique physiological processes makes lichens growth particularly susceptible to climatic variations, pollution and other environmental factors and liable to changes at genetic, individual, population and community levels. A lichen community is thus an assemblage of these atypical species living together (Nash 2008).

4.5 Bioindicator Lichen Communities

Fig. 4.16 Lichen communities: (**a**) *Calcioid*; (**b**) *Alectoroid* and *Usnioid*; (**c**) *Cyanophycean*; (**d**) *Lobarian* community

Forest lichen communities respond to primary climate variables such as precipitation and temperature and to geographical gradients such as elevation and latitude that integrate climate factors (Nash 2008; Will-Wolf et al. 2002, 2006).

The main groups of lichen bioindicator communities known from India and their predictive ability are described as follows:

Calcioid community: These 'pinhead' lichens are indicators of old growth forests. Many of these species are dependent on snags and old trees with stable rough bark (Fig. 4.16a).

Indicator value: Indicates undisturbed old forest ecosystem

Alectoroid and *Usnioid* community: The tufted and pendulous fruticose lichens including genera *Sulcaria*, *Bryoria*, *Ramalina* and *Usnea* (Fig. 4.16b).

Indicator value: Useful as indicators both of for older forest with better air quality

Cyanophycean community: The variation in diversity and abundance of epiphytic cyanolichens appears useful as an indicator of forest ecosystem function (Fig. 4.16c).

Fig. 4.17 Lichen communities: (**a**) *Xanthoparmelioid*; (**b**) *Graphidioid* and *Pyrenuloid*; (**c**) *Lecanorioid*; (**d**) *Parmelioid* community

Indicator value: Cyanophycean lichens play an important role in forest nutrient cycle and indicate forest age and continuity (McCune 1993).

Lobarian community: The Lobarian Group comprised of *Lobaria*, *Pseudocyphellaria*, *Peltigera* and *Sticta*. They are a distinctive group of lichen species associated with the bark of the trunks of old trees, which are not found in any of the other young stands (Fig. 4.16d).

Indicator value: They are sensitive to air quality and reliable indicators of species rich old forest with long forest continuity (Gauslaa 1995).

Xanthoparmelioid community: According to Eldridge and Koen (1998), the yellow foliose morphological group comprising of foliose lichen species of *Xanthoparmelia* forms this community (Fig. 4.17a).

Indicator value: Presence of these lichens is consistently correlated with stable productive landscape, i.e. landscapes with no accelerated erosion or least trampling by animals and trekking by humans.

Graphidioid and *Pyrenuloid* community: The growth of Graphidaceous (*Graphis*, *Opergrapha*,

4.5 Bioindicator Lichen Communities

Fig. 4.18 Lichen communities: (**a**) *Pertusorioid*; (**b**) *Lecideoid*; (**c**) *Leprarioid*; (**d**) *Physcioid* community

Scareographya, Phaeographis) and Pyrenocarpous (*Anthracothecium, Pyrenula, Lithothelium, Porina*) influenced by the nature of bark. Both groups mostly prefer to grow on a smooth bark tree in evergreen forest/regenerated forest (Fig. 4.17b).

Indicator value: Indicator of young and regenerated forest

Lecanorioid community: The group is comprised of *Lecanora*, *Lecidella* and *Biatora*. They prefer to grow on trees in thinned-out, regenerated or disturbed forest with more open area to receive more light and wind (Fig. 4.17c).

Indicator value: The *Lecanorioid* group indicates well-illuminated environmental condition of the forest with considerable exposure of light and wind.

Parmelioid community: Members of this group are comprised of mostly the species of lichen genera *Bulbothrix*, *Flavoparmelia*, *Parmotrema*, *Parmelia*, *Punctelia* and other genera of Parmeliaceae. The forest with closed canopy and less sunlight supports few species of Parmelioid genera to grow, while the open thinned-out forest with more sunlight exhibits dominance of Parmelioid lichens (Fig. 4.17d).

Fig. 4.19 Lichen communities: (**a**) *Teloschistacean*; (**b**) *Lichinioid*; (**c**) *Peltuloid*; (**d**) *Dimorphic* community

Indicator value: Indicator of thinned-out forest

Pertusorioid community: The group includes species of lichen genus *Pertusaria*. The *Shorea robusta* tree forests in the dry deciduous forest appear excellent host for this group of lichens to colonise (Fig. 4.18a).

Indicator value: Indicates old tree forest with rough-barked trees

Lecideoid community: The member of the group such as *Lecidea*, *Protoblastedia*, *Haematomma*, *Bacidia*, *Buellia* and *Schadonia* colonised mostly on bark of deciduous trees in sheltered and well-lit exposed sides (Fig. 4.18b).

Indicator value: Indicates exposed illuminated area

Leprarioid community: The species of *Chrysothrix*, *Cryptothecia* and *Lepararia* are the common lichens of the *Leprarioid* group, which forms powdery thallus on the substrates. The species of *Chrysothrix* appears first after forest fire (Fig. 4.18c).

Indicator value: Indicates moist and dry vertical slopes, rough-barked trees of moist and dry habitats

Physcioid community: The lichen species of *Physcia*, *Pyxine*, *Dirinaria*, *Heterodermia*, *Phaeophyscia* and *Rinodina* belongs to this

group. They have ability to grow on varied substrates in both moist and dry habitats (Fig. 4.18d).
Indicator value: The Physcioid lichens are considered as the pollution-tolerant lichens. Their presence indicates nitrophilous environment (Van Herk et al. 2002).

Teloschistacean community: The species of *Caloplaca, Letroutia, Brigantiaea* and *Xanthoria* having yellow thallus and apothecia belongs to this group. The members of this group have an ability to grow both on exposed and sheltered rocks. The dark orange pigment present on the upper cortex of the thallus acts as a filter and protects the lichens from high UV radiation (Fig. 4.19a).
Indicator value: Indicator of high UV irradiance

Lichinioid community: The genera of the lichen family Lichninaceae mostly having cyanobacteria as their photobiont belongs to this group. Members of this group prefer dry rocks and barks having higher concentration of calcium (Fig. 4.19b).
Indicator value: Indicates presence of calcareous substrates in the habitats

Peltuloid community: The species of lichen genera *Peltula* belongs to this group of lichens (Fig. 4.19c).
Indicator value: The presence of the species of this group indicates a stable rock substratum.

Dimorphic community: Species of the genera *Cladonia, Cladina* and *Stereocaulon* forms this community (Fig. 4.19d).
Indicator value: Indicates undisturbed soil ecosystem

Thus, lichen communities provide valuable information about the natural-/anthropogenic-induced changes in the microclimate and land-use changes due to human activity. Lichens have also been used to resolve environmental issues involving management of natural resources such as the effects of fragmentation and habitat alteration, the structure and management of forested stands, the ecological continuity on space and time of the natural or semi-natural forests, effects of development on biodiversity, the effectiveness of conservation practices for rare or endangered species and the protection of genetic resources. Owing to their excellence as sentinel organisms, lichens are being utilised in different countries to indicate high-value forests for conservation and to identify important biodiversity-rich sites which require management of forest resources (Will-Wolf 2010). The significant correlations found between stand age and lichen species richness in several forests substantiates the importance of old or died trees, and related factors, as a habitat for lichens.

References

Awasthi DD (1988) A key to the macrolichens of India and Nepal. J Hatt Bot Lab 65:207–303

Awasthi DD (1991) A key to the microlichens of India Nepal and Sri Lanka. Bibl Lichenol 40:1–337

Awasthi DD (2000) A hand book of lichens. Bishan Singh Mahendra Pal Singh, Dehradun

Awasthi DD (2007) A Compendium of the Macrolichens from India, Nepal and Sri Lanka. Bishan Singh Mahendra Pal Singh, Dehradun

Bajpai R, Upreti DK (2011) New records of lichens from Mahabaleshwar and koyna areas of Satara district, Maharashtra, India. Geophytology 40(1–2): 61–68

Bajpai R, Upreti DK, Dwivedi SK (2008) Diversity and distribution of lichens on some major monuments of Madhya Pradesh, India. Geophytology 37:23–29

Bajpai R, Upreti DK, Dwivedi SK, Nayaka S (2009) Lichen as quantitative biomonitors of atmospheric heavy metals deposition in Central India. J Atmos Chem 63:235–246

Bajpai R, Nayaka S, Upreti DK (2011) Distribution of lichens on four trees in east and south districts of Sikkim. Biozone 3(1 &2):406–419

Bajpai R, Shukla V, Upreti DK (2013) Impact assessment of anthropogenic activities on air quality, using lichen *Remototrachyna awasthii* as biomonitors. Int J Environ Sci Technol. doi:10.1007/s13762-012-0156-1

Bajpai R, Shukla V, Upreti DK (2013) Impact assessment of anthropogenic activities on air quality, using lichen *Remototrachyna awasthii* as biomonitors. Int J Environ Sci Technol. doi:10.1007/s13762-012-0156-1

Dubey U (2009) Assessment of lichen diversity and distribution for prospecting the ecological and economic potential of lichens in and around Along town, West Siang district, Arunachal Pradesh. PhD thesis, Assam University, Silchar

Dubey U, Upreti DK, Rout J (2007) Lichen flora of Along Town, West Siang district, Arunachal Pradesh. Phytotaxonomy 7:21–26

Eldridge DJ, Koen TB (1998) Cover and floristics of microphytic soil crusts in relation to indices of landscape health. Plant Ecol 137:101–114

Garty J, Tomer S, Levin T, Lehra H (2003) Lichens as biomonitors around a coal-fired power station in Israel. Environ Res 91:186–198

Gauslaa Y (1995) The Lobarion, an epiphytic community of ancient forests, threatened by acid rain. Lichenologist 27:59–76

Goswami DB (1995) Distribution of population tribes in India. Vanyajati 43(4):23–26

Goward T, McCune B, Meidinger D (1994) The lichens of British Columbia Illustrated keys, Part I – Foliose and Squamulose species, Special report series 8. Ministry of Forests Research Program, Victoria

Hale ME (1983) The biology of lichens, 3rd edn. Edward Arnold, London

Ingle KK, Bajpai R, Upreti DK, Trivedi S (2012) Lichen distribution pattern in Bhopal city, with reference to air pollution monitoring. Indian J Environ Sci 1692:75–80

Jagadeesh Ram TAM (2006) Investigation on the lichen flora of Sundarbans biosphere reserve, West Bengal. PhD thesis, Gauhati University, Assam

Jagadeesh Ram TAM, Sinha GP (2002) Phytodiversity of Sundarbans Biosphere Reserve with special reference to lichens. Geophytolgy 32(1–2):35–38

Joshi S (2009) Diversity of lichens in Pidari and Milam regions of Kumaun Himalaya. PhD thesis, Kumaun University, Nainital

Joshi Y, Upreti DK (2007) *Caloplaca awasthii*, a new lichen species from India. Bot J Linn Soc 155(1):149–152

Joshi S, Upreti DK, Das P (2011) Lichen diversity assessment in Pindari glacier valley of Uttarakhand, India. Geophytology 41(1–2):25–41

Joshi S, Upreti DK, Das P, Nayaka S (2012) Lichenometry: a technique to date natural hazards. Science India- www.earthscienceindia.info, Popular Issue, V(II):1–16

Kantvillas G, Jarman S (2004) Lichens and bryophytes on Eucalyptus obliqua in Tasmania: management implications in production forests. Biol Conserv 117:359–373

Kholia H, Mishra GK, Upreti DK, Tiwari L (2011) Distribution of lichens on fallen twigs of *Quercus leucotrichophora* and *Quercus semecarpifolia* in and around Nainital city, Uttarakhand, India. Geophytology 41(1–2):61–73

Kumar B, Upreti DK (2008) An account of lichens on fallen twigs of three Quercus species in Chopta forest of Garhwal Himalayas, India. Ann For 16(1):92–98

Logesh AR, Upreti DK, Kalaiselvam M, Nayaka S, Kathiresan K (2012) Lichen flora of Pichavaram and Muthupet mangrove (Southern coast of India). Mycosphere 3(5):884–888. doi:10.5943/mycosphere/3/6/1

McCune B (1993) Gradients in epiphyte biomass in three *Pseudotsuga-Tsuga* forests of different ages in western Oregon and Washington. Bryologist 96:405–411

Mishra GK (2012) Distribution and ecology of lichens in Kumaun Himalaya Uttarakhand PhD thesis, Kumaun University, Nainital

Mishra GK, Joshi Y, Upreti DK, Punetha N, Dwivedi A (2010) Enumeration of lichens from Pithoragarh district of Uttarakhand, India. Geophytology 39(1–2):23–39

Mohabe S (2011) Taxonomic and ecological studies on lichens of Vindhyan region, Madhya Pradesh. PhD thesis, Barkatullah University, Bhopal

Mohabe S, Upreti DK, Trivedi S, Mishra GK (2010) Lichen flora of Rewa and Katni district of Madhya Pradesh. Phytotaxonomy 10:122–126

Mohan MS, Hariharan GN (1999) Lichen distribution pattern in Pichavaram – a preliminary study to indicate forest disturbance in mangroves of south India. In: Mukerji KG, Chamola BP, Upreti DK, Upadhyay RK (eds) Biology of lichens. Aravali Books International, New Delhi

Nash TH III (ed) (2008) Lichen biology, 2nd edn. Cambridge University Press, Cambridge, p 486,

Nayaka S, Upreti DK (2011) An inventory of lichens in Uttar Pradesh through bibliographic compilation. In: National conference on forest biodiversity: earth's living treasure, 22 May 2011 organised by Uttar Pradesh State Biodiversity Board, Lucknow (U.P.)

Nayaka S, Upreti DK (2012) An overview of lichen diversity and conservation in western Ghats, India. In: The 7th international association for lichenology symposium. Lichens: from genome to ecosystems in a changing world, 9–13 January 2012, Bangkok, Thailand

Nayaka S, Upreti DK, Pathak S, Samuel C (2004) Lichen of Bondla and Bhagwan Mahavir Wildlife Sanctuaries, Goa. Biol Memoirs 30(2):115–119

Nayaka S, Satya, Upreti DK (2007) Lichen diversity in Achanakmar wildlife sanctuary, core zone area of proposed Amarkantak Biosphere reserve, Chattisgarh. J Econ Taxon Bot 31(1):133–142

Nayaka S, Upreti DK, Punjani B, Dubey U, Rawal J (2010) New records and notes on some interesting lichens of family Roccellaceae from India. Phytotaxonomy 10:127–133

Negi HR, Upreti DK (2000) Species diversity and relative abundance of lichens in Rumbak catchment of Hemis National Park in Ladakh. Curr Sci 78(9):1105–1112

Pathak S, Nayaka S, Upreti DK, Singh SM, Samuel C (2004) preliminary observation on lichen flora of Cotigao Wildlife Sanctuary, Goa, India. Phytotaxonomy 4:104–106

Rani M, Shukla V, Upreti DK, Rajwar GS (2011) Periodical monitoring with lichen, *Phaeophyscia hispidula* (Ach.) Moberg in Dehradun city, Uttarakhand, India. Environmentalist 31:376–381. doi:10.1007/s10669-011-9349-2

Rawat S (2009) Studies on medicinally important lichens and their conservation in some forest sites of Chamoli district, Uttarakhnad, India. PhD thesis, Babasaheb Bhimrao Ambedkar University, Lucknow

Rawat S, Upreti DK, Singh RP (2011) Estimation of epiphytic lichen litter fall biomass in three temperate forests of Chamoli district, Uttarakhand, India. Trop Ecol 52(2):193–200

Rout J, Kar A, Upreti DK (2004) Lichens of Sessa Orchid Sanctuary, West Kameng, Arunachal Pradesh. Phytotaxonomy 4:38–40

Rout J, Das P, Upreti DK (2010) Epiphytic lichen diversity in a reserve forest in south Assam, north India. Trop Ecol 51(2):281–288

Satya, Upreti DK (2009) Correlation among carbon, nitrogen, sulphur and physiological parameters of *Rinodina sophodes* found at Kanpur city, India. J Hazard Mater 169:1088–1092. doi:10.1016/j/jhazmat.2009.04.063

Satya, Upreti DK (2011) Lichen bioindicator communities in Achanakmar biosphere reserve, Madhya Pradesh and Chhattisgarh. In: Vyas D, Paliwal GS, Khare PK, Gupta RK (eds) Microbial biotechnology and ecology. Daya Publishing House, New Delhi, pp 669–682

References

Satya, Upreti DK, Patel DK (2012) *Rinodina sophodes*(Ach.) Massal.: a bioaccumulator of polycyclic aromatic hydrocarbons (PAHs) in Kanpur city, India. Environ Monit Assess 184:229–238

Saxena S, Upreti DK, Sharma N (2007) Heavy metal accumulation in lichens growing in north side of Lucknow city. J Environ Biol 28(1):45–51

Sheik MA, Upreti DK, Raina AK (2006a) Lichen diversity in Jammu and Kashmir, India. Geophytology 36:69–85

Sheik MA, Upreti DK, Raina AK (2006b) An enumeration of lichens from three districts of Jammu and Kashmir, India. J Appl Biosci 32(2):189–191

Shukla V (2007) Lichens as bioindicator of air pollution. Final technical report. Science and Society Division, Department of Science and Technology New Delhi. Project No. SSD/SS/063/2003

Shukla V (2012) Physiological response and mechanism of metal tolerance in lichens of Garhwal Himalayas. Final technical report. Scientific and Engineering Research Council, Department of Science and Technology, New Delhi. Project No. SR/FT/LS-028/2008

Shukla V, Upreti DK (2007a) Lichen diversity in and around Badrinath, Chamoli district (Uttarakhand). Phytotaxonomy 7:78–82

Shukla V, Upreti DK (2007b) Physiological response of the lichen *Phaeophyscia hispidula* (Ach.) Essl. to the urban environment of Pauri and Srinagar (Garhwal), Himalayas. Environ Pollut 150:295–299. doi:10.1016/j.envpol.2007.02.010

Shukla V, Upreti DK (2007c) Heavy metal accumulation in *Phaeophyscia hispidula* en route to Badrinath, Uttaranchal, India. Environ Monit Assess 131:365–369. doi:10.1007/s10661-006-9481-5

Shukla V, Upreti DK (2008) Effect of metallic pollutants on the physiology of lichen, *Pyxine subcinerea* Stirton in Garhwal Himalayas. Environ Monit Assess 141:237–243. doi:10.1007/s10661-007-9891-z

Shukla V, Upreti DK (2009) Polycyclic Aromatic Hydrocarbon (PAH) accumulation in lichen, *Phaeophyscia hispidula* of Dehradun city, Garhwal Himalayas. Environ Monit Assess 149(1–4):1–7

Shukla V, Upreti DK (2011) Changing lichen diversity in and around urban settlements of Garhwal Himalayas due to increasing anthropogenic activities. Environ Monit Assess 174(1–4):439–444. doi:10.1007/s10661-010-1468-6

Shukla V, Upreti DK, Patel DK, Tripathi R (2010) Accumulation of polycyclic aromatic hydrocarbons in some lichens of Garhwal Himalayas, India. Int J Environ Waste Manag 5(1/2):104–113

Shukla V, Patel DK, Upreti DK, Yunus M (2012a) Lichens to distinguish urban from industrial PAHs. Environ Chem Lett 10:159–164. doi:10.1007/s10311-0110336-0

Shukla V, Upreti DK, Patel DK, Yunus M (2012b) Lichens reveal air PAH fractionation in the Himalaya. Environ Chem Lett. doi:10.1007/s10311-012-0372-4

Shukla V, Patel DK, Upreti DK, Yunus M, Prasad S (2013) A comparison of heavy metals in lichen (*Pyxine subcinerea*), mango bark and soil. Int J Environ Sci Technol 10:37–46. doi:10.1007/s13762-012-0075-1

Shyam KR, Thajuddin N, Upreti DK (2011) Diversity of lichens in Kollihills of Tamil Nadu, India. Int J Biodivers Conserv 3(2):36–39

Singh KP (1999) Lichens of Eastern Himalayan region. In: Mukerji KG, Upreti DK, Upadhyay (eds) Biology of lichens. Aravali Books International, New Delhi, pp 153–204

Singh KP, Sinha GP (2010) Indian lichens an annotated checklist. Botanical Survey of India/Ministry of Environment and Forest, Kolkata

Singha AB (2012) Lichen flora on Betel nut host plant: distribution, diversity and assessment as air quality indicator in Cachar district, Southern Assam, North East India

St. Clair LL (1989) Report concerning establishment of a lichen biomonitoring program and baseline for the Jarbidge Wilderness Area, Hunbolt national Forest, Naveda. U.S. Forest Service technical report, pp 15

St. Clair BS, St. Clair LL, Mangelson FN, Weber JD (2002a) Influence of growth form on the accumulation of airborne copper by lichens. Atmos Environ 36:5637–5644

St. Clair BS, St. Clair LL, Weber JD, Mangelson FN, Eggett LD (2002b) Element accumulation patterns in foliose and fruticose lichens from rock and bark substrates in Arizona. Bryologist 105:415–421

Riga-Karandinos NA, Karandinos GM (1998) Assessment of air pollution from a lignite power plant in the plain of megalopolis (Greece) using as a biomonitors three species of lichens; impact on some biochemical parameters of lichens. Sci Total Environ 215:167–183

Upreti DK (1996) Lichen on *Shorea robusta* in Jharsuguda district, Orissa, India. Flora Fauna 2(2):159–161

Upreti DK (1998) Diversity of lichens in India. In: Agarwal SK, Kaushik JP, Kaul KK, Jain AK (eds) Perspective in environment. A.P.H. Publishing Corporation, New Delhi, pp 71–79

Upreti DK (2001) Taxonomic, pollution monitoring and ethnolichenological studies on Indian lichens. Phytomorphology 51:477–497

Upreti DK, Chatterjee S (1999) Distribution of lichens on Quercus and Pinus trees in Pithoragarh district, Kumaun Himalayas-India. Trop Ecol 40(1):41–49

Upreti DK, Nayaka S (2000) An enumeration of lichens from Himachal Pradesh. In: Chauhan DK (ed) Recent trends in biology, Prof. DD Nautiyal commemoration vol. Botany Department, Allahabad University, Allahabad, India, pp 15–31

Upreti DK, Negi HS (1995) Lichens of Nanda Devi Biosphere Reserve, Uttar Pradesh, India, I. J Econ Taxon Bot 19(3):627–636

Upreti DK, Nayaka S, Bajpai A (2005a) Do lichens still grow in Kolkata city? Curr Sci 88(3):338–339

Upreti DK, Nayaka S, Satya (2005b) Enumeration of lichens from Madhya Pradesh and Chattisgarh, India. J Appl Biosci 31(1):55–63

Van Herk CM, Aptroot A, van Dobben HF (2002) Long-term monitoring in the Netherlands suggests that lichens respond to global warming. Lichenologist 34:141–154

Will-Wolf S (2010) Analyzing lichen indicator data in the forest inventory and analysis program. General technical report, PNW-GTR-818. U.S. Department of Agriculture, Forest Service, Pacific Northwest Research Station, Portland, p 62

Will-Wolf S, Neitlich P, Esseen P-A (2002) Monitoring biodiversity and ecosystem function: forests. In: Nimis PL, Scheidegger C, Wolseley P (eds) Monitoring with lichens–monitoring lichens, NATO Science Series. Kluwer Academic Publishers, The Hague, pp 203–222

Will-Wolf S, Geiser LH, Neitlich P, Reis A (2006) Comparison of lichen community composition with environmental variables at regional and subregional geographic scales. J Veg Sci 17:171–184

Wolseley PA, Moncrieff C, Aguirre-Hudson B (1994) Lichens as indicators of environmental stability and change in the tropical forests of Thailand. Glob Ecol Biogeogr Lett 1:116–123

Yadav V (2005) Lichen flora of Himachal Pradesh. PhD thesis, Lucknow University, Lucknow

Ecosystem Monitoring

Monitoring the quality and sustainability of the ecosystem with lichens has been studied worldwide. Three major categories of assessment that have been identified so far for the role of lichens in ecosystem monitoring include air quality, climate and biodiversity. Both natural and manmade disturbances/disasters are responsible for imbalance in the ecosystem.

With increasing economic growth, environmental contamination, especially air pollution, is resulting in environmental degradation in the developing nations of Asia, especially India. In order to attain sustainable economic development, monitoring and eradication of environmental problems is important. The highest priority issues include monitoring of the quality of air, water and soil, deforestation and degradation of the natural environment.

Lichens are very useful for monitoring spatial and/or temporal deposition patterns of pollutants as they allow accumulation of pollutant throughout its thalli, and concentrations of pollutants in lichen thalli may be directly correlated with environmental levels of these elements. Lichens also meet other characteristics of the ideal sentinel organism: they are long lived, having wide geographical distribution, and accumulate and retain many trace elements to concentrations that highly exceed their physiological requirements. The details of the factors affecting the ecosystem, natural as well as anthropogenic, and role of lichens in ecosystem monitoring have been discussed.

5.1 Introdcution

Monitoring the ecological effects of contaminants released as a result of natural and/or man-made processes in the ecosystems, to detect changes in environmental quality, is best referred as *ecological monitoring*, as pollutants are not only localised to their source of origin but also contaminate global environment. The widespread dispersing nature of pollutants makes spatio-temporal monitoring of pollutants pertinent to ensure sustainable development either using air samplers or bioindicator species (Seaward 1974).

Natural sources of atmospheric pollutants generally include dust emissions, living and dead organisms, lightning and volcanoes. Over Asia, mineral dust is the major natural aerosol because of the vast desert regions. Other gases and aerosols are mainly the result of anthropogenic emissions, or airborne in situ production, rather than natural emissions. As the most important natural pollutant, dust aerosols play an important role in the climate system by affecting the radiation budget (Tegen and Lacis 1996; Sokolik et al. 1998), biogeochemical cycles (Martin and Fitzwater 1988; Martin 1991; Archer and Johnson 2000) and atmospheric chemistry (Dentener et al. 1996; Dickerson et al. 1997; Martin et al. 2003). Moreover, they have important consequences on surface air quality (Prospero 1999). Dust aerosols originating from East Asia, one of the major dust emission regions in the world, may influence the

Table 5.1 Methods for ecosystem monitoring utilising lichens

	Techniques	
1	Transplant	Transplanting healthy lichens into a polluted area and measuring changes in thallus physiology and elemental composition
2	Lichen zone mapping	To indicate the severity of pollution with reference to distance from the source, as reflected by change in composition of lichen diversity
3	Sampling individual species	Measurement of contaminants accumulated within the thallus

After Garty (2001)

ecological cycle of the North Pacific Ocean and the air quality over North America (Prospero 1999; Wuebbles et al. 2007).

A number of traditional studies dealing with atmospheric contamination are available but most of them have been limited to problems of high cost and the difficulty of carrying out extensive sampling, in terms of both time and space. There is, thus, an ever-increasing interest in using indirect monitoring methods such as analysis of organisms that are biomonitoring (Garty 2001). Biomonitoring involves the use of organisms and biomaterials to obtain information on certain characteristics of the ecosystem. According to Markert et al. (2003), biomonitoring is a composite phenomenon comprised of several interrelated terms.

Bioindicator	provides information on the environment or the quality of environmental changes
Biomonitor	provides quantitative information on the quality of the environment
Reaction indicators	are organisms which are sensitive to air pollutants and are utilised in studying the effects of pollutants on species composition and on physiological and ecological functioning
Accumulation indicators	are organisms which readily accumulate a range of pollutants without being harmed by the excessive concentration of the pollutants
Passive biomonitors	occurs naturally in the area
Active biomonitors	are transplanted into the research area for a specific period of time

Lichens are undoubted reliable bioindicators in the monitoring of ecosystem changes as Litmus test for ecosystem health (Hawksworth 1971). Biomonitoring provides relevant information about ecosystem health either from changes in the behaviour of the monitor organism (species composition and/or richness, physiological and/or ecological performance, morphology) or from the concentrations of specific substances in the monitoring organisms (Table 5.1). Thus, lichens can be used for the quantitative and qualitative determination of natural and human-generated environmental factors.

Biomonitors which are mainly used for qualitative determination of contaminants can be classified as being sensitive or accumulative. Sensitive biomonitors may be of the optical type and are used as integrators of the stress caused by contaminants and as preventive alarm systems. They are based upon either optical effect as morphological changes in abundance behaviour related to the environment and/or physical and chemical aspects as alteration in the activity of different enzymes systems as well as in photosynthetic or respiratory activities, while accumulative bioindicators have the ability to store contaminants in their tissues and are used for the integrated measurement of concentration for such contaminants in the environment, and those species involved are called accumulator (indicator) species. According to Wolterbeek et al. (2003), indicator species are recognised as 'Universal' and 'Local'. The term Universal is restricted to those species which are found exclusively on substrates containing high concentrations of pollutants for which the species is proposed as an indicator species. Local indicators are species which are associated with pollutant-bearing substrates in certain geographical areas but which also grow elsewhere in non-mineralised areas.

Various monitor materials have been applied in trace element air monitoring programmes, such as lichens, mosses, ferns, grasses, tree and pine needles. The mechanisms of trace element uptake and retention are still not sufficiently known. It is necessary to select out appropriate organisms which are suitable for the study purposes.

Lichens are natural sensors of our changing environment: the sensitivity of particular species and communities to a very board spectrum of environmental condition, both natural and unnatural, is widely acknowledged. Nevertheless, lichens undoubtedly represent one of the most successful forms of symbiosis in nature. They are to be found worldwide, exploiting not only all manner of natural, usually stable, micro- and macro-environments, but in many cases adapting to extreme conditions, including some brought about by human disturbance. Lichens are therefore used increasingly in evaluating threatened habitats, in environmental impact assessment and in monitoring environmental alterations, particularly those resulting from a disturbingly large and growing number of anthropogenic pollutants.

That lichens are sensitive to air pollution is, of course, a generalisation that requires cautious interpretation and limited extrapolation. The distribution patterns of a lichen species may well reflect varying levels of an air pollutant, but variation in its distribution may also be caused by a variety of abiotic or biotic factors as well. Furthermore, all lichens are not equally sensitive to all air pollutants; rather, different lichen species exhibit differential sensitivity to specific air pollutants. Sensitive species may become locally extirpated when a pollutant is present, but at least some tolerant species are likely to persist. This differential sensitivity is, however, very useful when interpreting air pollution effects. The absorption of metals in lichens involves some mechanisms like intercellular absorption through an exchange process, intracellular accumulation and entrapment of particles that contain metals.

Thus, lichens have a long history of use as monitoring of environmental pollution (Nimis 1990). A number of authors have advocated the use of biological monitoring to assess and understand the status of trends within natural ecosystem and focused on heavy metals, PAHs accumulation and their interactions due to natural and man-made disturbances/disasters (Nriagu and Pacyna 1988; Jeran et al. 2002; Godinho et al. 2008; Upreti and Pandey 2000; Shukla and Upreti 2007a, b). In areas that lack naturally growing lichens due to high levels of air contamination, lichen transplantation method, first introduced by Brodo (1961), is being applied as a standard method to study airborne metal and sulphur pollutants (Conti and Cecchetti 2001; Budka et al. 2004; Baptista et al. 2008).

There are large amounts of studies with respect to the effects of air pollution to epiphytic lichens and the use of lichens as bioindicators (Geebelen and Hoffman 2001; Paoli et al. 2011; Davies et al. 2007; Giordani 2007) as well as biomonitors (Nash 1976; Herzig et al. 1989; Gombert et al. 2002; Frati et al. 2006; Tretiach et al. 2007; Guidotti et al. 2009). Most of the studies have been carried out in urban areas where air pollution is caused by a number of factors. However, comparatively little comprehensive and precise data about biomonitoring with lichens from India is available.

5.2 Natural and Human Disturbances/Disasters

Nature is the major source of elements of inorganic and organic origin. Various categories of pollutants having natural and man-made origin include heavy metals, metalloids, polycyclic aromatic hydrocarbons (PAHs) and radionuclides. Natural processes (volcanic eruption) result in release of elements but nature has sinks which consume those elements via mineral cycling as in the case of metals. In lower concentration, especially metals except few are not toxic, they facilitate physiological functioning of the organism, but when its concentration exceeds permitted levels and shows phytotoxicity, then it is considered as pollutant. Radioactive substances have biological half-life which facilitates decay of these chemicals. Best example of source and sink is

CO$_2$ emission (potential greenhouse gas); in natural conditions CO$_2$ emitted by plants is consumed by other organisms maintaining the environmental levels roughly constant, but when human activity also releases these chemicals in the atmosphere, it results in imbalance in the natural cycling of the elements which enter either food chain via biomagnification and gases released remain in the atmosphere and act as potential greenhouse gases, resulting in elevation of global temperature and glacier retreat phenomenon.

Nitrogen oxides (NOx), volatile organic compounds (VOCs), polycyclic aromatic hydrocarbons (PAHs), particles and metals are some of the pollutants considered to be of ecological significance (Bignal et al. 2007, 2008). In addition, effects of ammonia (NH$_3$) on vegetation at roadside verges (Truscott et al. 2005) and on bark pH that influences lichen vegetation around a pig stock farm in Italy (Frati et al. 2006) and toxicity of nitric acid (HNO$_3$) to the lichen *Ramalina menziesii* have been recognised. Sulphur dioxide (SO$_2$) was once regarded as the most notorious pollutant affecting lichens, but the rapid reduction in SO$_2$ has been remarkable in the industrialised world today (Nieboer et al. 1977).

The relation of sulphur dioxide (SO$_2$) to lichens has been broadly studied throughout the world as this pollutant was once regarded as the most harmful compound to lichens. A wide range of methods is used to analyse the physical properties of lichens such as chlorophyll, sulphur isotope composition, sugar content, spectral reflectance, membrane proteins, moisture content and ethylene content (Conti and Cecchetti 2001). These studies have proved the deterioration of physical structures of lichens being exposed to sulphur compounds (LeBlanc et al. 1974; Shirazi et al. 1996).

Many recent studies relating to the effects of air pollution on lichens focus on NOx and there are numerous results showing the significant correlation between lichens and the pollutant. Nitrogen is an essential element for life, being involved in the synthesis of protein and nucleic acids (Nash 2008). However, an excess amount of nitrate deposition can deteriorate the symbiotic relationship. For instance, NOx has a strong effect on (Davies et al. 2007; Aragón et al. 2010), its community composition, frequency and dispersal (Larsen et al. 2007). In addition, lichen population declines in high NOx content (van Dobben et al. 2001; Giordani 2007). At the molecular level, Tretiach et al. (2007) showed that a large amount of NOx can damage the photobionts of transplanted *Flavoparmelia caperata*, hindering photosynthesis. This is probably due to the increased reactive oxygen species (ROS). They reported that a high concentration of NO$_2$ in the cells forms nitrous and nitric acids, which acidifies the cytoplasm and results in protein denaturation and deamination of amino acids and nucleic acid.

Despite those negative effects, some lichen species such as *Lecanora dispersa* and *Phaeophyscia orbicularis* are NOx tolerant (Davies et al. 2007). Even though Nash (1976) confirmed in a laboratory experiment the phytotoxic effect of NO$_2$ on lichens that were fumigated with 4 ppm (7,520 µg m^{-3}) for 6 h, he suggested that the pollutant would probably not be harmful to lichens since the NO$_2$ concentration detected in natural environment was usually less than 1 ppm. Thus, the effect of NOx on lichens seems controversial and unclear.

5.2.1 Heavy Metal

Heavy metals are chemical elements that exhibit metallic properties. Many different definitions of the term heavy metal have been proposed, based on density, atomic number, atomic weight, chemical properties or toxicity (Prasad 1997). Toxicity of metal is mainly dependent on the oxidation state of the element which is responsible for its bioavailability to the plants (Table 5.2).

Heavy metals are natural constituents of the earth's crust. Most of the metals are toxic in nature; their stable and non-biodegradable character allows their entry into the food chain and pose harmful effects to the organism in contact. Anthropogenic activities have drastically altered the biochemical and geochemical cycles and balance of some heavy metals. The principal man-made sources of heavy metals are point sources related mainly to industrial activities, e.g.

Table 5.2 Oxidation state, toxicity and bioavailability of various metals present in the ecosystem due to natural and/or anthropogenic processes

Elements	Redox state	Major sources	Toxicity and bioavailability
Arsenic (As)	III, V	Natural	Arsenite (III) is commonly the dominant species in moderate to strongly anoxic soil environments and is much more toxic, soluble and mobile than the oxidised form, arsenate (V) (Aide 2005)
Nitrogen (N)	III, 0, III, V	Natural	An essential nutrient but odd nitrogen compounds like PAN are toxic. Bioavailability to the food chain is facilitated by nitrogen-fixing bacteria in lichens and root nodules of higher plants
Carbon (C)	IV to IV	Natural/anthropogenic	Carbocatenation properties makes them main constituents of hydrocarbons
Chromium (Cr)	III, VI	Anthropogenic	Only +3 and +6 states are stable under most conditions of the surface environment. Cr(VI) is both highly soluble and toxic to plants and animals, yet Cr(III) is relatively insoluble and an essential micronutrient (Fendorf 1995; Negra et al. 2005)
Copper (Cu)	I, II	Anthropogenic	Copper mobility is decreased by sorption to mineral surfaces. Cu 2+ sorbs strongly to mineral surfaces over a wide range of pH values (Gasparatos 2012)
Sulphur (S)	II to VI	Anthropogenic	In II oxidation state it forms phytotoxic gas SO_2 and in VI state it forms SF_6 a potential green house gas
Iron (Fe)	II, III	Natural	
Manganese (Mn)	II, IV	Natural	Mn (IV), which is the most stable in neutral to slightly alkaline conditions, and Mn (II), which is stable in reducing conditions (Post 1999)
Nickel (Ni)		Anthropogenic	Various oxidation states but only Ni (II) is stable in the pH and redox conditions found in the soil environment (Essington 2004; McGrath 1995)
Lead (Pb)	II	Anthropogenic	Lead is a widespread pollutant in the soil environment with a long residence time compared with most other inorganic pollutants (Gasparatos 2012)

mines, foundries and smelters, and combustion byproducts and traffic, etc. Relatively volatile heavy metals and having lower density get associated with the particulate matter which may be widely dispersed throughout to longer distances, often being deposited thousands of miles from the site of initial release (Bari et al. 2001; Goyal and Seaward 1982). In general, the smaller and lighter a particle is, the longer it will stay in the air. Larger particles, greater than 10 μm in diameter tend to settle to the ground by gravity in a matter of hours, whereas the smallest particles (less than 1 μm in diameter) can stay in the atmosphere for weeks and are mostly removed by precipitation (Buccolieri et al. 2006).

A dynamic equilibrium exists between atmospheric nutrient/pollutant accumulation and loss that can make lichen tissue analysis a sensitive tool for the detection of changes in air quality of many pollutants (Farmer et al. 1991; Baddeley et al. 1972). In an entire lifespan, lichens undergo multiple wetting and drying cycles during a day. When hydrated, nutrients and contaminants are absorbed over the entire surface of the lichen. During dehydration, nutrients and many contaminants concentrate by absorption to cell walls, cloistering inside organelles or crystallising between cells (Nieboer et al. 1978). Lichens posses an ability to accumulate sulphur, nitrogen and metals from atmospheric sources better than plants (Berry and Wallace 1981; Brown et al. 1994; Beckett and Brown 1984). Macronutrients, such as nitrogen, sulphur, potassium, magnesium and calcium, are comparatively mobile and easily leached, and therefore, measurable changes in tissue concentrations can occur over weeks or months with seasonal changes in deposition (Bargagli 1998; Boonpragob et al. 1989), while trace and toxic metals such as cadmium, lead, and zinc are more tightly bound or sequestered within lichens and therefore more slowly released (Garty 2001). However, metals can stay in the environment for 20 years or more after their deposition, and elevated levels in lichens reveal this. If air quality improves, levels of metals will decrease over time, and changes in air quality can be detected in lichen tissue over a period of years (Bargagli and Nimis 2002). While it may take decades to return to background levels, changes may be observable from 1 year to the next as new growth takes place and metals are leached from older tissues. For spatio and temporal air quality assessment purposes, collection of a large-enough sample size, comprised of many individuals, should be sufficient to determine the average tissue concentration for that population.

As both natural and anthropogenic sources contribute to the bioaccumulation of metals in an organism, identifying the relative contribution from each source becomes a complex task. Puckett (1988) reported a method of calculating enrichment factors (EFs) to compare the concentration of metal within a plant with potential sources in the environment. The equation EF calculation is:

$$EF = \frac{x \text{ / reference element in lichen}}{x \text{ / reference element in crustal rock}}$$

Enrichment factors (EFs) are calculated to know the origin of metal either lithogenic and/or airborne. EFs studies indicate that Fe and Mn have a significant crustal origin and that the lower-concentration heavy metals Cd, Cu, Pb and Zn are mainly of anthropogenic origin. Accordingly to geochemical calculations, Al, Fe, and Mn have a significant crustal origin, while, Cd, Cu, Pb and Zn are of anthropogenic origin. Moreover, Fe resulted predominant in the coarse particle fraction, while Ni, Pb, V and Zn were predominant in the fine particle fraction (Buccolieri et al. 2006).

The major sources of heavy metal pollution in urban areas of India are anthropogenic, while contamination from natural sources predominates in the rural areas. Anthropogenic sources of pollution include those associated with fossil fuel (vehicular activity) and coal combustion, industrial effluents, solid waste disposal, fertilisers and mining and metal processing. At present, the impact of these pollutants is confined mostly to the urban centres with large populations, high traffic density and consumer-oriented industries. Natural sources of pollution include weathering of mineral deposits, forest/bush fires and windblown dusts. The heavy metals which are monitored by lichens include: aluminium

(Al), arsenic (As), cadmium (Cd), chromium (Cr), copper (Cu), iron (Fe), nickel (Ni), lead (Pb), manganese (Mn), strontium (Sr), tin (Sn), titanium (Ti), vanadium (V) and zinc (Zn).

As developing countries of Asia become industrialised and urbanised, heavy metal pollution is likely to reach disturbing levels. Lichens are frequently exposed to excess metals which they may tolerate as a result of detoxification mechanisms. Considerable amounts of heavy metals are immobilised by cell wall components, and the main processes for maintaining metal homeostasis in lichens, including transport of heavy metals across membranes, are chelation and sequestration (Backor and Loppi 2009). There is a conspicuous lack of data on the nature and extent of metal pollution either at local or regional levels, particularly to assist in the understanding of metal cycling in the environment. No systematic studies are being carried out to examine the dynamics of tropical ecosystems.

Although most countries in the sub-region recognise the need to combat pollution, environmental controls are either non-existent or inadequate. Most industries discharge effluents into the environment without any prior treatment, and the manufacture of 'pollution-intensive' products is being shifted to the developing countries where strict controls do not exist. The establishment of comprehensive monitoring systems and information gathering should be given priority by governments of the developing countries (for lichen biomonitoring data, see Sect. 5.2.1.5).

5.2.2 Arsenic (Metalloids)

Arsenic is a naturally occurring toxic element and its toxicity, mobility and bioavailability in soil are highly dependent on pH and redox potential (Aide 2005). Arsenic in air is found in particulate forms as inorganic As, and the dust of industrial periphery contains huge amount of As element (Menard et al. 1987). Methylated arsenic is a minor component in the air of suburban, urban and industrial areas and that the major inorganic portion is a variable mixture of the trivalent As(III) and pentavalent As(V) forms, the latter being predominant. From both biological and toxicological aspects, arsenic compounds can be classified into three major groups: inorganic arsenic compounds, organic arsenic compounds and arsenic gas. Two inorganic forms of As, arsenite (As[OH]$_3$) and arsenate (H$_2$·AsO$_4^-$ and HAsO$_4^{2-}$), are the main species in soils. Arsenite is commonly the dominant species in moderate to strongly anoxic soil environments and is much more toxic, soluble and mobile than the oxidised form, arsenate (Aide 2005). Common organic arsenic compounds are arsanilic acid, methylarsonic acid, dimethylarsinic acid and arsenobetaine.

Arsenic is released in the atmosphere from both natural and anthropogenic sources. The dominance of the anthropogenic factors is made obvious by high levels of As recorded near coal-based power plant and other industrial sites. The principal natural source is volcanic activity with minor contributions by exudates from vegetation and windblown dusts. Man-made emissions to air arise from the smelting of metals; the combustion of fuels, especially of low-grade brown coal; and the use of pesticides. According to Niriagu and Azcue (1990), arsenic is widely used in agriculture (manufacturing of pesticides), livestock (preservatives), medicines, electronic industries and mostly metallurgy. It is also released unintentionally as a result of many human activities such as smelting or roasting of any sulphide-containing mineral and, with combustion of fossil fuels, releases due to rapid leaching of exposed wastes from mining and ore processing activities, due to the manufacture of arsenicals and due to the greatly accelerated erosion of the land. The areas in the neighbourhood of the industrial complex, mining and vehicular activities exhibit significant increase in the concentration of arsenic. According to US EPA, the mean level of As ranges from <1 to 3 ng/m^3 in remote areas and from 20 to 30 ng/m^3 in urban areas.

Asia, especially Southeast Asia, has problem of As contamination of ground water as serious threat to human health. Recent studies show that more than 100 million people in Bangladesh, West Bengal (India), Vietnam, China and other South Asian countries drink and cook with

Table 5.3 Arsenic concentration (in µg g^{-1}) in different lichen species collected from diverse regions of India

S. No.	Species	Sites	Conc. (mini-max)	References
1.	*Caloplaca subsoluta* (Nyl.) Zahlbr.	Mandav, Madhya Pradesh	0.46–19	Bajpai et al. (2009a, b, 2010b)
2.	*Diploschistes candidissimus* (Kr.) Zahlbr.	Mandav, Madhya Pradesh	1.24–17.34	Bajpai et al. (2009a, b, 2010b)
3.	*Lepraria lobificans* Nyl.	Mandav, Madhya Pradesh	28.63–51.20	Bajpai et al. (2009a, b, 2010b)
4.	*P. praesorediosum* (Nyl.) Hale	Mandav, Madhya Pradesh	12.2–42.12	Bajpai et al. (2009a, b, 2010b)
5.	*P. euploca* (Ach.) Poelt in pisut	Mandav, Madhya Pradesh	5.87–10.52	Bajpai et al. (2009a, b, 2010b)
6.	*P. hispidula* (Ach.) Essl.	Mandav, Madhya Pradesh	10.98–51.95	Bajpai et al. (2009a, b)
7.	*P. hispidula* (Ach.) Essl.	Rewa, Madhya Pradesh	0.00–19.60	Bajpai et al. (2011)
8.	*Phylliscum indicum* Upreti	Mandav, Madhya Pradesh	8.11–20.99	Bajpai et al. (2009a, b, 2010b)
9.	*P. cocoes* (Sw.) Nyl.	NTPC, Uttar Pradesh	8.9–77	Bajpai et al. (2010a, b)
10.	*P. cocoes* (Sw.) Nyl.	Katni, Madhya Pradesh	BDL–33.4	Bajpai et al. (2011)
11.	*P. cocoes* (Sw.) Nyl.	West Bengal	5.20–48.10	Bajpai and Upreti (2012)
12.	*Remototrachyna awasthii*	Mahabaleshwar city, Maharashtra (2011)	0.18–3.96	Bajpai et al. (2013)
13.	*R. awasthii*	Mahabaleshwar city, Maharashtra (2012)	0.36–4.11	Bajpai et al. (2013)

arsenic-contaminated water, which can cause skin lesions, internal cancers, respiratory illnesses, cardiovascular diseases and neurological problems (Cheng et al. 2005). In Bangladesh, about one-third of the wells among nearly five million tested are considered unsafe (van Geen et al. 2005). This is due to the excessive use of tube wells as safe surface waters become scarce. When the tube wells pump ground water through geological layers rich with arsenic, the toxic metal is leached and accumulates in the wells, especially when the infiltrated waters become polluted (Cheng et al. 2005).

In India the states of West Bengal, Madhya Pradesh, Bihar and some parts of Uttar Pradesh are facing a lot of skin problem due to arsenic in water pollution. Few studies of arsenic accumulation in hydrophytes and ferns are available. Singh et al. (2006) examined the metabolic adaptation of *Pteris vittata* L., an arsenic hyperaccumulator fern, in different concentrations of arsenic solution and evaluated their tolerance capacity. Srivastava et al. (2007) reported metabolic adaptations at >315 µg g^{-1} dry weight of *Hydrilla verticillata* (L.f.) Royle, exposed to different concentrations of arsenic. Mishra et al. (2008) studied the phytochelating activities of *Ceratophyllum demersum* L., against 76 µg g^{-1} dry weight of arsenic concentration. In India most of the studies of heavy metal accumulation in lichens are focused on metals originating from vehicular and industrial activities like Pb, Cu, Fe and Zn, and little attention is being paid on other trace elements such as Hg, Mn and Ag as well as metalloids (Shukla and Upreti 2012).

Biomonitoring of As content in lichens has been carried out in Europe and in America, but in India recently biomonitoring studies employing lichens have been carried out to study the arsenic accumulation in lichen species having different growth form and growing naturally at different parts of India as well as explore a suitable As accumulator species (Table 5.3).

From the levels of As estimated in lichens of different parts of the country, it is clear that As represents one of the most abundant metalloid in the air and easily accumulated by different growth forms of lichens. The lichen thallus exhibits significant increase in the concentration of arsenic than their substrates.

Rate of absorption and accumulation of heavy metals is dependent on morphological feature of lichen thalli in addition to the kind and intensity of emission sources. The uptake of arsenic by a particular organism depends on the bioavailability of arsenic (which depends on its chemical form and environmental conditions) and the characteristics of the organism itself and of its substratum (Garty 2001; Loppi and Pirintsos 2003).

West Bengal is widely known for higher arsenic-contaminated state in India. As has been found accumulated in higher concentration in lichen thallus and lesser in substratum. Accumulation of pollutants is a continous process for lichens, as in the pollutants accumulate pollutants cummulatively over the year with least possibilities of leaching out from the thallus. However, pollutants deposited on substratum can be washed out with rainwater or blown off by wind. According to Deb et al. (2002), the concentrations of As in different areas shows the accumulation sequences as industrial > heavy traffic > commercial > residential.

Wind and its direction are probable agents for dispersion of elements, away from the source. Dispersion of metals depends on the gravity of a particular metal along with speed and direction of wind. Correlation coefficient indicated the dispersion of As in all the directions, however poor towards south and east affirmed the role of prevailing wind in bioaccumulation (Garty 2001).

A foliose lichen (*Pyxine cocoes*) growing luxuriantly in As-contaminated sites of W. B. in Hooghly and Nadia was analysed for arsenic. The mean arsenic concentration in lichen thallus ranged between 6.3 ± 1.5 and 48.1 ± 2.1 µg g^{-1} dry weight, whereas in substratum it was quite low and ranged between 0.8 ± 0.1 and 2.3 ± 0.9 µg g^{-1} dry weight. Maximum As was observed in samples collected from the immediate surroundings of Regional Rice Research Station. The higher concentration of As in thallus as compared to substrates clearly indicates that the As accumulation is airborne and not taken up from the substratum.

Arsenic is widely used in manufacturing of pesticides, weedicides and fertilisers which are capable to contaminate the atmosphere (Al and Blowes 1999). The agricultural land exhibited maximum As concentration (48.1 ± 2.1 µg g^{-1} dry weight) probably due to the frequent use of pesticides and fertiliser in the paddy fields.

Detailed studies on effect of As on transplanted lichen thalli of *P. cocoes* were carried out with different arsenate concentrations of 10, 25, 50, 75, 100 and 200 µM. The thalli were sprayed every alternate days. The thalli harvested on 10, 20, 30 and 45 days exhibit changes in photosynthetic pigments, chlorophyll fluorescence, protein content and antioxidant enzymes. The quantity of photosynthetic pigments exhibited a decreasing trend till 20 days but increased from 30 days onwards. Concomitantly, chlorophyll fluorescence also showed a decreasing trend with increasing arsenic treatment duration as well as concentration. The higher concentration of arsenate was found to be deleterious to the photosynthesis of lichen as the chlorophyll fluorescence and the amount of pigments decreased significantly. The protein content of lichen increased uninterruptedly as the concentration of arsenate as well as duration of treatment increased. The enzymatic activities of superoxide dismutase and ascorbate peroxidase increased initially at lower concentration of arsenate but declined at higher concentrations and longer duration of treatment. The catalase activity was found to be most susceptible to arsenate stress as its activity started declining from the very beginning of the experiment (Bajpai et al. 2012).

It is evident that most of the As metalloid associated with anthropogenic sources deposited directly over the lichen surface. The recorded significant difference in As concentrations among exposure areas may further emphasise the acceptance of lichen to monitor As from the atmosphere. In Indian context more studies are required to determine the concentration of As in lichens around pollution sources in different phytogeographical areas of the country. It is also evident that past mining records in the area, vehicular emission and use of agricultural pesticides played a significant role in the release of As in the environment (Bajpai et al. 2009a).

Among the different growth forms of lichens, the foliose lichen *Pyxine cocoes* appears more suitable species for carrying out arsenic accumulation studies in India. The data obtained on various parameters in transplant study and passive biomonitoring studies clearly shows the effectiveness of *P. cocoes* as biomonitor.

5.2.3 Polycyclic Aromatic Hydrocarbons (PAHs)

PAHs and their homologues are synthesised by the incomplete combustion of organic material arising, partly, from natural combustion and majority due to anthropogenic emissions. In nature, PAHs may be formed three ways: (a) high temperature pyrolysis of organic materials, (b) low to moderate temperature diagenesis of sedimentary organic material to form fossil fuel, and (c) direct biosynthesis by microbes and plants (Ravindra et al. 2008). Forest fires, prairie fires and agricultural burning contribute the largest volumes of PAHs from a natural source to the atmosphere. The actual amount of PAHs and particulates emitted from these sources varies with the type of organic material burned, type of fire (heading fire vs. backing fire), nature of the blaze (wild vs. prescribed; flaming vs. smouldering) and intensity of the fire. PAHs from fires tend to sorb to suspended particulates and eventually enter the terrestrial and aquatic ecosystems as atmospheric fallout (Baumard et al. 1998). Incomplete combustion of organic matter at high temperature is one of the major anthropogenic sources of environmental PAHs. The production of PAHs during pyrolysis (i.e. partial breakdown of complex organic molecules during combustion to lower molecular weight) is the major anthropogenic contribution of PAHs to an ecosystem (Xu et al. 2006; Baek et al. 1991; Zhang and Tao 2008). As the efficiency of energy utilisation has improved, emissions of PAHs in developed countries have decreased significantly in the past decades (Pacyna et al. 2003). However, PAH deposition on the Greenland ice sheet indicates that global PAH emissions have been constant from the beginning of the industrial period up to the early 1990s (Masclet et al. 1995), suggesting that PAH emissions from developing countries have been increasing due to rapid population growth and the associated energy demand. Because of this close relationship between PAH emissions and energy consumption, a strong correlation is anticipated between PAH emissions and some social and economic parameters. Moreover, a strong correlation exists between atmospheric PAH concentrations and population (Hafner et al. 2005). PAH emission inventories have been developed for several countries (the US and UK) and regions (the former USSR, Europe and North America) (Pacyna et al. 2003; Tsibulsky et al. 2001; US EPA 1998; Wenborn et al. 1999; Galarneau et al. 2007; Van der Gon et al. 2007). China has the only PAH emission inventory for a developing country, with km^2 resolution and dynamic PAH emission changes from 1950 to 2004 (Zhang and Tao 2008, 2009).

Srogi (2007) has discussed in detail the effect of PAHs exposure on different components of the ecosystem together with assessment of the risks and hazards of PAH concentrations for the ecosystem (Figs. 5.1 and 5.2) as well as on its limitations. Being semivolatile organic pollutants, PAHs exists in gas and/or particulate phase depending upon the vapour pressure of the individual PAHs remaining gas and particulate phase (Table 5.4).

The PAH concentration varies significantly in various rural and urban environments and is mainly influenced by vehicular and domestic emissions. Due to the persistent nature of PAHs, they have an ability to get transported to long distances far away from their origin mainly in polar regions via regular process of volatilisation and condensation termed as 'Hoping' (Fernandez et al. 1999).

In recent years, PAHs studies have attracted attention in air quality monitoring studies mainly due to its carcinogenic and mutagenic properties (Table 5.4). Among the different PAHs, 5- and 6-ringed PAHs are known to be potential carcinogens (Fig. 5.1); benzo(*a*)

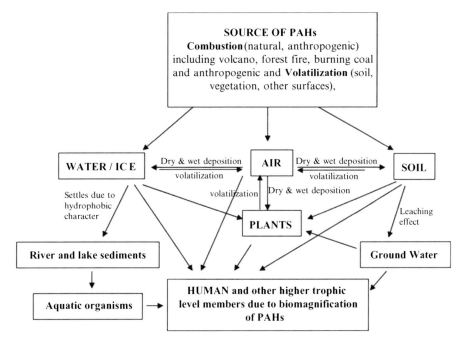

Fig. 5.1 Fate of organic pollutants after being emitted from the source resulting in the contamination of the entire ecosystem

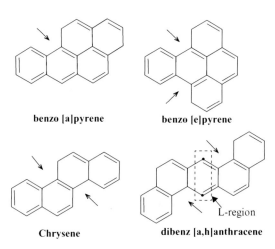

Fig. 5.2 Mechanism of gas phase degradation of naphthalene resulting in the formation of 2-nitronaphthalene; reaction is being initiated by a hydroxyl radical (Adapted from Bunce et al. 1997 and Sasaki et al. 1997)

pyrene (B(a)P) has been identified as being highly carcinogenic (Park et al. 2002). In view of health concern, monitoring the level of particle-bound PAHs in urban areas needs urgent attention (Chetwittayachan et al. 2002). Both petrol- and diesel-fuelled vehicles produce PAHs and nitro-PAHs (Ravindra et al. 2006). Nitro-PAHs are more potent carcinogen which are being produced by gas phase degradation of naphthalene resulting in the formation of 2-nitronaphthalene (Fig. 5.3).

It has been observed that 70–75 % of the carbon in coal is in aromatic form; the 6-membered ring aromatics are dominant with a small 5-membered ring fraction present as well and dominance of particular PAHs indicates its source of origin (Table 5.5) (Ravindra et al. 2008). PAHs such as benz(a)anthracene, benzo(a)pyrene, benzo(e) pyrene, dibenzo(c,d,m)pyrene, perylene and phenanthrene have been identified in coal samples.

Terrestrial sources of PAHs include the non-anthropogenic burning of forests, woodland and moorland due to lightning strikes. In nature, PAHs may be formed in three ways: (1) high-temperature pyrolysis of organic materials, (2) low to moderate temperature diagenesis of sedimentary organic material to form fossil fuels and (3) direct biosynthesis by microbes and plants (Ravindra et al. 2008).

In the last decade there has been increased interest in quantification of organic contaminants

Table 5.4 PAHs monitored using lichens including 16 US EPA priority PAHs, their phase distribution and related health risk

S. No.	PAHs	Particle/gas phase distribution	Carcinogenicity
1	Naphthalene	Gas phase	Nitro substituted form potential carcinogen
2	Acenaphthylene	Gas phase	
3	Acenaphthene	Gas phase	
4	Fluorene	Gas phase	
5	Phenanthrene	Particle gas phase	
6	Anthracene	Particle gas phase	
7	Fluoranthene	Particle gas phase	
8	Pyrene	Particle gas phase	
9	Benz[a]anthracene	Particle phase	√
10	Chrysene	Particle phase	√
11	Benzo[b]fluoranthene	Particle phase	√
12	Benzo[k]fluoranthene	Particle phase	
13	Benzo[a]pyrene	Particle phase	√√
14	Benzo[e]pyrene	Particle phase	
15	Dibenz[a,h]anthracene	Particle phase	√
16	Benzo[g,h,i]perylene	Particle phase	
17	Indeno[1,2,3-c,d]pyrene	Particle phase	√

√√ Potential carcinogen

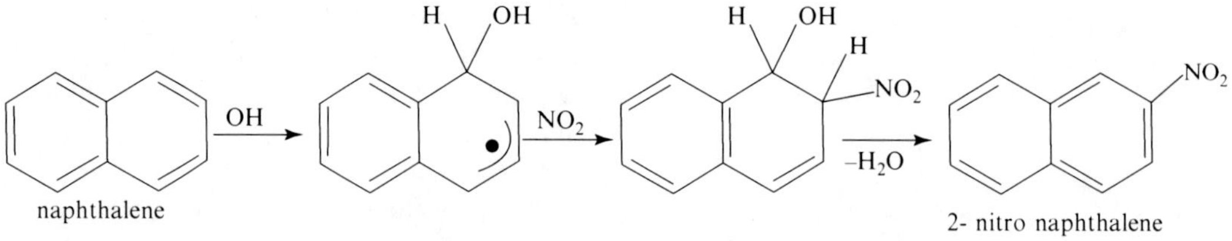

Fig. 5.3 Molecular structures of some PAHs having bay region (depicted with *arrows*) along with L-region having location of highest electron density (depicted by *dots*) in dibenz[a,h]anthracene, responsible for carcinogenicity in PAHs (Modified from Jerina et al. 1978 and Flesher et al. 2002)

in mountain regions, especially in Europe and America (Daly and Wania 2005). Several studies in Europe have demonstrated that chemicals emitted in low latitudes may be transported to higher latitudes as part of the moving air mass, where due to cooler temperatures they condense resulting in deposition of PAHs in high-altitude ecosystems and on ice (Mackay and Wania 1995), while few studies have been conducted in the Indian Himalayas. Detailed effect of PAHs on ecosystem and health risk requires extensive sampling and modelling for interpretation of the results.

Apart from assessment of PAHs and its influence around source of pollution, long-range atmospheric transport and deposition of PAHs is another area of concern, particularly biosphere reserves which sustain high endemism of species and high-altitude ecosystems, especially snow-capped peak, glaciers and pristine alpine forest in the Himalayan region, which are exposed to long-range dispersal of pollutants, as persistent organic pollutants are volatile and can evaporate into atmosphere and are transported to high-altitude regions.

Globally various studies have been conducted using indicator species including lichens (Table 5.6) as a reliable environmental tool for spatio-temporal monitoring of PAHs to identify

Table 5.5 Individual PAHs and their source of origin

S. No.	PAHs	Source of origin	References
1	Chrysene and benzo[k]fluoranthene	Coal combustion	Khalili et al. (1995) and Smith and Harrison (1998)
2	Benzo[g,h,i]pyrene, coronene and phenanthrene	Motor vehicle emission	Smith and Harrison (1998)
3	Low-molecular-weight PAHs; fluoranthene and pyrene	Diesel trucks	Miguel et al. (1998)
4	High-molecular-weight PAHs especially B[a]P and dibenz[a,h]anthracene	Light duty vehicle	Miguel et al. (1998)
5	Phenanthrene, fluoranthene and pyrene	Vehicular activity (Salting of road during winter)	Harrison et al. (1996)
6	Phenanthrene, fluoranthene and pyrene	Emission from incineration	Smith and Harrison (1998)
7	Fluorine, fluoranthene and pyrene along with moderate level of benzo[b]fluoranthene and indeno[1,2,3]pyrene	Oil combustion	Harrison et al. (1996)

emission sources, dispersal and atmospheric deposition (Augusto et al. 2009, 2010; Guidotti et al. 2003; Garćia et al. 2009; Shukla et al. 2012a).

In India PAHs accumulation studies (Table 5.7) with lichens have been recently initiated in the Himalayan region of Uttarakhand (Shukla and Upreti 2009). The PAHs accumulation in lichens of different localities of Dehradun city and on the way to Badrinath was estimated recently (Shukla and Upreti 2009; Shukla et al. 2010). The first baseline data on the distribution and origin of polycyclic aromatic hydrocarbons (PAHs) in *Phaeophyscia hispidula* collected from Dehradun city and other urban settlements of Uttarakhand exhibit the presence of 13 types of PAHs (naphthalene (0.14–5.65 ppm), acenaphthylene (0.89–22.13 ppm), fluorene + acenaphthylene (0.07–3.38 ppm), phenanthrene (0.06–6.47 ppm), anthracene (0.01–0.38 ppm), fluoranthene (0.01–3.58 ppm), pyrene (0.13–14.46 ppm), benzo(*a*)anthracene + chrysene (0.01–0.13 ppm), benzo(*k*)fluoranthene (0.01–0.03 ppm), benzo(*b*)fluoranthene (0.02–0.09 ppm), benzo(*a*)pyrene (0.00–0.03 ppm), dibenzo(*a, h*)anthracene (0.17–0.31 ppm), indino (1,2,3–cd ppm) and pyrene + benzo(ghi)perylene (0.00–0.20 ppm). The PAHs were of mixed origin, a major characteristic of urban environment. Significantly higher concentration of phenanthrene, pyrene and acenaphthylene indicates road traffic as major source of PAH pollution. Probable mechanism of bioaccumulation may be attributed to the donor–acceptor complex and has been reported to be formed between polycyclic aromatic hydrocarbons (carcinogenic and noncarcinogenic) and compounds of biological importance (Harvey and Halonen 1968). Therefore, PAHs (hydrophobic in nature) readily combine with these organic moiety to form adduct. The higher accumulation of 2- and 3-ring PAH in lichens may be because most of the species contains depsides and depsidones with active –OH sites, which facilitate adduct formation. *Phaeophyscia* and *Pyxine* have skyrin triterpine and lichenoxanthone (having hydroxyl group) which readily combine with most of the PAHs.

According to Domeňo et al. (2006) lichens could be used as good bioindicators for air PAHs quantification. Twelve out of the sixteen PAHs studied were found in lichen *Xanthoria parietina* samples with concentration ranging from 25 to 40 ng g^{-1}. The highest concentrations in lichens *Xanthoria parietina* were found for dibenzo(*a,h*)anthracene and benzo(*k*)fluoranthene, followed by benzo(*a*)anthracene, chrysene and fluorene. The reason of non-detection in lichens of other PAHs (five or more rings in their structure)

Table 5.6 Studies reporting concentration of PAHs contaminants in different parts of the world

S. No.	Chemical	Sample	Location	Date	Concentration	References
1.	PAHs associated with PM	Air sample	São Paulo city, Brazil (S.A.)	Aug.–Sep. 2000	0.065–31.2 ng m^{-3}	Vasconcellos et al. (2003)
2.	PAHs associated with PM	Air sample	Hong Kong	1993–1995	0.41–48 ng m^{-3}	Zheng and Fang (2000)
3.	PAHs	Lichen biomonitoring *Pseudevernia furfuracea*	Rieti, Italy	Nov. 1999–July 2001	36–375 µg kg^{-1}	Guidotti et al. (2003)
4.	PAHs	*Poa trivialis* (grass) and soil	Nancy (France)	March 1998		Bryselbout et al. (2000)
5.	PAHs	Salt marsh plant, *Spartina alterniflora*	Dover, New Hampshire (U.S.A)	2003	BDL–71 µg g^{-1}	Watts et al. (2006)
6.	PAHs	SPM Sediment Road dust	Yangtze estuarine (China)	Feb. 2006–Aug. 2006	3- and 4-ring PAH in higher concentration	Ou et al. (2010)
7.	PAHs	Soil	Tokushima (Japan)	NM	1–147 µg kg^{-1}	Korenaga et al. (2000)
8.	PAHs	Air	Tehran (Iran)	Apr. 2004–Mar. 2005	18.71–3085 ng m^{-3}	Halek et al. (2010)
9.	PAHs	Lichen (*Parmelia sulcata, Evernia prunastri, Ramalina farinacea, Pseudevernia, Usnea* sp. *Lobaria pulmonaria, Xanthoria parietina, Hypogymnia physodes*)	Aragón valley	2004	692–6,420 ng g^{-1}	Blasco et al. (2008)

Table 5.7 Total PAHs concentration and individual PAH concentration in different lichen species collected from different regions of India

Species	PAH	Naph	Acy	Fl and Ace	Phen	Anthr	Fluo	Pyr	B(a)A & Chry	B(k)F	B(b)F	B(a)P	D(a,h)A	IP & B(g,h,i)P	ΣPAH	References
Acarospora bullata	Mana (Uttarakhand)	9.42	4.42	BDL	16.18	BDL	BDL	BDL	0.05	BDL	BDL	BDL	BDL	BDL	30.07	Shukla et al. (2010)
Acarospora praeradiosa	Mana (Uttarakhand)	2.91	0.21	BDL	16.75	BDL	BDL	BDL	BDL	BDL	BDL	0.05	BDL	0.09	22.98	Shukla et al. (2010)
Dermatocarpon vellereum	Joshimath (Uttarakhand)	10.55	16.31	BDL	6.4	BDL	0.36	BDL	0.05	BDL	BDL	BDL	BDL	0.05	33.72	Shukla et al. (2010)
D. vellereum	Rudraprayag (Uttarakhand)	BDL–4.74	BDL–0.29	BDL–0.3	BDL–0.2	BDL–0.06	BDL–0.16	BDL–0.009	0.002–0.68	BDL–0.013	BDL–0.002	BDL–0.03	BDL–0.09	BDL–0.04	0.14–4.96	Shukla et al. (2013b)
Dimelaena oreina	Mana (Uttarakhand)	18	BDL	BDL	BDL	BDL	BDL	BDL	BDL	BDL	BDL	BDL	BDL	BDL	18	Shukla et al. (2010)
Heterodermia angustiloba	Badrinath (Uttarakhand)	6.65	18.6	BDL	7.73	BDL	BDL	BDL	BDL	BDL	BDL	BDL	BDL	BDL	32.98	Shukla et al. (2010)
Lepraria lobificans	Rishikesh (Uttarakhand)	0.2–41.97	BDL–1.98	BDL–0.29	BDL–0.11	0.003–1.06	BDL–8.67	BDL–0.04	BDL–0.013	0.006–0.012	BDL–0.009	BDL–0.003	BDL–0.09	BDL–0.17	0.5–43.1	Shukla (2012)
Phaeophyscia. Hispidula	Badrinath (Uttarakhand)	0.01–2.5	0.26–1.51	BDL–2.16	BDL–2.08	BDL–0.08	BDL–1.2	BDL	0.008–2.6	BDL	BDL–0.01	BDL–0.06	BDL–0.16	BDL–0.01	0.68–7.7	Shukla et al. (2010)
P. hispidula	Dehradun (Uttarakhand)	BDL–5.66	BDL–22.14	BDL–3.38	BDL–6.47	BDL–0.38	BDL–3.59	0.09–14.47	BDL–0.132	BDL–0.03	BDL–0.08	BDL–0.04	BDL–0.18	BDL–0.2	3.38–25.01	Shukla and Upreti (2009)
P. hispidula	Dehradun (Uttarakhand)	BDL–0.24	BDL–0.94	1.57–1.69	0.26–0.39	BDL–0.01	0.4–1.85	0.15–2.6	0.14–0.18	0.01–0.02	0.02	BDL–0.003	BDL	BDL	5.1–5.3	Shukla et al. (2010)
Phaeophyscia orbicularis	Srinagar (Uttarakhand)	0.42	ND	0.9	0.56	0.01	0.47	ND	ND	0.11	0.13	ND	0.04	0.013	2.653	Shukla et al. (2010)
Pyxine subcinerea	Haridwar (Uttarakhand)	0.025–2.9	0.03–42.0	BDL–0.63	0.01–0.75	BDL–0.19	0.043–185.8	BDL–0.009	BDL–0.0012	BDL–0.002	BDL–0.02	BDL–0.01	BDL–0.01	BDL–0.009	1.25–187.3	Shukla et al. (2013a)
Remototrhyna awasthii	Mahabaleshwar (Maharashtra)	BDL–5.67	0.07–14.47	BDL–3.75	BDL–4.34	BDL–0.16	0.05–3.21	0.06–10.48	BDL–0.32	BDL–1.39	BDL–1.37	BDL–1.97	BDL–0.59	BDL–0.35	0.193–54.78	Bajpai et al. (2013)
Rinodna sophodes	Kanpur city (Uttar Pradesh)	BDL–0.32	0.01–0.099	BDL–0.16	0.027–0.06	BDL–0.07	BDL–0.06	BDL–0.05	BDL–0.13	ND	ND	ND	ND	ND	0.19–0.49	Satya et al. (2012)

present in the atmosphere in high concentrations may be of being almost exclusively adsorbed on suspended particulate matter. Concerning the origin of the PAHs found in the lichen, benzo(*a*) pyrene is usually emitted from catalyst and non-catalyst automobiles. Benzo(*a*)anthracene and chrysene are often resulted from the combustion of both diesel and natural gas. In both cases the origin suggests the traffic road as a major source of these compounds, which fits to other studies in which benzo(*a*)pyrene and dibenzo(*a,h*)anthracene indicate traffic emission and identify traffic as the main source of urban PAH emission.

Apart from quantification of PAHs (Table 5.7) lichen biomonitoring has been successfully employed to monitor the spatial behaviour of PAHs along with changes in the PAHs in the land use class and related health risk has been estimated (Augusto et al. 2009, 2010; Shukla et al. 2010, 2012a, b; Bajpai et al. 2013). In the high-altitude Himalayan ecosystem, impact of air PAHs fraction has been observed which widely influence the spatial behaviour of PAHs in the area. In a study carried out in Central Garhwal Himalayas, it was observed that the bioaccumulation of 2- and 3-ringed PAHs was higher in samples from higher altitude, while bioaccumulation of fluoranthene (4 ringed PAH), having high spatial continuity, showed higher concentration in samples from higher altitude, while PAHs with 5 and 6 rings were confined to the lower altitude at the base of the valley justifying its particulate bound nature (Shukla et al. 2012a, b).

In India PAHs profile in lichens considerably varies from site to site. Diagnostic molecular ratio has been applied to the biomonitoring data and the results were found to be in conformity with the pollution source, dominant mode of transport. As in Haridwar city, commercial and tourist activity encourages more and more diesel-driven vehicles and has been affirmed by diagnostic ratios at industrial and city centre, an important holy pilgrimage, having combustion being predominant source (Shukla et al. 2012a).

Metallic content (originating mainly due to vehicular activity) bioaccumulated in lichen correlated with its PAH concentration to trace the source of PAH in the air of Haridwar city. Lichen thalli of *Pyxine subcinerea*, collected from 12 different localities of Haridwar city were analysed. The total metal concentration of four metals (chromium, copper, lead and cadmium) ranged between 369.05 and 78.3 µg g^{-1}, while concentration of 16 PAHs ranged between 1.25 and 187.3 µg g^{-1}.

Statistical correlation studies revealed significant positive correlation between anthracene and chromium ($r=0.6413$, $P<0.05$) and cadmium with pyrene ($r=0.6542$, $P<0.05$). Naphthalene, acenaphthene, fluorene, acenaphthylene, anthracene and fluoranthene are reported to be main constituent of diesel vehicle exhaust which is in conformity with the present analysis as lead (indicator of petrol engine exhaust) had negative correlation with all these PAHs (Shukla and Upreti 2011a, b).

Growth from of lichens may also play a significant role in the accumulation of PAHs. The saxicolous, crustose and squamulose species growing on rocks mostly accumulated uniform concentration of low-molecular-weight 2- and 3-ringed compounds. The higher vehicular activities or excessive usage of wood and coal in a particular area is responsible for higher concentration of PAHs (Blasco et al. 2011).

Studies carried out till now in India establishes the utility of *Phaeophyscia hispidula* as an excellent biomonitoring organism in monitoring of PAHs from foothill to the sub-temperate area of Garhwal Himalayas and may be effectively utilised in the other part of the country with transplant studies.

5.2.4 Radionuclides

Radionuclides occur naturally as trace elements in rocks and due to radioactive decay of Uranium-238 and Thorium-232 resulting in release of large amount of energy in the form of ionising radiation which may be alpha, beta or gamma radiations (Table 5.8). When these ionising radiations strike a living organism, it may injure cellular integrity of the organism. Consequently it may lead to cancer and other health problems.

Table 5.8 Common radionuclides, its sources and related health impact

Contaminant	Sources	Health impact
Radium-226	Natural	Carcinogen
Radium-228	Natural	Carcinogen
Radon-222	Natural	Carcinogen
Uranium	Natural	Kidney toxicity, carcinogen
Adjusted gross alpha emitters	Natural and man made	Carcinogen
Gross beta and photon emitter	Natural and man made	Carcinogen

Natural radioactivity arises mainly from primordial radionuclides, like 40-K, and the radionuclides 238-U and 232-Th and their fission products. These radionuclides are present at trace levels in all rocks and ground formations with concentrations varying within a wide range in different geological settings. Soils will have concentration of natural radionuclides determined by their concentrations in the parent rock from which the soils originate and also by various geological processes. Physicochemical properties of soils also have an important role to play in the concentration, distribution and behaviour of radionuclide in soils (Baeza et al. 1995; Belvermis et al. 2010). Silicic igneous rocks like granites are considered to be important sources of uranium mobilisation as they contain higher uranium and thorium content. They are also associated with uranium deposits and contain significant amounts of labile uranium (Ivanovich and Harmon 1982). Hence, the levels of natural environmental radioactivity and the associated external exposure due to gamma radiation are observed to be at different levels in the soils of different regions in the world (UNSCEAR 1993; Patra et al. 2013).

Man-made source including the exploitation of nuclear energy for military and peaceful purposes has considerably increased the risks of higher doses of ionising radiations being received by different components of the ecosystem. The release and effect of nuclear radiation has received particular attention after the tragic accident in April 26, 1986, at the Chernobyl Nuclear Power Plant (NPP) in Ukraine, and another large-scale accident, explosion of the deposit of nuclear wastes in Kyshtym in Eastern Urals in 1957, has also been investigated (Nikipelov et al. 1990; Sawidis 1988). The Fukushima nuclear disaster in the year 2011 was a result of the location of the NPP in the high seismic zone in Japan which resulted in the release of a large amount of nuclear radiation.

It has been observed that the impact of nuclear accident is maximum within the radius of 6–8 km; the fallout of products of nuclear fuel is more or less homogenous and rather dense. At greater distances the fallout of radioactive particles of nuclear fuel was influenced by atmospheric processes and by physical peculiarities of thrown-out particles; their sedimentation was patchy and accumulation in lichens is heterogenous (Biazrov 1994).

As nuclear accidents result in considerable release of fission products of the nuclear fuel into the atmosphere in the area, therefore, application of lichen biomonitoring related with the ecological consequences of nuclear catastrophes is related to the understanding of the spatial distribution of a number of radionuclides in the thalli of different lichen species and elucidation of both general patterns and regional features of the accumulation of radionuclides in the thalli of lichens from the nuclear accidents both spatially and temporally (Biazrov 1994).

A number of studies carried out before and after the Chernobyl accident, have demonstrated that lichens and mosses can accumulate high amounts of radioactivity (Hviden and Lillegraven 1961; Eckl et al. 1986; Papastefanou et al. 1988, 1992; Seaward et al. 1988; Hanson 1967). Interest in the behaviour of radionuclides in natural ecosystems was first developed during the 1960s, at the time of weapons testing in the atmosphere, because of the observed transfer of Cs-137 along the lichen–reindeer–man food chain (Hanson 1967).

The most critical food chain in the world for concentrating airborne radionuclides is the lichen–caribou–human food chain. Lichens accumulate atmospheric radionuclides more efficiently than other vegetation due to their lack of roots, large surface area and longevity. Uptake from the substrate is minimal compared with the uptake from wet or dry deposition. Lichens are the main winter forage for caribou, which in turn, are a main dietary staple for many northern Canadians. Thus, airborne radionuclides, particularly cesium-137 (137Cs), lead-210 (210Pb) and polonium-210 (210Po), are transferred efficiently through this simple food chain to people, elevating their radiological dose (Thomas and Gate 1999).

Lichens have been frequently used to monitor spatial patterns in radioactive deposition over wide areas (Feige et al. 1990). Terricolous lichens as well as epiphytic lichens may be effectively utilised as biomonitors (Sloof and Wolterbeek 1992). Radiocesium content in Finnish lichens was 5–10 times higher than in higher plants before 1960, but after 1965, there was a rapid decrease of nuclear weapon tests and subsequently radionuclide fall out. There was no apparent decrease in concentration in lichens (Salo and Miettinen 1964; Tuominen and Jaakkola 1973).

Radiocesium uptake is generally highest in terricolous and lowest in epiphytic lichens, epilithic species ranging in between which depends on several factors, such as the inclination of the thallus and its hydration physiology (Kwapulinski et al. 1985a, b). Guillitte et al. (1994) found that lichens with a horizontal thallus occurring on tree branches were twice as contaminated as those with a vertical thallus growing on tree trunks. Most of the radiocesium is deposited at the thallus surface, whereas uptake from the soil seems to be negligible; only 2 % of the soil radiocesium can penetrate into terricolous lichen thalli. Hanson and Eberhardt (1971) found a seasonal cycle of radiocesium in lichens, with maximum values in summer and a minimum in midwinter. The distribution of radiocesium in lichen thalli was the object of several studies, starting from the early 1960s. In fruticose lichens the apical parts of the thalli contain 2–14 times more cesium than the basal parts (Paakola and Miettinen 1963; Hanson 1967). The mobility of radiocesium inside the thallus was studied by Nevstrueva et al. (1967); results indicate that Cs and Sr are rather mobile within the thallus, Cs being less leachable. *Cladonia stellaris* was periodically monitored in Sweden from 1986 to 1990 which showed that there was a slight downward movement of radiocesium through the lichen carpets; however, some 70–80 % of radiocesium still resided in the upper 3 cm (Kreuzer and Schauer 1972; Mattsson 1974). According to Hanson and Eberhardt (1971), the concentrations of radiocesium are relatively stable in the upper parts of terricolous lichens, but the radionuclide is apparently cycled between the lower portions of the lichen mats and the humus layer. Feige et al. (1990) studied the radionuclide profile of *Cetraria islandica* and *Cladonia arbuscula*: the radionuclides are almost uniformly distributed throughout the thalli, although the upper parts of *Cladonia arbuscula* appear to be more radioactive than the lower parts. In *Cetraria islandica*, the apothecia tend to accumulate more radionuclides than the rest of the thallus. Some of the pictures show also the presence of locations with higher concentration corresponding to products of nuclear fusion or to highly radioactive particles deriving from the Chernobyl accident and trapped inside the thalli. The same authors have also tried to wash the lichens in deionised water: after a week only 8 % of the radionuclides were removed, and after 2 weeks the removal interested only 3 % of the remaining radioactivity (Nimis 1996).

Morphological differences between species may play an important role in their capacity to intercept and retain radiocesium. Kwapulinski et al. (1985a, b) found species-specific differences in four species of *Umbilicaria* collected in Poland. Sloof and Wolterbeek (1992) studied radiocesium accumulation in a foliose lichen, *Xanthoria parietina*, and expressed the activity on a weight and on an area basis and found large variations between parts of the thallus with and without fruit bodies, whereas the average radiocesium activity expressed per surface area was almost constant. In general, lichens, especially foliose and fruticose species, have a high surface area to mass ratio; this property is often reported as one of the main reasons for their relatively high capacity to accumulate heavy metals and radionuclides (Seaward et al. 1988; Nimis et al. 1993). Like in higher plants, uptake and release of cesium in lichens may be affected by the chemically related and physiologically important elements, potassium, sodium and, in a lesser degree, calcium (Tuominen and Jaakkola 1973). This factor, however, seems to be important only on a physiological level. Much less studied are the physiological mechanisms underlying radiocesium uptake by lichens. According to Tuominen and Jaakkola (1973) some process of cationic exchange should be involved. However, Handley and Overstreet (1968) demonstrated that the fixation of radiocesium in lichen thalli does not depend on their physiological activity, being mostly a passive phenomenon.

According to Subbotina and Timofeeff (1961), however, radiocesium ions were still strongly bound and difficult to remove from partially decomposed thalli. This would suggest that the ions are transported into the thallus and bound to cytoplasmic molecules through processes of active translocation. There is some evidence that lichens are more resistant than other organisms to high radioactivity: according to Biazrov (1994), lichen thalli measured near Chernobyl showed extremely high radioactivity values, but these did not cause any visually discernible anomalies in the development of lichen thalli, confirming the data on the high resistance of lichens to radioactive irradiation earlier presented by Brodo (1964). The biological half-time of radiocesium in lichens is very variable, depending on the species and especially on precipitation (Tuominen and Jaakkola 1973). The literature values range from 2.7 to 17 years. The effective half-life of radiocesium in carpets of *Cladonia* was estimated differently by different authors: from 5 to 8 years, to 17 + 4 years, and 7–8 years in the upper 3 cm and about 8–10 years in the whole carpet (Ellis and Smith 1987; Lidén and Gustavsson 1967). Martin and Koranda (1971) gave a biological half-time of ca. 8 years in interior Alaska and of 3–3.7 years in coastal areas. These differences might be due to differences in precipitation between the humid coastal areas and the relatively dry internal regions. Lidén and Gustavsson (1967) suggested that as time elapses from the moment of deposition, the effective half-life of radiocesium for lichens will increase. In Canada, after the cessation of nuclear weapons' testing in 1962, the cesium deposited as fallout was available to agricultural plants for only a few years (Bird 1966, 1968); further north, the fallout was not lost as quickly; lichens, mosses and vascular cushion plants between 60° and 70°N demonstrated significant available Cs-137 in the 1980s, long after it had disappeared from the more contaminated regions further south (Hutchinson-Benson et al. 1985; Meyerhof and Marshall 1990). According to Hanson (1967), the biological half-life period in *Cladonia stellaris* is of 3–6 years when deposition has happened in the liquid form, of 1–13 years when it has occurred in the gaseous form. Different formulas to calculate the removal half-times in lichens were proposed (e.g. Gaare 1990; Sloof and Wolterbeek 1992). However, a generalisation is probably difficult: different factors affect the actual half-life of radiocesium in lichens; some of them depend on features of the lichen itself, such as growth rates, genetic variability and density of fructifications; others depend on characteristics of the station, such as microclimatic variability, leaching of the substrata and geographic situation. The sampling techniques, as well, may have an influence on the estimates: different values might be obtained if sampling the upper vs. the lower parts of the thalli.

The estimation of the residence time of long-lived radionuclides in lichens and mosses is important for mineral cycling studies in natural ecosystems, especially in case of edible species, because of their important role in the food chain (Iurian et al. 2011).

Biological half-life, also termed as ecological half-life, residence time or environmental half-life, refers to the time it takes to reduce the amount of a deposited element to half its initial value by natural processes. The effective half-life is the time taken for the amount of a specified radionuclide in the body to decrease to half of its initial value as a result of both radioactive decay and natural elimination.

Lichens are living accumulators of natural and man-made radionuclides and heavy metals (Table 5.9) (Eckl et al. 1986; Seaward 1992; Jeran et al. 1995; Boileau et al. 1982). Gorham (1959) reported for the first time that lichens are

Table 5.9 Levels of radionuclide content in different lichen species

S. No	Lichen		Radionuclide content								References
			Ra-226	Pb-210	Mn-54	Zn-65	Sr-90	Cs-137	Ce-Pr-144	U-238	
1	*Cetraria nivalis* (pCi gm^{-1} dry weight)	Greenland	–	–	0.8	1.0	4.9	12.0	13.2	–	Hanson (1971)
2	*Alectoria ochroleuca*	Greenland	–	–	0.07	2.3	2.6	13.3	17.6	–	
3	*Cetraria delisei*	Greenland	–	–	1.9	0.9	1.4	25.0	11.5	–	
4	*Stereocaulon paschale*	Greenland	–	–	N.D.	1.6	4.7	20.0	14.9	–	
5	*Cladina stellaris* (Bq kg^{-1} dry weight)	Polar Urals	–	–	–	–	90–150	190–450	–	–	Nifontova (2000)
6	*Cladina stellaris*	Northern Urals	–	–	–	–	50–100	120–310	–	–	
7	*Flavocetraria nivalis*	Polar Urals	–	–	–	–	50–160	100–430	–	–	
8	*Flavocetraria nivalis*	Northern Urals	–	–	–	–	70–90	220–230	–	–	
9	*Hypogymnia physodes* (Bq kg^{-1} dry weight)	Slovenia	6–279	175–1904	–	–	–	–	–	0.14–6.16	Jeran et al. (1995)

much more efficient in accumulating radiocesium than higher plants, hence representing suitable bioindicators of the radioactive fallout. A good correlation is known to exist between total radiocesium content in lichens and total estimated deposition, which led to the use of these organisms as biomonitors of radioactive compounds (Hanson 1967; Nimis 1996). The large surface area of lichens, relative to their mass, is one of the main reasons for their relatively high capacity to accumulate radionuclides or other elements, like heavy metals. Handley and Overstreet (1968) demonstrated that the fixation of radiocesium in lichen thallium does not depend on their physiological activity, being mostly a passive phenomenon. In general, epigeic lichens accumulated higher Cs-137 concentrations than epiphytic lichens. The cesium content in the epiphytic lichens is mostly due to the absorption of the airborne radionuclides, unlike the epigeic lichens which can accumulate the Cs-137 from air but also through the exchange of cesium atoms between the lithogenic substrate and lichen thallus. Moreover, epiphytic lichens are somehow protected from any kind of contamination, by the crown of the tree on which they grow.

Radionuclide concentrations, e. g. in the thalli of lichens, exceed by far the content of the same substances in various organs of vascular plants (Biazrov and Adamova 1990; Adamova and Biazrov 1991). As was revealed experimentally, these organisms are capable of retaining high dosages of ionising radiation (1,000 R/24 h during 22 months) without detrimental effects (Brodo 1964).

Cesium-137 concentrations in lichen and moss samples have been studied for calculations of natural depuration rates. The natural depuration rates are estimated at biological half-lives. The biological half-lives of 137Cs in a lichen and moss samples (*Xanthoria parietina* and *Leucodon immersus*) are estimated to be 58.6 and 10.9 months, respectively. The result supports the view that radioactivity monitoring in lichens can be a more useful monitor than mosses to determine the lasting effect of radioactive fallout (Topcuoğlu et al. 1995).

It has been studied that high nuclear irradiation does not cause any visually discernable anomalies in the development of lichen thalli which imparts high resistance to lichens to radioactive irradiation and makes them a bioindicator of nuclear fallout (Brodo 1964).

Specific activity of Cs-137, K-40 and Be-7 in four lichen species were chosen for the calculation of biological half-times: the epiphytic lichens *Pseudevernia furfuracea* (from *Betula pubescens*) and *Hypogymnia physodes* (from *Betula verrucosa*), specific for the mountainous and subalpine regions, and the epigeic lichens *Cladonia squamosa* and *Cladonia fimbriata*.

Natural samples of lichen *Peltigera membranacea* were tested for uranium sorption mechanism. Thalli were incubated in solutions containing 100 ppm U for up to 24 h at pH values from 2 to 10. U the pH range 4–5. Maximum U uptake by *P. membranacea* averaged nearly 42,000 ppm which represented the highest concentration of biosorbed U of any lichen reported. Electron probe microanalysis (EPM) revealed that U uptake is spatially heterogeneous within the lichen body, and U attains very high local concentrations on scattered areas of the upper cortex. Energy dispersive spectroscopic (EDS) analysis revealed that strong U uptake correlates with phosphate signal intensity, suggesting involvement of biomass-derived phosphate ligands or surface functional groups in the uptake process (Haas et al. 1998).

Differences in radionuclide content have been observed between the lichen species growing on different mountain rocks and slopes and at different elevations. These differences depend both on the structural and functional characteristics of lichen species and on specific climatic and ecological conditions at the site of their growth (Seaward 1988; Nifontova 2000).

However, in India no such study has been carried out yet but it has prospects as India has large deposits of thorium in the southern coast of the country as well as use of nuclear energy in power generation is also starting. Therefore, in view of the studies carried out worldwide, study of the spatio-temporal behaviour of these radionuclides, lichen biomonitoring may be explored in India as sentinels of radionuclides.

5.2.5 Climate Change

Many chemical compounds present in the earth's atmosphere act as greenhouse gases. Some of them occur in nature (water vapour, carbon dioxide, methane and nitrous oxide), while others are exclusively human-made (like gases used for aerosols). These gases allow sunlight to enter the atmosphere freely. When sunlight strikes the earth's surface, some of it is reflected back towards space as infrared radiation (heat); it gets trapped by these compounds. In order to maintain temperature equilibrium over time, the amount of energy sent from the sun to the earth's surface should be about the same as the amount of energy radiated back into space, leaving the temperature of the earth's surface roughly constant. But the presence of greenhouse gases which absorb this infrared radiation and trap the heat in the atmosphere causes imbalance resulting in global temperature rise. Levels of several important greenhouse gases have increased by about 25 % since large-scale industrialisation began around 150 years ago. During the past 20 years, burning of fossil fuel is the source of anthropogenic carbon dioxide emissions. Concentrations of carbon dioxide in the atmosphere are naturally regulated by carbon cycle. The movement (flux) of carbon between the atmosphere, land and oceans is dominated by natural processes via plant photosynthesis. These natural processes can absorb net 6.1 billion metric tonnes of anthropogenic carbon dioxide emissions produced each year (measured in carbon equivalent terms); an estimated 3.2 billion metric tonnes is added to the atmosphere annually. The earth's positive imbalance between emission and absorption results due to continuing increase in greenhouse gases in the atmosphere due to excessive anthropogenic contribution. World carbon dioxide emissions was expected to increase by 1.9 % annually between 2001 and 2025 mainly in the developing world where emerging economies, such as China and India, where fossil fuel is used for energy generation. In the developing countries emissions are expected to grow at 2.7 % annually between 2001 and 2025 and surpass emissions of industrialised countries near 2018 (http://www.eia.gov/environment.html; US Energy Information Administration 1998).

Global-warming potential (GWP) is a relative measure of how much heat a greenhouse gas traps in the atmosphere. It compares the amount of heat trapped by a certain mass of the gas to the amount of heat trapped by a similar mass of carbon dioxide. A GWP is calculated over a specific time interval, commonly 20, 100 or 500 years. GWP is expressed as a factor of carbon dioxide (whose GWP is standardised to 1). For example, the 20-year GWP of methane is 72 (IPCC 2007). The GWP depends on the following factors:

1. The absorption of irradiation by the compound
2. The spectral location of its absorbing wavelengths
3. The atmospheric lifetime of the compound

The combustion of fuels mainly releases SO_2, NOx, CO and ozone. NOx is easily oxidised to HNO_3 (the resulting lifetime of NOx is approximately 1 day); it cannot be directly transported over long distances. Additionally, HNO_3 in the troposphere is removed quickly by deposition and is not an effective reservoir for NOx. However, research in the past decades has shown that peroxyacetyl nitrate (PAN) is a more efficient reservoir for NOx in long-range transport (Hov 1984; Staudt et al. 2003).

Ozone is a major pollutant in the intercontinental transport research not only due to its adverse effects on air quality and climate over the downwind regions but also because of the complexity in the photochemistry involved in its production and destruction along the transport process. Recent surface ozone measurements in Asia in comparison with earlier measurements indicate that Asian ozone concentrations have increased significantly in the last few decades (Intergovernmental Panel on Climate Change IPCC 2001), due to increases in anthropogenic emissions of ozone precursors in Asia (for details see in Sect. 5.2.5).

Carbon monoxide (CO) a product of incomplete combustion, can be effectively transported globally due to its long lifetime of 1–3 months in the troposphere (Staudt et al. 2001; Liu et al. 2003). A major source of the springtime Asian CO outflow is biomass burning in Southeast Asia, extending from northeast India to southern China and maximising in Burma and Thailand (Heald et al.

2003). The air plumes from biomass burning are transported over the Pacific at lower latitudes than typical of other Asian anthropogenic pollutants (Heald et al. 2006). The CO, however, oxidises very fast and forms CO_2, which though is not noxious but is one of the major contributors of greenhouse effect. This implies a reduction of CO, hence CO_2 emissions, can only be achieved by improving the engine efficiency or by using fuels containing lower concentration of carbon such as natural gas.

The compressed natural gas (CNG) is a clean-burning alternative fuel for vehicles (Kathuria 2002) with a significant potential for reducing harmful emissions especially fine particles. It has been observed that diesel combustion emits 84 gram per kilometre (g/km) of such components as compared to only 11 g/km in CNG. The levels of greenhouse gases emitted from natural gas exhaust are 12 % lower than diesel engine exhaust when the entire life cycle of the fuel is considered. It has also been found that one CNG bus achieves emission reduction equivalent to removing 85–94 cars from the road. The emission benefits of replacing conventional diesel with CNG in buses lies on the fact that it results in reduction (%) of CO, NOx and PM to 84, 58 and 97 %, respectively (Kathuria 2004).

Another potential greenhouse gas (GHG), methane (CH_4) is involved in a number of chemical and physical processes in the earth's atmosphere. In the global CH_4 cycle, substantial amount of CH_4 is consumed by biological processes. The only known biological sink for atmospheric CH_4 is its oxidation in aerobic soils by methanotrophs or methane-oxidising bacteria (MOB), which can contribute up to 15 % to the total global CH_4 reduction. Methanotrophs, Gram-negative bacteria that utilise CH_4 as their sole source of carbon and energy, play a crucial role in reducing global CH_4 load due its CH_4 consumption characteristics (Singh 2011).

Although the complexity of the natural system sets fundamental limits to predictive modelling, the approach is useful in obtaining a first approximate estimate for the potentially dramatic impact of climate change on biodiversity. Models that can elucidate the correlations between climate and biophysical processes and thus increase ability to predict the consequences of the effects of climate change on distribution of species and their habitats should be developed (Sutherland et al. 2006). Better information on the climate sensitivity of species is essential to be able to detect responses of individual species to climate change, to assess critical levels and to develop anticipatory strategies. Thus, there is a need to find appropriate indicators to identify the effects of climate change, to verify results of modelling and to determine the response of species (Sutherland et al. 2006; Bässler et al. 2010).

The effects of climate change can be best monitored in alpine and montane ecosystems, as clearly described by Grabherr et al. (1994). Mountain species are unusually sensitive to the climate and are threatened by climate change (Pauli et al. 2007; Thuiller et al. 2005) because they lose parts of their range. The effects of warming are worsened by the disproportionately rapid decrease in available land surface area with increasing altitude. Species in low mountain ranges are limited in how they can adjust their ranges in response to increasing temperature. Currently, the most relevant physical and temporal scales of ecological investigation are local (Walther et al. 2002). At this local scale, there is a need to consider a variety of taxonomic groups (Ellis et al. 2007). At a regional scale, a strong impact of global warming on various taxonomic groups is expected, with some species becoming extinct. The species–environment relationship for high-montane species is expected to be less complex and seems to be dominated mainly by the effect of low temperatures for several taxonomic groups. Thus, it is assumed that climate warming will lead to a sensitive behaviour, specifically to changes in distribution in the form of decreasing probability of occurrence. These selected species are hence good indicators at a regional scale, suitable for long-term monitoring designed to validate results from modelling and to determine the response of species to climate change. It is assumed that all high-montane species will have the same response driven by the same environmental factors, which would make most of these species suitable as cross-taxon climate-sensitive indicators. Such indicator species might reduce the potentially high costs of climate change monitoring of species inhabiting more complex systems at lower altitudes (Chapin and Körner 1994).

Lichens are considered as sensitive indicators of global warming, as the spread of several thermophilous epiphytes in north-western Central Europe has been attributed to late twentieth-century warming (Hauck 2009). Occurrence of range expansions especially in the western-most parts of temperate Europe with its mild climate supports the hypothesis that these recent changes in the distribution or regional frequency of lichen species are driven by rising temperatures. In the recent years some thermophilous lichen species strongly increased in frequency, such as *Candelaria concolor*, *Flavoparmelia caperata*, *Hyperphyscia adglutinata*, *Hypotrachyna* sp., *Parmotrema perlatum* and *Punctelia borreri* (Søchting 2004; Aptroot and van Herk 2007). Others lichen species have invaded Western Europe from outside (*Heterodermia obscurata*, *Physcia tribacioides*) (Wolfskeel and van Herk 2000). Some lichens from cold environments with arctic alpine or boreal-montane distribution patterns have declined at lowland sites and on isolated mountains of temperate Western and Central Europe (Aptroot and van Herk 2007).

Biological monitoring (with epilithic lichens) of the local consequences of anticipated global climate change has been studied in Israel. The study was based on standardised protocol which included sampling scheme, including lichen measurement along transects on flat calcareous rocks, and construction of a trend detection index (TDI). TDI is a sum of lichen species cover with coefficients chosen so as to ensure maximum ability to detect global climate trends. Coefficients were estimated along an altitudinal gradient from 500 to 1,000 MSL. The gradient study demonstrated that the TDI index is performed better than other integrated indices. Measuring, for instance, a 100 transects in 50 plots (two-transect-per-plot scheme) allows one to detect a climate-driven change in the epilithic lichen community corresponding to a 0.8 C shift in annual mean temperature. Such resolution appears sufficient in view of global warming of 2.5 C considered by the Intergovernmental Panel on Climate Change as a realistic prediction for the end of the next century (Insarov et al. 1999; Insarov 2010).

Change in the lichen diversity in relation to climate change has been well worked out (Hauck 2009; Aptroot and van Herk 2007), but in India few such studies have been carried out (Joshi et al. 2011, 2008) which revealed terricolous lichens respond to global warming as their distribution has been restricted due to warming phenomenon. In the Himalayas (Pindari region) green algae-containing lichens exhibit an increase in number than cyanolichen based on comparison of the past published account of lichens three decades earlier as well as there has been increase in trentepohiliod lichen species and decrease in soil- and rock-inhabiting lichens (Joshi and Upreti 2008).

Another aspect related with climate change is increased UV-B radiation due to ozone depletion caused by Chloro Fluoro Carbons (CFCs). This phenomenon is quite pronounced in the polar region. Spring time ozone depletion in the polar region is due to release of chlorofluorocarbons in the earth atmosphere and is a serious cause of concern among environmentalists. Flora of Antarctic is mostly dominated by cryptogamic plants with limited distribution mostly confined to Sub-Antarctic region. Cryptogams being photoautotrophic plant, for light requirements, are exposed to extreme seasonal fluctuation in photosynthetically active radiation (PAR) and ultraviolet (UV) radiation. Antarctic cryptogams are known to withstand the enhanced UV radiation by synthesis of screening compounds (UV-B-absorbing pigments and anthocyanin compounds). A major part of the UV-absorbing compounds appeared to be constitutive in lichens which are usnic acid, perlatolic acid and fumarphotocetraric acid which is particularly induced by UV-B. Secondary metabolites such as phenolics, parietin and melanin also enhance the plant defence, by different molecular targets in specific solar irradiance and potential for increased antioxidative protection to UV-induced vulnerability (Singh et al. 2011).

Photoprotective potential of desiccation-induced curling in the light-susceptible old forest lichen,

Lobaria pulmonaria, also provides evidence for morphological adaptations in lichens to tolerate high incident radiation (Barták et al. 2006).

Although the use of indicator species remains contentious, it can be useful if some requirements are fulfilled (Carignan and Villard 2002; Niemi and McDonald 2004). Correlation of changes at community or individual levels of lichens correlates better with changes in the land-use class and air quality rather than rising temperature. Therefore, these aspects need special consideration while utilising changes in lichen diversity in predicting the impact of climate change on biodiversity (Hauck 2009).

5.2.6 Assessment of Paleoclimatic Conditions (Lichenometry)

Quantitative estimation of paleoclimate is fundamental to the reconstruction of past environmental and biotic change and provides a baseline for predicting the effects of future regional and global climate change (Liu and Colinvaux 1988; Wilf 1997; Behling 1998). Some of the most widely used methods employ biological proxies such as pollen, diatoms or plant megafossils.

Accelerated rate of melting of glaciers due to global warming is a worldwide phenomenon, especially in tropical region. Due to temperature rise glaciers are melting rapidly causing the shrinking, subsidence and retreating of glaciers with the result of expansion and formation of glacial lakes to the stage of potential glacial lake outburst floods (GLOFs). The glacier retreat phenomena has been taking place rapidly in recent decades, with the common and widespread fear of too much water (GLOFs) and too little water (glacier retreat). Temperature, precipitation and humidity have changed significantly over the last half century (Vuille et al. 2008). Studies show a temperature increase of around 0.6–0.2 °C since 1900 (Lozan et al. 2001). In a study carried out in Northern Pakistan, increased seismic activity has been correlated with rise in temperature which results due to isostatic rebound of earth caused by loss in mass of glaciers, termed as 'unloading phenomenon' (Usman et al. 2011).

Himalayan region is one of the most dynamic, fragile and complex mountain ranges in the world due to tectonic activity and a rich diversity of climates, hydrology and ecology. The high Himalayan region is the fresh water source of largest river systems in Asia, on which over 1.3 billion peoples are dependent. The melting of snow and ice from these glaciers and snowmelt runoff from the mountains is of course just a part of the water supplies of these rivers. The instabilities can impact people severely, especially those residing within or near the mountains (Kulkarni 2007).

Lichenometric techniques is very useful in dating moraine ridges on recent glacier forelands in alpine regions, as once attached to substratum, the position of lichen thallus during the entire lifespan does not change; therefore, the age of lichen is an alternate for the minimum exposure time of a substrate to the atmosphere and sunlight. Lord William Hamilton (1730–1803), a naturalist, first applied botany on geological dating problems and tried to relate the density and type of vegetation cover with the age of lava flows of Vesuvius. The basic concept of lichenometry is based on the similar approach. The use of lichens growth for relative dating of the surfaces was first proposed by the botanist Knut Faegri in the 1930s which was further expanded by the Austrian botanist Roland Beschel in the 1950s (Beschel 1950; Joshi and Upreti 2008). Lichenometry has majority of applications from dating glacier moraines, landslides and fluvial deposits to calibrating the age through the formation of old monuments, buildings and other archaeological structures (Innes 1985). The applications of lichenometry based on different palaeoclimatic events, reconstructions and man-made artefacts used lichenometry in reconstructing the Holocene environment from deltaic deposits in northeast Greenland. Gob et al. (2003) used the technique in Figarella river catchment in France. The dating was performed on lichens present on terrace pebbles to determine the period of their deposition or terrace formation and incision of the river. According to the study the Figarella underwent three major incision phases. The highest level of 20–25 m did not have any

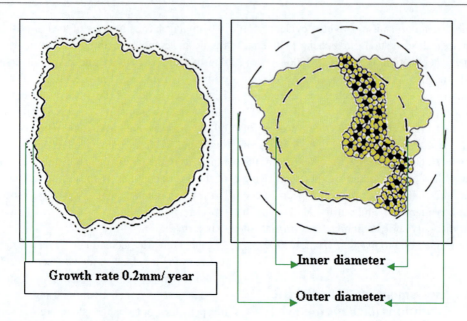

Fig. 5.4 Lichenometric studies are based on annual growth rate of radially growing lichens as in *Rhizocarpon geographicum*

lichen colonisation and was inhabited by high, dense scrub. The maximum lichen size below 12 m of terrace was 11 cm that envisaged the age of the terrace formation about 1,800 years. This leads us to conclude that the river began to incise and transform the pebble sheets in the terrace about 2,000 years ago. The third level was at 3 m and represented by largest thalli size of 7 cm that indicates 400 years old terrace level. The study also highlighted the relation of boulder transportation with palaeofloods. The presence of largest lichen thalli on riverbed boulders indicated the longevity of their stable conditions. That in turns gave an idea of the last or ancient flood occurrences (Joshi et al. 2012).

An interesting and rather unnoticeable application of lichenometry is to date the prehistoric eruption of volcanoes in a particular landscape. On account of high tolerance, lichens can survive in extreme conditions of heat. Many lichens such as *Caloplaca crosbyae, Dirinaria aegialita, D. applanata, Candelaria concolor, Ramalina umbilicata, Hyperphyscia adglutinata, Syncesia* and *Xanthoparmelia* sp. are known to grow on magma (Jorge-Villar and Edwards 2009).

Dating range depends on the specific species and environmental factors. In temperate environments foliose form is expected to survive about 150 years, while crustose forms can provide dating 400–600 years, and at high latitudes, dating may exceed 1,000 years (Winchester 2004). Absolute dating is based on the size of the largest surviving lichen. Therefore, reference to their specific details should be taken as minimum approximations only. Other factors leading to lichen mortality and renewed colonisation are competition for growing space on the rock surface, vegetation rate, weathering and geomorphic changes (Joshi 2009). In alpine environments, growth of *Rhizocarpon geographicum* is very slow, i.e. 0.2 mm year^{-1} (Hansen 2008).

In India, lichenometry has been initiated in the initial years of the last decade. But lack of valuable information in the study sites in favour of palaeoclimatic conditions and surfaces to date is the major problem in the application of the technique in India (Joshi and Upreti 2008). In a study carried out in Pindari glacier, lichen, *Rhizocarpon geographicum*, with a known growth rate of 0.2 mm year^{-1} was selected to date recent glaciations activity in the area (Fig. 5.4).

Advantage of applying lichenometry in reconstruction of paleoclimate with some restrains lies on simple methodology, where the other surface dating tools such as radiocarbon dating, dendrochronology and weathering techniques face difficulties. Species identification in the field, influence of environmental factors on growth rate, nature and timing of colonisation (colonisation delay), absence of reproducibility in many published sampling designs, the supposed inadequacy of a single parameter as an index of age and difficulties associated with the methodology of growth rate determination are some of the drawbacks in the implementation of lichenometry. However, as the technique is widely used in relative or approximate dating, its role in tectonic, geomorphic, geo-chronological studies and in other landform evolutions underlies the importance of the technique.

5.2.7 Loss of Biodiversity

Both natural and man-made disasters are responsible for the loss of biodiversity. Natural catastrophes include volcanic eruptions, hurricanes, heavy rains and floods, while economic growth is also causing decline in biodiversity. In most of South Asia, the percentage of land area in which nature is protected is low compared to that in the developed world. Most of the protected areas in India and Pakistan are only partially protected. Of all the major South Asian countries, as far as the area afforded nature protection is concerned, Sri Lanka is most environmentally protected. In the case of China, a much higher proportion of its land area than in India is protected and more than 80 % of its protected area is totally protected compared to India's 24 % (Alauddin 2004).

Rapid loss of forest area results in loss of flora and fauna and leads to loss of carbon sequestration potential. Tropical forest ecosystems contain the world's greatest diversity of flora and fauna (Sporn et al. 2010). The world's tropical forests disappeared at the rate of 15 million hectares per year (equivalent to 40 % of the Japanese archipelago) during the latter half of the twentieth century. During the 10-year period from 1990 to 2000, Indonesia's forest decreased by 1.2 % (from 118.1 million to 105 million ha), Malaysia's by 1.2 % (from 21.7 to 19.3 million ha), Myanmar's by 1.5 % (from 39.6 to 34.4 million ha), the Philippines' by 1.5 % (from 6.7 to 5.8 million ha) and Thailand's by 0.7 % (from 15.9 to 14.8 million ha). The deforestation in northern China and Mongolia is mainly due to overgrazing, which is aggravating the yellow dust storm phenomenon, while deforestation in Siberia is due to commercial logging (Japan Environmental Council 2005).

Increasing demand for wood and hence commercial logging is the main cause of the forest fires as burning is the cheapest way to clear land for agricultural and construction purposes (Fig. 5.5). In Southeast Asian countries like Indonesia, Malaysia, Singapore, Brunei and Thailand, forest fires are executed to expand palm oil plantations. Since 1977 Indonesian forest fire smoke has become a regular environmental event. The smoke gets worse in dry weather, especially when coupled with the *El Niño* phenomenon (Japan Environmental Council 2005).

Boreal forest fires also result in emissions of NOx and PAN which can enhance the formation of O_3 in the Arctic. Current estimates of the emission ratios for NOx and PAN (relative to CO) from boreal forest fires are highly uncertain and based on few studies (Singh et al. 1996, 2000a, b).

Another negative aspect of urbanisation and population rise is increase in demand of food grains which ultimately leads to conversion of forest area into cropland. Throughout most of Asia, the area under cropland has increased. Cropland area in Bangladesh and China has decreased in the early 1990s compared to the early 1980s. In South Asia, Pakistan has experienced the highest increases in cropland, while among the countries of East and Southeast Asia, Malaysia recorded the highest increase (~47 %) followed by Indonesia (20 %). Developed countries with the exception of Australia and Germany have recorded declines in cropland area. Land under permanent pasture has virtually remained unchanged for Indian subcontinent, while India and Nepal have recorded declines in their respec-

Fig. 5.5 Clearing of forest for construction, preparation of crop land and selective logging of trees has resulted in forest decline and loss of biodiversity

tive areas. All the Asian countries have experienced decline in the natural forest cover. Pakistan, Philippines and Thailand have registered much faster decline (more than 3 % per annum) than the rest (Alauddin 2004).

Naturally growing plant communities are reported to provide useful information regarding ambient air quality. Reduction in species diversity, selective disappearance of sensitive species and visible injury to the plant structure are some of the circumstantial evidences which indicate deterioration in the air quality of the area (Rai et al. 2011).

Secondary forest and plantations have lower lichen species richness compared with the primary forest as secondary rain forests lack the Thelotremataceae flora of the virgin forests. The lack of old stands and monospecific character of the plantations has led to a strong depletion and alteration of the lichen flora, with some species becoming dominant as Physciaceae members are predominant on *Mangifera indica* plantation in the majority of locations in India. Terricolous lichen (lichen growing on soil) indicates undisturbed area, Usneoid community indicates pollution-free area, while Physcioid lichen communities indicate polluted or highly polluted areas. Reduction in the thallus with closeness to the pollution source is also an indicator which can be utilised to understand the effect of point source (especially thermal power plants and industrial set ups) on air quality of the area (Chaphekar 2000; Singh et al. 1994). Changes due to forest destruction have severe impact at community level but it has rarely been documented. Among the different communities, foliicolous communities are more prone to microclimatic changes based on their substrate specificity and sensitivity. Sipman (1997) observed that clearing of forest caused foliicolous lichen species to become discoloured and moribund. Some foliicolous lichens reappear in secondary/regenerated forest but its frequency is lower as compared to primary forests.

Epigeic moss (*Hypnum cupressiforme* Hedw.) and epigeic lichen (*Cladonia rangiformis* Hoffm.) samples were collected simultaneously in the Thrace region, Turkey, where mosses were found at all sampling sites; the lichen could be collected only at 25 of the sites, presumably because lichens are more sensitive than mosses with respect to air pollution and climatic variations. All elements showed higher accumulation in the moss than in the lichen, whereas element intercorrelations were generally higher in the lichen (Coskun et al. 2009).

Lichens have been used for monitoring local hot spots of pollution, and regional patterns of pollutants which indicates the uptake of metals from the substrates, interspecies differences, and a comparison of the data with other bioindicator species provides the effectiveness of lichens as biomonitors (Garty 2001; Gombert et al. 2004).

Cyanolichens (blue-green algae-containing lichens) are useful as an indicator of forest ecosystem function in temperate and boreal forests. Cyanolichens are important in forest nutrient cycle, as these species are sensitive to both pollution and forest age continuity (Mc Cune 1993; Neitlich and Will-Wolf 2000; Sillett and Neitlich 1996). In India in an ecologically studied area of Pindari, glacier area exhibits less cyanophycean lichens than Milan glacier area. *Lobaria* and *Sticta* are sensitive to air quality as well as reliable indicators of species-rich old forest with long forest continuity (Kondratyuk and Coppins 1998; Kuusianen 1996a, b; Sillett et al. 2000; Joshi 2009).

Lichens are recognised as being very sensitive to air pollution, and in recent decades, several qualitative or quantitative methods have been proposed for assessing environmental quality of urban areas on the basis of lichen data (Seaward 1989). The correlation between SO_2 emissions and the nature of lichen communities led Hawksworth and Rose (1970) to develop a bioindication scale for the qualitative estimation of mean winter sulphur dioxide levels in England and Wales using epiphytic lichens. Two separate scales were prepared, one for lichens on moderately acid bark and the other for lichens on basic or nutrient-enriched bark. Several mapping studies based on this method were established (Hawksworth 1973; Belandria and Asta 1986).

Van Haluwyn and Lerond (1986) proposed a qualitative method based on lichenosociology with a 7-point scale based on easily recognisable species. Quantitative methods permit the calculation of a pollution index with a mathematical formula based on different parameters relative to the epiphytic flora. One of these is the Index of Poleotolerance (IP) proposed by Trass (1973):

$$IP = \sum_{1}^{n} a \times \frac{c}{C}$$

where n = the number of species, a = the degree of tolerance of each species on a scale of 1–10 determined by field experience, c = the corresponding level of covering and C = the overall degree of cover of all species. However, the best known is the method of De Sloover and Le Blanc (1968) or IAP (Index of Atmospheric Purity):

$$IP = \frac{1}{10}\sum_{1}^{n} Q \times f$$

Q = the resistance factor or ecological index of each species, and f = the frequency coverage score of each species.

Ammann et al. (1987); Herzig et al. (1989); Herzig and Urech (1991) tested 20 different IAP formulas, comparing IAP values with direct measurements of eight air pollutants (SO_2, NO_2, Pb, Cu, Cd, Zn, Cl and dust), and found the best correlation with the formula

$$IP = \sum_{1}^{n} F$$

(model based on the sum of the lichen species frequencies). The IAP approach to bioindication has recently been reviewed by Kricke and Loppi (2002). Asta and Rolley (1999) reviewed the formula of IAP, where f represents a cover value ranked from 1 to 5 only. Other quantitative methods have been established in different countries. In Germany the VDI (1995) method is based on the calculation of the frequency of species within a sampling ladder of 10 quadrats (each 10 = 10 cm) placed against the trunk. In Italy, the methodology proposed for monitoring the effects of air pollution by phytotoxic gases (SO_2 and NOx) based on a measure of biodiversity calculated as the sum of frequencies of epiphytic species within a sampling ladder of 10 quadrats (each 15 = 10 cm) (Nimis et al. 1990; Nimis 1999) has been widely applied there (Giordani et al. 2002; Loppi et al. 2002). The most recent methodology, strongly standardised to provide easier comparisons throughout Europe, is not related to any pollutant, but can be considered as an indicator of general environmental quality is LDV (Lichen Diversity Value (Asta et al. 2002)).

The LDV method determines the actual state of lichen diversity before or after long-term exposure to air pollution and/or to other types of environmental stress. The interpretation of geographic patterns and temporal trends of lichen diversity in terms of pollution, eutrophication, climatic change, etc. may be assisted by using ecological indicator values and a numerical analysis of a matrix of species and relevés (Asta et al. 2002).

LDV provides a rapid, low-cost method to define zones of different environmental qualities. It provides information on the long-term effects of air pollutants, eutrophication, anthropisation and climatic change on sensitive organisms. It can be applied in the vicinity of an emission source to prove the existence of air pollution and to identify its impact, or, on a larger scale, to detect hot spots of environmental stress. Repeated monitoring at the same sites enables assessment of the effects of environmental change. Data quality largely depends on the uniformity of growth conditions: the more uniform, the more reliable are the results. A high degree of standardisation in sampling procedures is therefore necessary.

The Lichen Diversity Value (LDV) involves two steps: the first step in calculating the LDV of a sampling unit (j) is to sum the frequencies of all lichen species found on each tree (i) within the unit. Since substantial differences in lichen growth may be expected in different sides of the trunks, the frequencies have to be summed separately for each aspect. Thus, for each tree there are four Sums of Frequencies (tree i: SFiN, SFiE, SFis, SFiiw).

Next, for each aspect the arithmetic mean of the Sums of Frequencies (MSF) for sampling unit j is calculated

$$MSFNj = \frac{\begin{pmatrix} SF1Nj + SF2Nj + SF3Nj \\ +SF4Nj + \cdots + SFnNj \end{pmatrix}}{n}$$

where MSF is the mean of the Sums of Frequencies of all the sampled trees of unit j; SF is the Sum of Frequencies of all lichen species found at one aspect of tree i; N, E, S and W are north, east, south and west; n is the number of trees sampled in unit and j = the Lichen Diversity Value of a sampling unit j, LDVj is the sum of the MSFs of each aspect

$$LDVj = (MSFNj + MSFEj + MSFSj + MSFWj)$$

Study on the diversity and distribution of lichens in and around Nainital city (Kumaun Himalayas) has been carried out by Kholia et al. (2011) and enumerated 105 species of lichens belonging to 48 genera and 21 families. The distribution pattern of lichens distinctly differentiated three zones (core: high pollution, intermediate: struggling zone and normal zone: good for lichen growth (Fig. 5.6)).

As lichens are well-documented indicators of air pollution, it was expected that lichen distribution and abundance would be negatively affected by air pollution (Hauck 2008). The results showed a gradient of increasing lichen cover and diversity with increasing distance from pollution discharge points. Transects upwind of these points also showed greater cover, diversity and height than those downwind of the pollution points. Comparisons between the sites revealed significantly higher cover, diversity and height at one control site. In the other control site only cover was significantly higher than that of the polluted area (Loppi and Frati 2006).

The analysis of air pollution level with metals using lichens and further statistical analysis includes calculation of the background concentrations and the contamination factors indicate extreme contaminations in the surroundings of

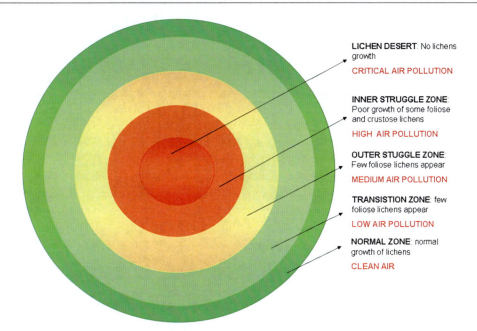

Fig. 5.6 Five classification zones corresponding to degree of injury to lichen flora and level of total air pollution (Herzig et al. 1989)

point sources. The comparison of the distribution maps for metal concentrations enables the identification of the pollution sources (State et al. 2012).

Epiphytic lichen diversity is impaired by air pollution and environmental stress. The frequency of occurrence of lichen species on a defined portion of tree bark is used as an estimate of diversity and as a parameter to estimate the degree of environmental stress. Lichen biomonitoring provides a rapid, low-cost method to define zones of different environmental qualities. It provides information on the long-term effects of air pollutants, eutrophication and climatic change on sensitive organisms. It can be applied in the vicinity of an emission source to prove the existence of air pollution and to identify its impact, or, on a larger scale, to detect hot spots of environmental stress. Repeated monitoring at the same sites enables assessment of the effects of environmental change. Data quality largely depends on the uniformity of growth conditions: the more uniform, the more reliable are the results. A high degree of standardisation in sampling procedures is therefore necessary (Pinho et al. 2004).

Sernander (1926) recognised the disappearance of lichens from cities due to increasing pollution and conducted the systematic mapping and recognised three distinct zonations in which 'Lichen desert' is the city centre where the tree trunks were devoid of lichens. Struggle zone is comprised of areas outside the city centre with tree trunks poorly colonised with lichens followed by the 'normal zone' where lichen communities on the tree trunks were well established. Subsequently, the large number of similar city maps showed that these zonations were well correlated with the degree of pollution, the size of the urbanisation area and the prevailing winds.

Lichen diversity is adversely affected by point source. Lichen thallus size, thallus number and frequency of occurrence, along with diversity of lichens at three levels (species, generic and family) are considered as variables to see the community composition across the distance from a point source (paper mill) in Assam. Result showed that the number of lichen thallus per tree in study area ranged from 3 to 16, while thallus area per tree varied from 20 to 256.48 cm^2. The number of species showed high positive correlation with the number of genera, families, thalli and thallus area. The number of thalli showed high positive correlation with area covered, number of thallus and thallus area per tree. Distance from the paper mill exhibited no significant correlation with either variable. Multivariate analysis showed two major groups and two subgroups of communities. Sites which are more polluted showed a decrease in the community variables. Fifteen out of seventeen sites were the most affected ones. Epiphytic lichen community study thus can be used to study levels of pollution impact around a source of pollution (Pulak et al. 2012).

The use of lichen in mapping lichen community changes with respect to changes in air quality has been carried out employing standardised protocols in Europe and America, where lichen biomonitoring is an integral part of environmental impact assessment programme (Pinho et al. 2004). But no such protocols are standardised in Asia; majority of the biomonitoring studies carried out till date is based on random sampling of lichens from an area of interest.

Foliose and fruticose epiphytic lichens are best suited for biomonitoring studies (Seaward 1993; van Dobben and ter Braak 1999). In urban environment, sulphur dioxide along with the NO_x gases (resulting from vehicular emissions) has antagonistic effect on lichen community at relatively high doses of gases in the environment (Balaguer et al. 1997). Changes in the composition of lichen diversity including frequency, density and abundance provide first-hand evidence on the alterations in air quality of an area due to air pollution or microclimatic changes (van Herk et al. 2002; Aptroot and van Herk 2007).

Since the last century there has been a considerable change/decline in the lichen biodiversity all around the world (Hauck 2009). It has been observed that rate of decline of lichen biodiversity in the Himalayan region, especially Garhwal Himalayas, is quite faster (Upreti and Nayaka 2008). There is a considerable increase in the abundance of thermophilous and poleotolerant lichens in the temperate climate of Garhwal Himalayas. (Shukla and Upreti 2011a, b; Shukla 2007). PCA analysis revealed that sites influenced with anthropogenic activity negatively contribute towards lichen diversity of the area (Shukla and Upreti 2011a, b).

In India grid plotting technique has been utilised to map lichen diversity in Lucknow city. In this study distribution of each species was plotted in 1 × 1 km grid in all direction. The distribution data of lichens collected from all the four areas, viz. north, east, west and south, provided four distinct zones, viz. Zone A, with no lichen growth, was the area within the centre of the city up to 5 km all around, Zone B showed presence of some calcareous lichens mostly in the areas with old historical buildings, Zone C had scarce growth of few crustose and foliose lichen in the localities with scattered mango trees, and Zone D showed normal growth of different epiphytic lichen taxa together with same foliicolous (leaf inhabiting) lichens, an indication of a more or less pollution-free environment (Saxena 2004).

Presence or absence of lichens has invariably been linked with environmental pollution and is used to estimate the range of pollution and pollutants from the source of emission. High or low lichen diversity is the result of various factors like certain types of air pollution, changes in forest management or stand structure, diversity of plant substrates available for colonisation, climate favourability and periodicity of fire (Jovan 2008). The widely used Shannon Index of general diversity or H index is a mimic of the so-called information theory formula that is hard to calculate factorials and combines the variety and evenness of components as one overall index of diversity (Odum 1996). It is calculated using the following formula:

$$H = -\sum \left(\frac{ni}{N}\right) \log \left(\frac{ni}{N}\right) \text{ or } -\sum (Pi) \log (Pi),$$

where ni = importance value for each species, N = total of importance values and Pi = importance probability for each species.

Giordano et al. (2004) studied the relation of Shannon Index with pollution in Italy and correlated biodiversity to the total number of species ($r = 0.88$). According to Odum (1996), species diversity tends to be low in physically controlled ecosystems (i.e. subjected to strong physicochemical limiting factors) and high in biologically controlled ecosystems. Wilhm (1967) demonstrated the changes in Shannon index of diversity (H) of the benthos downstream from a pollution outfall. Trivedi (1981) in a number of surveys has demonstrated that pollution produces striking changes in biotic community. Some species may be unable to survive and others may persist in reduced locations and certain other species may be able to attain greater abundance. Zullini and Peretti (1986) observed the significant decrease in Shannon diversity index of moss-inhabiting nematodes on an increase of the Pb content in the moss growing near the industrial area in Italy. Junshum et al. (2008) applied three biological indices, namely, algal genus pollution index, Saprobic index and Shannon index, to classify the water quality around a power plant in Thailand and concluded that the Shannon Index of diversity appeared to be much more applicable and interpretable for the classification of water quality into three categories (clean, moderately polluted, heavily polluted) in comparison to the other two.

In a study carried out in a paper mill area in Assam (India), Shannon diversity index (H) of lichen community around an industry has been used to determine the effect of air pollution in the surrounding areas. The Shannon index was mapped by plotting it with the help of kriging (interpolation technique) and a pollution gradient model was prepared. It was observed that higher polluted areas with low values of Shannon index are the regions around and nearer to the paper mill and town area and around stone crusher units. It is concluded that the Shannon index is a potential indicator to measure the effect of air pollution and can be used to delineate the pollution zones around any industrial area (Pulak et al. 2012).

The comparison of the lichen diversity with an earlier study carried out during 1960–1980s exhibits a distinct change in the Lucknow. In and around the city of Lucknow out of the 18 species recorded in the past, 14 species are common to the present study. It seems that the remaining 4 species (*Julella* sp., *Opergrapha herpetica*, *Peltula euploca* and *Phylliscum macrosporum*) of the former study might have become totally extinct from the area. The change in lichen communities in the district is mainly due to change in the environmental condition during the last 25 years. This indicates the replacement of the sensitive species of lichens with tolerant ones in the district (Saxena 2004).

Lichen flora of Kolkata revealed the exclusive occurrence of pollution-resistant species, *Parmelia caperata* (=*Flavoparmelia caperata*), on the roadside trees of Kolkata. The most probable reason for existence of resistant species was long-range dispersal of pollutants (caused by nearby factories) with wind (Das et al. 1986; Shukla and Upreti 2012).

In another study, Das et al. (2013) studied the impact of anthropogenic factors on abundance variability among lichen species in southern Assam. It was observed that the area with least anthropogenic pressure shows a general universal pattern of natural communities, i.e. a J-curve pattern, where majority of the species are rare and few are abundant. With a small change in anthropogenic pressure, there is little effect on rare species but the abundant species increased. With major changes both the rare and the abundant species decrease changing the overall community composition. The J-curve changes to a uni-modal curve; the moderately abundant species are found to be resilient against anthropogenic pressure levels. The community ecology of organisms has its root in evolutionary history, succession and biogeography. The studies on community ecology, therefore, may help in throwing significant light on these important aspects and can also be used as indicators of ecosystem health.

Lichen flora of Garden city, Bangalore, was explored by Nayaka et al. (2003). Significant change in the lichen diversity was observed in comparison to earlier study conducted 18 years ago. There were only four species common between two studies. Air quality of Pune city in Maharashtra Province was assessed by distribution of lichens in the city. It was observed that out of the 20 streets/sites of the Pune city surveyed, only 11 sites showed the presence of lichens Nayaka and Upreti (2005a).

An earlier enumeration of lichens of Indian Botanical Garden collected by Kurz in 1865 and described by Nylander in 1867 was compared by Upreti et al. (2005). It is interesting to note that in the last more than 140 years, the lichen flora of the area has been changed significantly as only 3 species out of 50 species (recorded earlier) were common between the two studies.

More systematic and standardised lichen diversity studies are required to be conducted in India in order to monitor the air quality and establish lichen biomonitoring, an integral part of environmental assessment programmes.

5.3 Solely Human Disturbances/Disasters

With increasing economic growth (industrialisation and urbanisation) (Fig. 5.7), environmental contamination, especially air pollution, is resulting in environmental degradation in the developing nations of Asia. In many Asian countries including India, environmental component of sustainable development is virtually ignored compared with economic benefits, resulting in deterioration of ecosystem affecting quality of air, water and soil. In the quest of rapid economic development, political and business support for environmental impact assessment (EIA) is given low priority (Foster 1993; Curran 2000). Countries that have achieved rapid industrialisation and economic development have done so at the cost of extensive environmental damage (Alauddin 2003; Li et al. 2009).

The relationship between economic growth and the environment has been controversial. Classical economic theory always indicated negative relation between economic growth and environmental quality. The empirical and theoretical literature on the Environmental Kuznets Curve (EKC) has suggested that the relationship between economic growth and the environment could be positive and hence growth is a prerequisite for environmental improvement (Fig. 5.8). The EKC depicts the empirical pattern that at relatively low levels of income per capita, pollution level (and intensity) initially increases with rising income but then reaches a maximum and falls thereafter. The dominant theoretical expla-

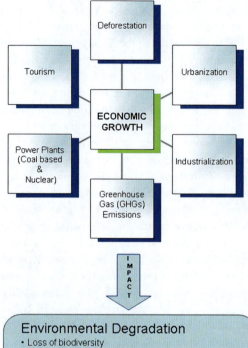

Fig. 5.7 Conceptual diagram showing the impact of various types of developmental activities resulting in environmental degradation

nation is that when GDP increases, the greater scale of production leads directly to more pollution, but, at a higher level of income per capita, the demand for health and environmental quality rises with income which can translate into environmental regulation, in which case there tend to be favourable shifts in the composition of output and in the eco-friendly techniques of production (Panayotou 2003). For example, the air in London, Tokyo and New York was far more polluted in the 1960s than it is today. This theoretical assumption has been proved by several lichen biomonitoring studies carried out in advanced countries of Europe, according to which the present levels of pollutants are lower in comparison to earlier data (Lisowska 2011; Crespo et al. 2004; Seaward 1997).

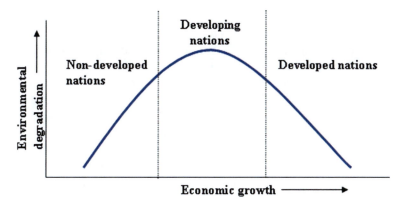

Fig. 5.8 Environmental Kuznets Curve (EKC) showing relation between environmental and economic developments (Modified from Panayotou 2003)

Asia is the world's largest and most populous continent and covers 8.6 % of the earth's total surface area (or 29.4 % of its land area). With over four billion people, it has more than 60 % of the world's current human population. As evident and measured by annual percentage change in GDP, many Asian countries, especially in East Asia, have experienced rapid economic growth (Alauddin 2003; Li et al. 2009).

In developing countries unplanned rapid expansion of economic activities, consideration for environmental conservation and many problems resulting due to degradation in quality of ambient environment such as clean air, safe drinking water and quality of food are being given low priority (Japan environmental council 2005). Statistical data provided by various international organisations, including the World Bank, have indicated that the annual cost of all aspects of air quality degradation is substantial and could constitute up to around 2 % of GDP in developed countries and more than 5 % in developing countries. These costs include mortality, chronic illness, hospital admissions, lower worker and agricultural productivity, IQ loss and reduction of visibility. Up to 800,000 premature deaths and up to one million prenatal deaths have been estimated as one consequence of air pollution globally (Abdalla 2006).

In the developing countries a causal relationship between air pollution and health effects has been reported. Increased cases of respiratory problems and low pulmonary function is due to exposure to respirable suspended particulate matter (RSPM) which remains suspended in the urban air and easily inhaled. Asthmatic populations are also susceptible to the impact of particulate and SO_2 exposure. Most evidence suggests that populations living in cities with high levels of air pollution in developing countries experience similar or greater adverse effects of air pollution. According to Smith et al. (1999) around 40–60 % of acute respiratory infection is due to environmental causes.

Association between mortality rate and particulate air pollution has long been studied. Dockery et al. (1993) related excess daily mortality from cancer and cardiopulmonary disease to several air pollutants, especially fine particulate matter ($PM_{2.5}$, particulate matter with aerodynamic diameter of equal to or less than 2.5 µm), in their prospective cohort study. Since then, many other epidemiological studies on the adverse human effects of air pollutants have been carried out, ranging from variations in physiological functions and subclinical symptoms (heart rate variability, peak expiratory flow rate, etc.) to manifest clinical diseases (asthma, chronic obstructive pulmonary disease, stroke, lung cancer, leukaemia, etc.), premature births and deaths (Delfino et al. 1998; Naeher et al. 1999; Laden et al. 2000; Suresh et al. 2000; Janssen et al. 2002; Calderón-Garcidueñas et al. 2003; Wilhelm and Ritz 2003; O'Neill et al. 2004; Preutthipan et al. 2004; Han and Naeher 2006).

Even though the current fossil fuel use in developing countries is half that of the developed countries, it was expected to increase by 120 % by the year 2010. If control measures are not implemented, it has been estimated that by the year 2020 more than 6.34 million deaths will occur in developing countries due to ambient concentrations of particulate air pollution (Mukhopadhyaya and Forssel 2005).

Present environmental crisis in Asia is mainly because of non-standardised environmental parameters for environmental regulation, and moreover environmental impact assessment (EIA) is overpowered by economic gains in terms of GDP. In absence of standardised EIA protocol, newly developing countries are susceptible to accumulate more pollution-emitting industries. As a result many countries in the East Asian region are very likely to accumulate pollution emitting industries as rich countries filter out such industries and transfer them to newly developing countries (Kim 1990). According to Kim (2006) without strict environmental control, it is very likely that East Asia will accumulate the worst pollution in the world.

Variation in the extent, regulatory form and practical application of EIA in different developing countries is dependent on resources, political and administrative systems, social systems as well as the level and nature of economic development (George 2000). Other than problem in the system, there are several prominent difficulties in developing countries in relation to EIA report preparation, prominent being the lack of trained human resources and of financial resources which often leads to the preparation of inadequate and irrelevant EIA reports in developing countries (Clark 1999) as well as baseline socio-economic and environmental data being inaccurate, difficult to obtain or nonexistent in developing countries (Wilbanks et al. 1993). Lohani et al. (1997) attributed lack of attention and commitment to follow-up as a serious shortcoming in Asian EIA practice. Projects in developing countries may change substantially between authorisation and implementation, and environmental controls may not be observed or monitored. There is relatively little information about the accuracy of developing country EIA predictions. The major limiting factors concerning the development of Asian EIA practice is the lack of effective communication of EIA results and recommendations to decision makers. In some eastern Asian countries, EIA begins after the construction commences and is used only to confirm that the environmental consequences of the project are acceptable (Brifett 1999; Wood 2003).

The worst impact of non-standardised economic development is the increasing global temperature and increased frequency of natural disasters.

Anthropogenic activities resulting due to human settlements result in three main sources of air pollution: (1) stationary or point, (2) mobile and (3) indoor. In developing countries, especially in the rural area, indoor air pollution from using open fires for cooking and heating is a serious problem, while industries, power plants, etc. are the cause of stationary air pollution. In urban areas, both developing and developed countries, it is predominantly mobile or vehicular pollution that contributes to overall air quality problem. Air pollutants emitting due to vehicular activity has local–regional–global impact, viz. local (e.g. smoke affecting visibility, ambient air, noise), regional (such as smog, acidification) and global (i.e. global warming) (Faiz et al. 1996).

5.3.1 Urbanisation, Expanding Cities and Industrialisation

5.3.1.1 Urbanisation and Expanding Cities

Many metropolitan cities in developing countries located in the Asia face serious air pollution problem. Asian megacities cover <2 % of the land area but emit >16 % of the total anthropogenic sulphur emissions of Asia. It has been estimated that urban sulphur emissions contribute over 30 % to the regional pollution levels in large parts of Asia. The average sulphur contribution of megacities over the western Pacific increased from <5 % in 1975 to >10 % in 2000 (Guttikunda et al. 2003). Urban air has high concentrations of sulphur dioxide and suspended particulate matter (SPM). In large cities where traffic is concentrated, air pollution is reaching serious levels, mainly caused by vehicles. The health impacts of airborne fine particulate matter (known as $PM_{2.5}$), and diesel exhaust particles (DEP), are matters of particular concern (Zhang et al. 2007).

The pollution from vehicles are due to discharges like CO, unburned hydrocarbons (HC), Pb compounds, NOx, soot and suspended particulate matter (SPM) mainly from exhaust pipes. A study reports that in Delhi one out of every ten school children suffers from asthma that is worsening due to vehicular pollution (CPCB 1999; Cropper et al. 1997).

The rapid urbanisation of many cities in South and Southeast Asia has increased the demand for bricks, which are typically supplied from brick kilns in peri-urban areas. Brick kilns, along with aluminium smelters, ceramic manufacture and phosphorus fertiliser factories, are the major sources responsible for atmospheric fluoride pollution (Weinstein and Davison 2003). Bricks are produced from soil (usually clay) that may contain fluoride at concentrations up to 500 ppm in brick kilns at temperatures ranging from 900 to 1,150 °C. At these temperatures, fluoride compounds are released into the atmosphere in the form of gaseous HF, silicon fluoride and particulate calcium fluoride, along with other pollutants such as sulphur dioxide (SO_2) (Ahmad et al. 2012).

As India and China develop into the world's manufacturing centre, air quality in this region is also deteriorating rapidly. The amount of air pollutants generated in the region is the world's largest. As the air pollution becomes regionalised, the soils and waters of the entire Northeast Asian region will also be affected (Kim 1993). China in particular, because of its rapid push to industrialise, is experiencing dramatic levels of aerosol pollution over a large portion of the country (Liu and Diamond 2005; Kim 1993, 2006).

By 1992, China was producing an estimated 14.4 million tonnes of dust a year and 16.85 million tonnes of sulphur dioxide; and solid wastes were increasing by 20 million tonnes annually. Many of the fastest-growing enterprises are based on high energy and high material consumption, while rural enterprises are among the heaviest polluters.

There has been exponential increase in anthropogenic NOx air pollution in Asia between the years 1975 and 2000 from ~10,000 to ~30,000 kt/year (Akimoto 2003). In India, air quality in national capital Delhi, data shows that of the total 3,000 metric tonnes of pollutants (Blackman and Harrington 2000) bleached out every day, close to two-third (66 %) is from vehicles. Similarly, the contribution of vehicles to urban air pollution is 52 % in Bombay and close to one-third in Calcutta (Button and Rietveld 1999). The worst thing about vehicular pollution is that it cannot be avoided as the emissions are emitted at the near-ground level where we breathe. Pollution from vehicles gets reflected in increased mortality and morbidity and is revealed through symptoms like cough, headache, nausea, irritation of eyes, various bronchial problems and visibility.

Anthropogenic aerosols, mainly black carbon soot, alter the regional atmospheric circulation, contributing to regional as well as global climate change (Dutkiewicz et al. 2009). Menon et al. (2002) and Rosenfeld et al. (2007) have suggested that reducing the amount of anthropogenic black carbon aerosols, in addition to having human health benefits, may help diminish the intensity of floods in south China and droughts and dust storms in north China. Similar considerations apply to India. India's air pollution, because it is also rich in black carbon, has reached the point where scientists fear it may have already altered the seasonal climate cycle of the monsoons.

5.3.1.2 Industrial Development

More than 80 % of energy is produced from coal, a fuel that emits a high amount of carbon and greenhouse gases and other toxic inorganic and organic pollutants. Fine particles or microscopic dust from coal or wood fires and unfiltered diesel engines are rated as one of the most lethal forms or air pollution caused by industry and ageing coal or oil-fired power stations. Industrial processes also release chemicals known as halocarbons and other long-lived gases, some of which trap heat in the atmosphere and lead to global warming.

India, a faster developing economy, is among the ten most industrialised countries in the world. It has the world's eighth largest economy. Since economic liberalisation beginning in 1991, India's economy grew by 5 % a year, on average, during 1992–1997 and at a higher rate after that. However, rapid economic and industrial growth is causing severe urban and industrial pollution. India's per capita carbon dioxide emissions were roughly 3,000 lb (1,360 kg) in 2007, which is less as compared to China and the USA, with 10,500 lb (4,763 kg) and 42,500 lb (19,278 kg), respectively. India has been ranked as the seventh most environmentally hazardous country in the world by a new ranking released recently. Brazil was found to be the worst on environmental indicators, whereas Singapore was the best. The USA was rated second worst and China was ranked third. According to Time magazine's list of most polluted cities in the world, New Delhi and Mumbai figure in top 10. Heavy reliance on coal for power generation has exacerbated India's environmental problems.

The burning of fossil fuels is the main cause of atmospheric air pollution and the main source of anthropogenic CO_2 emissions (IEA 1999).

About 90 % of total emissions of carbon monoxide (CO) in Gulf countries are due to transportation activities. In the Gulf Cooperation Council (GCC) countries total atmospheric emission loads are about 3.85 million tonnes per year, made of 28 % CO, 27 % SO_2 and 23 % particulates (UNEP, 1999). Recent studies have indicated that the Gulf countries emit about 50 % of the total of Arab countries' (254 million metric tonnes of carbon) emissions of CO_2.

5.3.1.3 Mercury Emission

Industrial and allied activities like coal combustion, waste incineration, metal mining, refining and manufacturing and chlorine-alkali production are the major anthropogenic sources of highly toxic metal, Mercury (Hg). Due to long residence time of Hg^0 in the atmosphere from 0.5 to 2 years and being able to travel long distances and deposited in remote places even 1,000 km away from the sources, mercury (Hg) is considered as a global pollutant. Anthropogenic activities emit both elemental Hg (Hg^0) with a long life in the atmosphere and reactive gaseous mercury (RGM) and particulate Hg, which are short lived in the air and deposited near the emission source. Natural processes, including volcanoes and geothermal activities; evasion from surficial soils, waterbodies, and vegetation surfaces; and wild fires, as well as the reemission of deposited mercury also result in the release of substantial amount of mercury in the atmosphere mainly in the form of Hg^0. Global oceanic emission is estimated to be 800–2,600 tonnes/a and global natural terrestrial emission is estimated to be 1,000–3,200 tonnes/a. Thus, through all sources there is a total natural mercury emission of 1,800–5,800 tonnes/a. The global anthropogenic Hg emission to the atmosphere was estimated to be 2,190 tonnes in 2000, of which two-thirds of the total emission of ca. 2,190 tonne of Hg came from combustion of fossil fuels. Asian countries contributed about 54 % (1,179 tonnes) to the global Hg emission from all anthropogenic sources worldwide in 2000. China leads the list of the ten countries with the highest Hg emissions from anthropogenic activities followed by India, Japan, Kazakhstan and Democratic People's Republic of North Korea. With more than 600 tonnes of Hg, China contributes about 28 % to the global mercury emission (Li et al. 2009).

Hg can be converted to toxic methylmercury (Me–Hg) and gets accumulated in the food chain. The outbreaks of severe mercury poisoning in Minamata, Japan, and Iraq in the last centaury had posed threat to the eco-environment system and human beings. In Japan, Mercury-contaminated effluent discharged into Minamata Bay from an acetaldehyde-producing factory. Me–Hg got bioaccumulated by fish and shellfish which when consumed by humans caused Minamata disease. Mercury poisoning killed more than 100 people and paralysed several thousands of people around Minamata Bay and the adjacent Yatsushiro

Table 5.10 Hg concentrations (in μg g^{-1} dry weight) in lichen thallus of different species from India

S. No.	Lichen species	Locality	Concentration	References
1	*Phaeophyscia hispidula* (Ach.) Essl.	Dehradun, Uttarakhand	0.00–47.00	Rani et al. (2011)
2	*Pyxine cocoes* (Sw.) Nyl.	North side Lucknow city	2.10–5.90	Saxena et al. (2007)
3	*Phaeophyscia orbicularis*	North side Lucknow city	3.4	Saxena et al. (2007)
4	*Lecanora leprosa*	North side Lucknow city	1.10–3.10	Saxena et al. (2007)
5	*Arthopyrenia nidulans*	North side Lucknow city	5.90	Saxena et al. (2007)
6	*Sphinctrina anglica*	North side Lucknow city	3.90	Saxena et al. (2007)
7	*Bacidia submedialis*	North side Lucknow city	7.50	Saxena et al. (2007)

Sea since 1956. In the early 1970s in Iraq, a major methylmercury poisoning resulted due to consumption of seed grain treated with a Me-Hg fungicide, which resulted in the death of 10,000 people and 100,000 had permanently brain damaged (Li et al. 2009). Some lichen biomonitoring related with Hg pollution has been carried out in India (Table 5.10).

5.3.1.4 Fluoride Emission

Hydrogen fluoride (HF) is one of the most phytotoxic air pollutants (Weinstein and Davison 2003). HF and other fluoride compounds in the atmosphere are deposited onto the vegetated surfaces either in gaseous or particulate form, while airborne gaseous fluorides can enter directly into the leaf through stomata. This fluoride then dissolves in the apoplast, altering the photosynthetic process, causing visible injury and ultimately affecting growth and yield. The impact of atmospheric fluoride pollution from various sources on crops has also been well documented (Brewer 1960; Mason et al. 1987; Moraes et al. 2002). Lee et al. (2003) reported damage to vegetation around brick and ceramic factories in Taiwan, and local effects on vegetation of fluoride emissions from aluminium factories and thermal power plants have been reported in India (Lal and Ambasht 1981; Pandey 1981, 1985; Narayan et al. 1994; Ahmad et al. 2012).

Other than fluoride emission the poorly regulated brick kilns also contribute significantly to local sulphur and black carbon emissions (Emberson et al. 2003; Iqbal and Oanh 2011).

5.3.1.5 Assessment of Urban and Industrial Pollution with Lichens

Lichens have been recognised and successfully utilised as biological indicators of air quality. They are among the most valuable and reliable biomonitors of atmospheric pollution. Primarily lichens were utilised to monitor gaseous pollution, namely, sulphur (SO_2) and nitrogen (NO_x, NH_3, NO_3, etc.) (Rao and LeBlanc 1967; Vestergaard et al. 1986). Lichens show high sensitivity towards sulphur dioxide because their efficient absorption systems result in rapid accumulation of sulphur when exposed to high levels of sulphur dioxide pollution (Wadleigh and Blake 1999). The algal partner (5–10 % of total thallus structure) is most affected by the sulphur dioxide as chlorophyll is irreversibly converted to phaeophytin, and thus, photosynthesis is inhibited (Upreti 1994). Lichens also absorb sulphur dioxide dissolved in water (Hawksworth and Rose 1970). Excessive levels of pollutants in the atmosphere, especially SO_2, have detrimental effect on the physiology and morphology of sensitive species and causes extinction of the species, which ultimately results in changed lichen diversity pattern (Haffner et al. 2001; Purvis 2000).

Photosynthesis is the core physiological function of an autotrophic organism and its functional state has been considered as an ideal tool to monitor the health and vitality of plants (Clark et al. 2000). Between gas exchange and chlorophyll fluorescence methods for measuring the photosynthetic performance in a plant, the latter technique has become more popular in the recent years. Introduction of highly user-friendly portable fluorometers further widened the scope of photosynthesis research, especially in in situ conditions. The technique was also found to be useful for studying samples such as lichens and bryophytes, whose structure otherwise makes them difficult to study with conventional gas exchange systems (Maxwell and Johnson 2000; Genty et al. 1989). The chlorophyll fluorescence technique provides a large amount of data with a minimum of expertise and time and without injury to the plants. The technique most frequently utilises the parameter Fv/Fm as a reliable indicator of the maximum photochemical quantum efficiency of photosystem II or photosynthetic performance of organism under investigation (Butler and Kitajima 1975). It is the only parameter that is not temperature sensitive and measured in dark-adapted samples. Fv/Fm is calculated from F0, the fluorescence when the reaction centre of PSII is fully open, and Fm, the maximum fluorescence when all the reaction centres are closed following a flash of saturation light. Fv (i.e. Fm−F0) is the maximum variable fluorescence in the state when all nonphotochemical processes are at a minimum (van Kooten and Snel 1990).

Photosynthetic performances of 82 lichens occurring in Western Himalayas were determined using chlorophyll fluorescence. Fv/Fm ranged from 0.023 to 0.655, with terricolous Cladonia subconistea at alpine region having maximum value. Photosynthetic performances of alpine lichens were found to be better than those of temperate due to the influence of favourable climate, wet soil and rock in the region. As the study was carried out during early summer, most of the lichens started experiencing stress which is evident by their Fv/Fm values. As many as ten chlorolichens (with green alga as photobiont) growing in temperate region are severely stressed and have values <0.1. The stress components in the study area are mostly water availability and high-intensity light. The cyanolichens (with blue-green alga as photobiont) have relatively lower Fv/Fm ranging from 0.075 to 0.315. On the basis of their Fv/Fm values, the lichens in the present study are classified into three categories: normal, moderate and severely stressed with values ranging from 0.5 to 0.76, 0.3 to 0.49 and 0.01 to 0.29, respectively (Nayaka et al. 2009).

In recent times, sensitivity to other pollutants has been explored. Lichens are adversely affected by short-term exposure to nitrogen oxides as low as 564 µg m^{-3} (0.3 ppm; Holopainen and Kärenlampi 1985). Most reports regarding lichen sensitivity to fluorine relate the physical damage of lichens to tissue concentrations or a specific point source of emissions rather than ambient levels. In general, visible damage to lichens begins when 30–80 ppm fluorine has been accumulated in lichen tissues (Perkins 1992; Gilbert 1971). In one fumigation study (Nash 1971), lichens exposed to ambient F at 4 mg m^{-3} (0.0049 ppm) accumulated F within their thalli and eventually surpassed the critical concentration of 30–80 ppm. Fluorine is associated with aluminium production and concentrations in vegetation may be elevated near this type of industrial facility. In addition to gaseous pollutants, lichens are sensitive to depositional compounds, particularly sulphuric and nitric acids, sulphites and bisulphites and other fertilising, acidifying or alkalinising pollutants such as H$^+$, NH$_3$ and NH$_4^+$. While sulphites, nitrites and bisulphites are directly toxic to lichens, acidic compounds affect lichens in three ways: direct toxicity of the H$^+$ ion, fertilisation by NO$_3^-$ and acidification of bark substrates (Farmer et al. 1992).

Effect of simulated acid rain and heavy metal deposition on the ultrastructure of the lichen *Bryoria fuscescens* (Gyeln.) Brodo and Hawksw. was studied. Algal and fungal components responded differently to pH, and there was an interaction with metal toxicity. The algal partner was the most sensitive to acid rain and heavy metal combinations and had more degenerate cells than the fungal partner. Damage was apparent in chloroplasts and mitochondria, where thylakoid and mitochondrial cristae were swollen. The fungal partner was the more sensitive to high concentrations of metal ions in non-acidic conditions, suggesting a synergistic interaction between the metals and acidity. The results suggest that acid wet deposition containing metal ions may reduce survival of lichens (Tarhanen 1998).

Lichen partners differ in the ability to sequester the heavy metals; mycobiont, comprising more than 90 % of total lichen biomass, accumulates most of the heavy metals from the environment. Sanità di Toppi et al. (2005) found that mycobiont hyphae, especially those forming the upper cortex of lichen thalli, were the main site of Cd accumulation which may be attributed to the presence of extracellular lichen substances produced by fungal hyphae. Significant positive correlation of Cd and Cr content and lichaxanthone has been observed in *Pyxine subcinerea* (Shukla 2012).

In the Netherlands, a number of studies have demonstrated that ammonia-based fertilisers alkalinise and enrich lichen substrates that in turn strongly influence lichen community composition and element content (van Herk and Aptroot 1999; van Dobben et al. 2001; van Dobben and ter Braak 1998, 1999).

In a mixed urban environment, pollutant mixes can have synergistic, additive or antagonistic effects on lichens, and individual species differ in their sensitivity to these pollutants and their response to pollutant mixes (Hyvärinen et al. 2000; Farmer et al. 1992). During the past 20 years, much data have been collected concerning metal tolerance and toxicity in lichens (Garty 2001).

Metals can be classified into three groups relative to their toxicity in lichens (Nieboer and Richardson 1981):
1. Class A metals: K^+, Ca^{2+} and Sr^{2+} are characterised by a strong preference for O_2-containing binding sites and are not toxic.
2. Ions in the B metals class: Ag^+, Hg^+ and Cu^+ tend to bind with N- and S-containing molecules and are extremely toxic to lichens even at low levels.
3. Borderline metals: Zn^{2+}, Ni^{2+}, Cu^{2+} and Pb^{2+} are intermediate to Class A and B metals. Borderline metals, especially those with class B properties (e.g. Pb^{2+}, Cu^{2+}), may be both detrimental by themselves and in combination with sulphur dioxide.

This provides a good rationale to monitor both metal- and sulphur/nitrogen-containing pollutants simultaneously if possible.

Adverse effects of metals include decreases in thallus size and fertility, bleaching and convolution of the thallus, restriction of lichens to the base of vegetation (Sigal and Nash 1983) and mortality of sensitive species. Microscopic and molecular effects include reduction in the number of algal cells in the thallus (Holopainen 1984), ultrastructural changes of the thallus (Hale 1983; Holopainen 1984), changes in chlorophyll fluorescence parameters (Gries et al. 1995), degradation of photosynthetic pigments (Garty 1993) and altered photosynthesis and respiration rates (Sanz et al. 1992). The first indications of air pollution damage from SO_2 are the inhibition of nitrogen fixation, increased electrolyte leakage and decreased photosynthesis and respiration followed by discoloration and death of the algae (Fields 1988). More resistant species tolerate regions with higher concentrations of these pollutants but may exhibit changes in internal and/or external morphology (Nash and Gries 1991). Elevation in the content of heavy metals in the thallus has also been documented in many cases (Garty 2001), but it is not always easy to establish what specific effect these elevated levels will have on lichen condition or viability. Tolerance to metals may be phenotypically acquired, but sensitivity of lichens to elevated tissue concentrations of metals varies greatly among species, populations and elements (Tyler 1989). The toxicity of metal ions in lichen tissue is the result of three main mechanisms: the blocking, modification or displacement of ions or molecules essential for plant function. Metal toxicity in lichens is evidenced by adverse effects on cell membrane integrity, chlorophyll content and integrity, photosynthesis and respiration, potential quantum yield of photosystem II, stress ethylene production, ultrastructure, spectral reflectance responses, drought resistance and synthesis of various enzymes, secondary metabolites and energy transfer molecules (Garty 2001).

The influence of pollution sources on the presence of trace elements and on lichen species community composition in the natural area has been carried out which revealed that lichen diversity negatively correlated with Cu, Pb and V. The study also underlined the value of combining the use of biomonitors, enrichment factors and lichen diversity for pollution assessment to reach a better overview of both trace elements' impact and the localisation of their sources (Achotegui-Castells et al. 2012).

In India lichen biomonitoring has been successfully employed for air quality monitoring in different regions of the country employing different bioindicator species (Table 5.11). Accumulation of metal (Al, Cd, Cu, Cr, Fe, Pb, Ni, Zn) pollutants in lichen thallus by passive as well as active principals are well known from different cities of the country such as Uttar Pradesh (Faizabad, Lucknow, Kanpur and Raebareli district), Madhya Pradesh (Dhar, Katni and Rewa district) and West Bengal (Hooghly and Nadia district), Maharashtra (Pune and Satara district) and Uttarakhand (Dehradun, Pauri district). The accumulation of different metals decrease with increasing distance from the city centre. The metals Cr, Cu and Pb were more at the higher vertical position (20–25 ft), whereas other metals (Zn, Fe) accumulated maximum at lower vertical position (4–5 ft).

Accumulation rate of heavy metals depends to a large extent on physical aspects such as thallus type and their morphological features (Garty et al. 1979). A number of atmospheric pollutants are recognised as a significant factor in the deterioration of cultural properties. Atmospheric deposition of heavy metals such as As, Al, Cd, Cr, Cu, Fe, Ni and Zn in four different growth forms of lichens exhibit diverse quantitative variations. The difference in metal concentrations were tested using an ANOVA test ($p<5$ %) between the species and between the sites. Among the different growth forms, most of the heavy metal exhibits their higher level of accumulation. The competitive uptake studies revealed the selectivity sequence as foliose > leprose > squamulose > crustose. In calcium and magnesium these selectivity sequence represented as crustose > squamulose > foliose > leprose forms. The morphology and anatomy of the thallus may play an important role in accumulation of pollutants.

All the sites in the city exhibited an enhanced level of metals, as most of the metals analysed except Ca and Mg have anthropogenic origin due to vehicular activities.

The *Phaeophyscia hispidula*, *Diploschistes candidissimum*, *Phylliscum indicum* and *Lepraria lobificans* belonging to foliose, crustose, squamulose and leprose growth forms, respectively, seem to be efficient bioaccumulators of inorganic pollutants. These species are good mitigator of atmospheric fallouts and can be utilised for air quality monitoring in the area.

The bioaccumulation factor (BAFs) was maximum in *Lepraria lobificans* for Ca, Mg and Al, whereas *Diploschistes candidissimum* has maximum BAF for As. Both the foliose lichens (*P. hispidula* and *P. praesorediosum*) have maximum BAFs for Fe. The bioaccumulation factor was zero for Zn, Ni, Cd and Cr because these metals were not detected in substratum.

The damage caused by the metallic pollutants in the lichen *Pyxine subcinerea* Stirton, by measurements of Chl a, Chl b, total Chl, carotenoid and protein and OD 435/415 ratio significantly, exhibits the changes in physiology. It was observed that Cu, Pb and Zn significantly affect the physiology of the lichen *P. subcinerea*. Multiple correlation analysis revealed significant correlation (<0.001) among the Fe, Ni, Cu, Zn and Pb metals analysed. Cd did not correlate with any other metals except Fe ($p<0.05$). Cu, Pb and Zn are the main constituents of the vehicular emissions and had significant positive correlation ($p<0.001$) with protein content, while the OD 435/415 ratio values decreased statistically ($p<0.001$) with increase in amount of Cu, Pb and Zn (Shukla and Upreti 2008).

According to Beckett and Brown (1983), Cd and Zn compete with one another for sites holding bivalent cations, which is consistent with the observation that the correlation coefficients between Cd and Zn, although not statistically significant, exhibit negative trends (Shukla et al. 2012a, b). Usually Zn, Cu and Fe are supposed to act antagonistically against Cd. Zn, Fe and Cu are negatively correlated with Cd. The significant positive correlation of protein with Cd indicates the synthesis of protein under Cd-stressed condition similar to expression of stress protein 70 (hsp 70) in the lichen photobiont *Trebouxia erici*

Table 5.11 Concentraion (in µg g^{-1} dry weight) of various lichens collected from different phytogeographical regions of India

	Species	Sites	Ni	Al	Ca	Cd	Cr	Cu	Fe	Pb	Mg	Mn	Zn	References
1.	*Acarospora gwynii*	Maitri Station (East Antarctica)				19.28–97.87	36.36–124.84	6,420–8,560					12.4–69	Upreti and Pandey (1999)
2.	*Arthopyrenia nidulans*	Lucknow city (North) (Uttar Pradesh)	BDL			BDL	137.5	21.7	5,183	15.6			219.7	Saxena et al. (2007)
3.	*Bacidia submedialis*	Lucknow city (North) (Uttar Pradesh)	BDL			BDL	127.4	66.6	5,283	5.9			96.7	Saxena et al. (2007)
4.	*Buellia grimmae*	Maitri Station (East Antarctica)				57.75	395	11,290					71.8	Upreti and Pandey (1999)
5.	*B. isidiza* (Nyl.) Hale	Bangalore city (Karnataka)					12.99	86.02	22,721	22			102.7	Nayaka et al. (2003)
6.	*B. pallida*	Maitri Station (East Antarctica)				54.76	138.88	10,070					51	Upreti and Pandey (1999)
7.	*Caloplaca subsoluta* (Nyl.) Zahlbr.	Mandav (Madhya Pradesh)	0.60–10	17.7–176	147.9–195	0.2–0.9	0.5–6.5	1.0–16	6.2–14.2		210–376.8		48.5–81.6	Bajpai et al. (2009a, b, 2010a)
8.	*Chrysothrix candelaria* (L.) Laundon	Bangalore city (Karnataka)					5.18–95.3	19.66–23.7	748–7,556	0.00–623.95			95.76–157.49	Nayaka et al. (2003)
9.	*Cryptothecia punctulata*	Fungicidal. South India			100			575.4	2.226					Nayaka et al. (2005)
10.	*Dimelaena oreina*	Badrinath (Uttarakhand)	6.9–13.3			0.67–2	BDL	BDL	8,348–21,780	7.9–158.5			22.1–48.6	Shukla (2007)
11.	*Diploschistes candidissimus* (Kr.) Zahlbr.	Mandav (Madhya Pradesh)	1.2–10.6	16.5–181.6	150.68–314.45	0.10–0.80	0.50–5.80	0.9–12.5	8.3–19.0		283.4–415.39		45.9–94.9	Bajpai et al. (2009a, b, 2010a)
12.	*D. gypsaceus*	Bhimbetka (Madhya Pradesh)			77.45									Bajpai et al. (2010c)
13.					87.66									Bajpai et al. (2010c)
14.	*Dirinaria aegialita* (Afz. In Ach.) Moore	Bangalore city (Karnataka)					0.00–34.57	8.99–16.06	6,887–7,358	BDL–46.4			98.60–122.39	Nayaka et al. (2003)
15.	*D. confluens* (Fr.) Awasthi	Faizabad city (High Way) (Uttar Pradesh)								0.17–6.0				Dubey et al. (1999)
	D. papillulifera (Nyl.) Awasthi													

(continued)

Table 5.11 (continued)

	Species	Sites	Ni	Al	Ca	Cd	Cr	Cu	Fe	Pb	Mg	Mn	Zn	References
16.	*D. consimilis* (Stirton) Awasthi	Bangalore city (Karnataka)					35.59	22.22	7081.00	149.15			198.14	Nayaka et al. (2003)
17.	*D. consimilis* (Stirton) Awasthi	Lucknow city (commercial/industrial) (Uttar Pradesh)					BDL–1740	BDL–13.65	BDL	BDL			BDL–36.60	Mishra et al. (2003)
18.	*D. consimilis* (Stirton) Awasthi	Lucknow city (residential sites) (Uttar Pradesh)					89.1–1,920	BDL–12.45					BDL–8.70	Bajpai et al. (2004)
19.	*Dermatocarpon vellereum*	Badrinath (Uttarakhand)	1.58–4.96			0.42–0.94	BDL–7.5	BDL–1.56	1,331–5,464	9.68–16.98			18–36.9	Shukla (2007)
20.	*Endocarpon subrosettum*				99.41									Bajpai et al. (2010c)
21.	*Graphis scripta* (L.) Ach.	Bangalore city (Karnataka)					BDL	10.06	863.00	BDL			384.55	Nayaka et al. (2003)
22.	*Heterodermia diademata* (Taylor) Awasthi	Bangalore city (Karnataka)					6.62–6.82	1.71–11.02	3,020–4,402	0.00–30.49			126.54–160.0	Nayaka et al. (2003)
23.	*Lecanora muralis*	Badrinath (Uttarakhand)	3.6–15.2			2.0–8.1	BDL–6.6	BDL–5.3	5,487–7,720	58.8–171			20.3–113.4	Shukla (2007)
24.	*Lecanora fuscobrunnea*	Maitri Station (East Antarctica)				25.35	59.95	9.820					15.86	Upreti and Pandey (1999)
25.	*L. cinereofusca* H. Magn.	Faizabad city (crowded places) (Uttar Pradesh)								BDL–1.05				Dubey et al. (1999)
26.	*L. expectans*	Maitri Station (East Antarctica)				60.38	53.29	9.420					37.44	Upreti and Pandey (1999)
27.	*L. leprosa*	Lucknow city (Uttar Pradesh)	2.20–5.40			0.30	0.00–25.60	8.30–13.20	1573.00–1748	6.0–12.80			53.2–66	Saxena et al. (2007)
28.	*L. leprosa* Fee	Bangalore city (Karnataka)					29.92	9.84	3,121	154			128.15	Nayaka et al. (2003)
29.	*L. perplexa* Brodo	Bangalore city (Karnataka)					7.96	7.37	265	199.32			531.5	Nayaka et al. (2003)
30.	*Lecidea cancriformis*	Maitri Station (East Antarctica)				20.72	75.41	4.950					44.71	Upreti and Pandey (1999)

31. *L. siplei*	Maitri Station (East Antarctica)				56.12	237.41	17.510	55.44	Upreti and Pandey (1999)
32. *Lepraria lobificans* Nyl.	Mandav (Madhya Pradesh)	1.53–17.05	2.15–124.82	11.68–23.87	BDL–3.05	17.12–978.99	12.46–41.4	319.8–3,196	6.0–17.79	110.49–165.67	Bajpai et al. (2009a, b)
33. *Parmotrema austrosinensis* (Zahlbr.) Hale	Bangalore city (Karnataka)					BDL	20.26–158.32	586–4,530	BDL	100.67–153.9	Nayaka et al. (2003)
34. *Parmotrema praesorediosa* (Nyl.) Hale	Bangalore city (Karnataka)					8.39–28.2	15.65–16.06	2,040–7,389	164.4–233.3	126.9–321.2	Nayaka et al. (2003)
35. *P. praesorediosum* (Nyl.) Hale				41.27						Bajpai et al. (2010c)
36. *P. praesorediosum* (Nyl.) Hale	Mandav (Madhya Pradesh)	0.70–10.50	25.4–285.1	98.57–109.31	0.30–2.0	15.50–1136.9	0.00–56.5	361.20–8172.5	27.85–61.3	140.20–218.50	Bajpai et al. (2009a, b, 2010a, b, c)
37. *Peltula euploca*				95.89						Bajpai et al. (2010c)
38. *P. euploca* (Ach.) Poelt in pisut	Mandav (Madhya Pradesh)	1.20–12.70	14.1–101.2	163.3–190.52	0.10–0.50	1.00–9.20	3.90–38.80	195.10–279.30	154.84–195.16	18.70–47.00	Bajpai et al. (2009a, b, 2010a, b, c)
39. *Pertusaria leucosorodes* Nyl.	Bangalore city (Karnataka)					3.04	5.84	570.00	31.92	79.86	Nayaka et al. (2003)
40. *Phaeophyscia hispidula* (Ach.) Essl.	Dehradun city (Uttarakhand)	7.22–17.39				103.79–1189.56	19.96–31.36	8348.37–12543.40	0.01–17.42	116.97–198.78	Shukla et al. (2006)
41. *P. hispidula* (Ach.) Essl.	Pauri and Srinagar (Uttarakhand)	54.00–67.90				79.65–151.89	24.02–35.76	4505.00–10923.00	231.90–425.90	84.99–141.80	Shukla and Upreti (2007b)
42. *P. hispidula* (Ach.) Essl.	(Mandav (Madhya Pradesh)	1.30–19.30	21.20–302.40	97.30–115.13	0.60–2.80	19.50–1091.80	14.70–88.90	985.10–7891.90	20.00–54.76	44.60–170.70	Bajpai et al. (2009a, b)
43. *P. hispidula* (Ach.) Essl.	Rewa (Madhya Pradesh)		92.70–561.80		0.00–6.80	0.00–35.20		176.50–419.40	0.00–11.70	103.10–214.60	Bajpai et al. (2011)
44. *P. hispidula* (Ach.) Essl.	Dehradun, (Uttarakhand)	52.50–1230.00			2.00–875.00	193.00–4410.00	42.20–780.00	1135.00–29456.00	124.00–6412.50	119.00–5910.00	Rani et al. (2011)
45. *P. hispidula* (Ach.) Essl.	Dehradun, (Uttarakhand)	7.90–24.20			0.06–33.60	2.68–22.00	0.90–21.30	136.00–234.00	6.35–52.40	16.10–69.60	Shukla et al. (2012a, b)

(continued)

Table 5.11 (continued)

Species	Sites	Ni	Al	Ca	Cd	Cr	Cu	Fe	Pb	Mg	Mn	Zn	References
46. *P. hispidula* (Ach.) Essl.	Badrinath (Uttarakhand)	11.1–18.1			0.5–1.9	BDL–5.5	BDL–5.3	4783–14335	38.8–52.2			43.5–70.5	Shukla (2007)
47. *P. orbicularis*	Lucknow city (Uttar Pradesh)	10.70			0.00	62.20	18.30	19374.00	4.80			70.90	Saxena et al. (2007)
48. *Phylliscum indicum* Upreti	Mandav (Madhya Pradesh)	0.60–9.40	2.50–103.10	229.86–320.70	0.10–0.60	0.90–7.30	5.30–33.10	196.70–306.70		102.70–168.78		16.80–37.50	Bajpai et al. (2009a, b, 2010a, b, c)
49. *Physcia caesia*	Maitri Station (East Antarctica)				106.66	100.00	10350.00					74.66	Upreti and Pandey (1999)
50. *P. tribacia* (Ach.) Nyl.	Bangalore city (Karnataka)					0.00	33.32	6683.00	191.12			276.47	Nayaka et al. (2003)
51. *Pyrenula nanospora* (A. Singh) Upreti	Bangalore city (Karnataka)					36.43	18.28	1506.00	175.90			231.01	Nayaka et al. (2003)
52. *Pyxine cocoes* (Sw.) Nyl.	Bangalore city (Karnataka)					9.53–10.31	16.30–19.32	9795–12,056	0.00–63.63			103.30–224.60	Nayaka et al. (2003)
53. *P. cocoes* (Sw.) Nyl.	Lucknow city (Uttar Pradesh)	0.00–9.60			0.00–0.40	0.00–34.40	10.20–21.70	3,255–5,183	3.30–10.60			57.60–63.40	Saxena et al. (2007)
54. *P. cocoes* (Sw.) Nyl.	Raebareli, NTPC, Uttar Pradesh	0.60–18.30	297–1,631			0.90–12.10	2.50–12.10	58.60–1498.40	0.50–9.30			7.80–59.60	Bajpai et al. (2010a, b)
55. *P. cocoes* (Sw.) Nyl.	Katni (Madhya Pradesh)		78.2–365.10		0.00–6.30	0.80–26.20		103.0–689.40	0.00–13.30			57.30–194.40	Bajpai et al. (2011)
56. *P. cocoes* (Sw.) Nyl.	Hooghly and Nadia (West Bengal)		98.80–1376.10		0.00–4.50	0.00–40.30	0.00–10.20	69.30–730.10	0.00–89.10			0.00–118.10	Bajpai and Upreti (2012)
57. *P. petricola* Nyl. In crombie	Bangalore city (Karnataka)					18.47–19.00	115.19–338.12	5538.00–9202.00	83.33–101.40			105.38–133.05	Nayaka et al. (2003)
58. *Pyxine subcinerea* Stirton	Haridwar city (Uttarakhand)	13.50–719.25			4.20–51.45	16.40–94.40	19.65–144.10	795.00–17280.00	17.25–157.80			158.00–1178.00	Shukla et al. (2013a, b)
59. *P. subcinerea* Stirton	Srinagar, Garhwal (Uttarakhand)	0.02–0.14			0.00–0.14	0.06–0.19	0.27–0.84	43.08–55.93	0.26–0.62			1.76–4.69	Shukla and Upreti (2008)
60. *P. subcinerea* Stirton	Rudraprayag (Uttarakhand)		66.7–369.1		BDL–56.5	30.1–246.9	18.4–101.8	38.4–150.5	8.4–44.7		BDL–274.9	27.2–189.4	Shukla (2012)
61. *Remototrachyna awasthii*	Mahabaleshwar city, Maharashtra		125.00–1295.00		0.11–1.83	0.92–3.01	89.1–966.40	0.00–18.80			7.30–31.00	15.06–45.90	Bajpai et al. (2013a)
62. *R. awasthii*	Mahabaleshwar city, Maharashtra		335.80–1295.16		0.16–1.90	0.97–3.53		334.97–989.83	0.38–19.57		4.62–37.00	16.60–49.46	Bajpai et al. (2013b)

63.	*Rhizocarpon flavum*	Maitri Station (East Antarctica)		51.08	96.56	10040.00		67.12	Upreti and Pandey (1999)
64.	*Rinodina olivaceobrunnea*	Maitri Station (East Antarctica)		46.01	82.33	14240.00		71.44	Upreti and Pandey (1999)
65.	*Sphinctrina anglica*	Lucknow city (Uttar Pradesh)	3.80	0.00	50.70	25.80	3251.00	3.10	Saxena et al. (2007)
66.	*Umbilicaria aprina* Nyl.	Maitri Station (East Antarctica)			7.40–15.63	55.00–138.00	5386.00–11386.00	6.33–25.75	Upreti and Pandey (1994)
67.	*U. decussata* (Vill.) Zahlbr.	Maitri Station (East Antarctica)			3.36–4.20	45.00–93.00	4966.00–12760.00	5.66–19.80	Upreti and Pandey (1994)
68.	*Usnea antarctica*	Maitri Station (East Antarctica)		122.39	166.66	8840.00		111.97	Upreti and Pandey (1999)
69.	*Xanthoria elegans*	Badrinath (Uttarakhand)	2.84–8.04	0.09–0.33	BDL	5.06–8.01	3.3–3.7	2.87–7.49	33.87–50.3	Shukla (2007)

(Bačkor et al. 2006). Metal tolerance in animal and plants (vascular and non-vascular) is known to be conferred by production of a special class of proteins called metallothioneins and phytochelatins (PCs). The PCs play a central role in the detoxification of several heavy metals, especially Cd (Prasad 1997).

The higher resistance of chlorolichens to heavy metals compared with cyanolichens may be attributed to the phytochelatin synthesis in lichens with *Trebouxia* algae (Branquinho et al. 1997). Similarly in the present study, *P. hispidula*, a trebouxioid lichen, may gain an ecological advantage from their ability to counter heavy metal with prompt phytochelatin synthesis. However, there are few researches on the biosynthesis and function of phytochelatins in lichens in response to heavy metal exposure (Branquinho 2001; Pawlik-Skowrońska et al. 2002). Heavy metals are known to induce 'hsc' synthesis (heat shock proteins) which ultimately prevents damage of the membranes against sublethal and lethal temperatures. In plants, thermo protection by heavy metals via heat-shock cognates and the role of heat-shock proteins in protecting membrane damage by functioning as molecular chaperones is a manifestation of co-stress (Prasad 1997), which seems to be true for *Phaeophyscia hispidula* (Shukla et al. 2012a, b).

Lichens take up nutrients from (a) the substratum on which it is attached (bark, as in the case of epiphytic lichens) and (b) the metal-enriched ambient atmosphere (particulates and dissolved ions) (Nieboer and Richardson 1981). The concentrations of trace elements in lichen thalli may be directly correlated with environmental levels of these elements (Loppi et al. 1998; Purvis et al. 2008; Shukla and Upreti 2007a). Therefore, lichen biomonitoring can be applied to assess the air quality of an area. However, bioaccumulation is dependent on various factors, one of them being the contribution of the substratum on which lichen colonises (Lodenius et al. 2010; Thormann 2006; Baker 1983; Markert et al. 2003; St. Clair et al. 2002a, b). In a study carried out for assessing metal contents of a lichen species (*Pyxine subcinerea* Stirton) and mango bark collected from Haridwar city (Uttarakhand), were compared with soil, sampled from beneath the tree from which lichens were collected. The metal content in lichen, bark and soil ranged from 1,573 to 18,793, 256 to 590 and 684 to 801 µg g^{-1}, respectively. This clearly indicates that lichens accumulated higher amounts of metal compared to bark or soil. Statistical analysis revealed that metal concentration in lichens did not show significant linear correlation with the bark or soil. Pearson's correlation coefficients revealed negative correlation of Pb ($r=-0.2245$) and Ni ($r=-0.0480$) contents between lichen and soil, which indicate direct atmospheric input of metals from ambient environment. Quantification and comparison of elemental concentration in lichens, its substratum and soil can provide valuable information about air quality in the collection area (Shukla et al. 2013a, b).

Assessment of concentration variation of heavy metal provides vital information on the spatial behaviour of those metals affecting the air quality. In a study, samples of *Pyxine subcinerea* Stirton (a lichen species) were collected from Rudraprayag valley (Uttarakhand) to investigate the metal profile accumulated in lichens. Multivariate statistical analysis was carried out to elucidate possible contribution of various sources of pollution, including the anthropogenic sources as well, on the heavy metal profile in lichens. Cluster analysis successfully grouped geogenic and anthropogenic inputs represented by Al and Mn and Cu, Cd, Pb and Zn, respectively. Principal component analysis did segregate sites based on the origin (major contributors): PC1 corresponds to major contribution of geogenic metals, while PC2 corresponds to anthropogenic loadings. PC1 is dominated by highly significant positive loadings of Al and Mn, and PC2 is dominated by significant loadings of Cr, Pb, Zn and Cd. The study shows that lichen biomonitoring data may be effectively utilised to distinguish source of heavy metals in air which bioaccumulates in lichens (anthropogenic and/or natural sources) (Shukla unpublished).

The origin of PAHs was also assessed using the Phe/Ant, Flu/Pyr, Ant/Ant + Phe, Flu/Flu + Pyr and Naph/Phen concentration ratios. The total concentration of 16 PAHs ranged from 3.38 to

25.01 µg g⁻¹ with an average concentration of 12.09 ± 9.38 (SD). The PAH ratios clearly indicate that PAHs were of mixed origin, a major characteristic of urban environment. Significantly higher concentration of phenanthrene, pyrene and acenaphthylene indicates road traffic as major source of PAH pollution in the city. The study establishes the utility of *P. hispidula* as an excellent biomonitoring organism in monitoring both PAH and metals from foothill to subtemperate area of the Garhwal Himalayas (Shukla and Upreti 2009).

In a long-term biomonitoring study carried out in Dehradun, capital city of Uttarakhand, it was observed that the total metal concentration was the highest at sites heavily affected by traffic like Mohkampur Railway Crossing, Haridwar Road (42,505 µg g⁻¹). Dela Ram Chowk, located in the centre of the city, also had higher metal concentration, 34,317 µg g⁻¹, with maximum concentration of Pb at 12,433 µg g⁻¹, while Nalapani forest area had minimum total metal concentration (1,873 µg g⁻¹) as well as minimum Pb level at 66.6 µg g⁻¹, indicating anthropogenic activity, mainly vehicular activity, responsible for the increase in metal concentration in the ambient environment. In comparison with the earlier years 2004 and 2006, air pollution as indicated by similar lichen shows a considerable increase in the total metal concentration (especially Pb) in the ambient air of Dehradun city, which may be attributed to exponential rise in the traffic activity in the last 5 years (Rani et al. 2011).

Phaeophyscia hispidula, a common foliose lichen, growing in its natural habitat, was analysed for the concentration of six heavy metals (Fe, Ni, Zn, Cr, Cu and Pb) from five different sites of Pauri city, Garhwal Himalayas, Uttaranchal, India. The concentration of metals is correlated with the vehicular activity and urbanisation. The total metal concentration is highest at Circuit House on Pauri-Devprayag Road, followed by Malli on Pauri-Srinagar Road, which experience heavy traffic throughout the year, while Kiyonkaleshwar area, having less vehicular activity, had minimum accumulation of metal. The statistical parameter, coefficient of variation % showed higher CV% for Fe and Cr but lower for Cu and Ni. The concentrations of most of the metals at different sites were statistically significant (0.01 level). There was high spatial variability in the total metal concentrations, at different sites, that ranged from 5,087.1 to 11,500.44 µg g⁻¹ with an average concentration of 8,220.966 ± 2,991.467 (SD) (Shukla and Upreti 2007b).

Pyxine cocoes a foliose lichen commonly growing on Mango trees in tropical regions of India is an excellent organism for determining the pollutants emitted from coal-based thermal power plant and accumulated in lichens after prolonged exposure. The diversity and distribution of lichens in and around such power plant act as useful tool to measure the extent of pollution in the area. The distributions of heavy metals from power plant showed positive correlation with distance for all directions. The speed of wind and direction plays a major role in dispersion of the metals. The accumulation of Al, Cr, Fe, Pb and Zn in the thallus suppressed the concentration of pigments (chlorophyll a, chlorophyll b, total chlorophyll); however, it enhanced the level of protein. Further the concentration of chlorophyll content in *P. cocoes* increased with decreasing the distance from the power plant, while protein carotenoid and phaeophytisation exhibit significant decrease.

The morphology, chemistry and anatomy of lichens play important role in accumulation of metals. Another common tropical lichen species *Phaeophyscia hispidula* belonging to the same lichen family (Physciaceae) as of *Pyxine* has distinct morphology and chemistry. A thick tuft of rhizinae (hair like structure) on the lower surface of the thallus in *Phaeophyscia hispidula* acts as a metal reservoir and thus exhibits higher accumulation of most of the metals than *Pyxine*. The crust-forming lichens attached tightly to the substrates through their whole lower surface have the highest accumulation of Al in the metal sequence, while the squamulose and foliose forms show Fe in the higher concentrations. The lichens have special affinity with iron and they accumulate iron in greater amount than other metals.

Chlorophyll degradation value is considered to be an appropriate index for evaluating the effects of heavy metal pollution in lichens. In the present investigation the value of chlorophyll degradation ranges between 0.579 and 0.995 and 1.4 and 0.80 for control site and a site with high levels of vehicular traffic, respectively, as reported by Kardish et al. (1987).

Coefficient of correlation between different physiological parameters shows that chlorophyll content is significantly correlated ($p<0.01$) with carotenoid content, while protein content is negatively correlated with all other physiological parameters (with Chl. b $r=-0.5491$, carotenoid $r=-0.5809$, OD $r=-0.5034$).

In many studies from different regions of the world, concentration of total chlorophyll was affected by traffic level (Kauppi 1980; Arb et al. 1990; Carreras et al. 1998; Shukla and Upreti 2007a). Specifically, lichen located at sampling sites with high traffic level had increased chlorophyll concentration. It might be inferred that the content of chlorophylls increased parallel to the level of pollutant emitted by traffic.

According to Beckett and Brown (1983), Cd and Zn compete with one another for sites holding bivalent cations. It is consistent with the observation that the correlation coefficients between Cd and Zn, although not statistically significant, exhibit negative trends. Usually Zn, Cu and Fe are supposed to act antagonistically against Cd. Zn, Fe and Cu are negatively correlated with Cd. The significant positive correlation of protein with Cd indicates the synthesis of protein under Cd-stressed condition similar to expression of stress protein 70 (hsp 70) in the lichen photobiont *Trebouxia erici* (Bačkor et al. 2006). Metal tolerance in animal and plants (vascular and non vascular) is known to be conferred by production of a special class of proteins called metallothioneins and phytochelatins (PCs). The PCs play a central role in the detoxification of several heavy metals, especially Cd (Prasad 1997).

The higher resistance of chlorolichens to heavy metals compared with cyanolichens may be attributed to the phytochelatin synthesis in lichens with *Trebouxia* algae (Branquinho et al. 1997). Similarly in the present study, *P. hispidula*, a trebouxioid lichen, may gain an ecological advantage from their ability to counter heavy metal with prompt phytochelatin synthesis. However, there are few researches on the biosynthesis and function of phytochelatins in lichens in response to heavy metal exposure (Branquinho 2001; Pawlik-Skowrońska et al. 2002). Heavy metals are known to induce 'hsc' synthesis (heat shock proteins) which ultimately prevents damage of the membranes against sublethal and lethal temperatures. In plants, thermo protection by heavy metals via heat-shock cognates and the role of heat-shock proteins in protecting membrane damage by functioning as molecular chaperones are a manifestation of co-stress (Prasad 1997), which seems to be true for *P. hispidula*.

The assessment of lichen diversity in Dehra Dun city clearly indicates that the members of lichen family Physciaceae, especially *P. hispidula*, grows luxuriantly in both busy sites in the centre of the city and periurban rural areas (Shukla and Upreti 2007b). Increase in Physciaceae in temperate regions has been associated with both increasing temperature and increasing availability of nutrients (Loppi and Pirintsos 2000; Saipunkaew et al. 2007; van Herk et al. 2002).

More than 70 % of the population of India lives in rural areas and agriculture is their major livelihood. In the recent years diesel, being cheaper than petrol, is the predominant fuel used (>70 % consumption) for various agriculture practices and automobiles. Diesel exhaust is known to produce higher concentration of carcinogenic nitro-PAHs in comparison to petrol-fuelled vehicles along with low-toxicity 2-, 3- and 4-ringed PAHs. Petrol engine exhaust gases tend to have higher concentrations of the 5- and 6-ringed PAHs (benzo(a)pyrene, benzo(g,h,i) perylene, indeno(1,2,3-cd)pyrene, coronene) which are more carcinogenic than the 2- and 3-ringed PAHs. Nitro-PAHs are known to be potential health hazardous compounds even if its concentration in diesel exhaust is quite lower. Further, non-toxic simple PAHs like naphthalene may undergo photochemical reactions to form highly toxic nitro-PAHs such 1-nitronaphthalene (C.P.C.B 2005). Thus, diesel in India is the major contributor of PAH emissions into the environment (C.P.C.B 2005).

In a study carried out in Haridwar city, metallic contents (originating mainly due to vehicular activity) bioaccumulated in lichens were correlated with their PAH concentration to trace the source of PAH in air. The total metal concentration of four metals (chromium, copper, lead and cadmium) ranged between 369.05 and 78.3 µg g^{-1}, while concentration of 16 PAHs ranged between 1.25 and 187.3 µg g^{-1}. Statistical correlation studies revealed significant positive correlation between anthracene and chromium ($r=0.6413$, $p<0.05$) and cadmium with pyrene ($r=0.6542$, $p<0.05$). Naphthalene, acenaphthene, fluorene, acenaphthylene, anthracene and fluoranthene are reported to be main constituent of diesel vehicle exhaust which is in conformity with the present analysis as lead (indicator of petrol engine exhaust) had negative correlation with all these PAHs. The results indicate that diesel-driven vehicles contribute more towards ambient PAHs level.

Gas-particulate phase partitioning of PAHs may be monitored by studying the PAH profile in lichens. Anthropogenic sources at large are responsible for the release of PAHs, which gets bioaccumulated in plants along various spatial scales, based on their physicochemical properties. Atmospheric PAHs are distributed both in gaseous and particulate forms. In the present study a lichen species, *Dermatocarpon vellereum* Zschacke, has been collected from different altitudes in and around Rudraprayag valley, located in Central Himalayan region of India to investigate the spatial distribution of PAHs in the valley. PAHs concentration recorded ranged from 0.136 to 4.96 µg g^{-1}. Variation in PAHs concentration (based on ring profile) at different altitudes in comparison to centre of the town (Rudraprayag) at an altitude of 760 m provided vital information regarding spatial behaviour of PAHs. Result revealed that the bioaccumulation of 2- and 3-ringed PAHs was higher in samples from higher altitude, while bioaccumulation of fluoranthene (4 ringed PAH), having high spatial continuity, showed higher concentration in samples from localities away from town centre. The result verifies association of fluoranthene with particulate matter resulting in its wider distribution as suspended particulate matter and thus remains in gas phase as well. PAHs with 5 and 6 rings were confined to the lower altitude at the base of the valley justifying its particulate bound nature (Shukla et al. 2013).

It is clear from the studies that of all the studies carried out till date, members of the Physciaceae family are hyperaccumulators of metals and few species are established bioindicator species like *Pyxine cocoes*, *Phaeophyscia hispidula* and *Rinodina sophodes*.

5.3.2 Power Plants

Extensive use of coal in the thermal power plants causes some of the worst air pollution and has been linked to respiratory diseases and lung cancer. The burning of low-quality coal that contains high levels of fluorides and heavy metals is the main source of air pollution in China. A study reported that the IQs of children aged 8–13 in villages with high occurrences of fluorosis from burning this coal were about ten points lower than those in other areas (Li et al. 1995). It is reported that more than 40 million Chinese people suffer from symptoms of fluorosis caused by burning coal, making it the most widespread epidemic in China (Lee 2005).

Fossil fuels are a source of sulphur dioxide (SO_2) and carbon dioxide (CO_2). These compounds contribute to acidification and climate change. As a result of rapid economic growth, the use of fossil fuels, and the consequent emission of air pollutants, has been increasing in Asia and may do so in the coming decades. As a result, SO_2 emissions may increase fast in the future, and critical loads for acidifying deposition may be exceeded for a range of ecosystems in large parts of Asia (Foell et al. 1995). In Europe and North America, countries have developed strategies to reduce acidification by emission control. In Asia, such policies have only recently received attention and focus mainly on technologies to control SO_2 emissions like fuel and fuel gas desulphurisation. Replacing fossil fuels by renewable energy sources may be an alternative to these technical measures. This may also reduce CO_2 emissions (Boudri et al. 2002).

Recent literature referring to lichen biomonitoring of power plants has dealt mostly with airborne elements emitted by power plants using fossil fuels. The majority of the investigations of power plant emissions, airborne pollutants and lichens as monitors were performed in temperate zones. Lichens applied as monitors near coal-fired power stations in Portugal, for example, were found to accumulate heavy metals such as Fe, Co, Cr and Sb originating from coal and ash particles drifting through the air and positioned on the thallus (Freitas 1994). Freitas (1995) analysed the comparative accumulation of Cr, Fe, Co, Zn, Se, Sb and Hg in two vascular plants and in the epiphytic lichen *Parmelia sulcata* in an industrial region occupied by a thermal coal-fired power station, a chemical plant and an oil refinery. Of the three organisms, the lichen was found to be the most effective bioaccumulator. The technique of biomonitoring with lichens was also applied to estimate the air quality in the La Spezia district, Italy, in relation to a coal-fired power plant and other industrial activities (Nimis et al. 1990). The lichen *Parmelia caperata* collected from *Olea sativa* trees was meant to biomonitor the SO_2 pollution in the study area. The applied index, based on the frequency of species within a sampling grid, showed a high statistical correlation with pollution data measured by recording gauges. The distribution of the lichen *P. caperata* was found to correspond best with the lichen index. The lichen *P. caperata* was used again as a bioindicator of heavy metal pollution in an additional study in La Spezia (Nimis et al. 1993). Data on the amounts of 13 metals in lichen thalli in 30 stations were compared with the data of lichens collected in other parts of Italy. Mn and Zn did not coincide with substantial pollution-related phenomena, whereas extreme pollution-related phenomena coincided with high levels of Pb and Cd in an area adjacent to an industrial zone. Within the industrial zone, phenomena related to evident deposition referred to Al, As, Cr, Fe, Ni, Ti and V. Additional studies reported on the use of lichens around coal-fired power stations in arid and semiarid areas. Nash and Sommerfield (1981) found, within a radius of a few kilometres of a power station in New Mexico, that lichens contained elevated concentrations of B, F, Li and Se relative to lichens in more remote sites. One lichen species was found to contain elevated concentrations of Ba, Cu, Mn and Mo in sites located in the vicinity of the station. Few studies have been conducted in India involving lichen, only heavy metals (Bajpai et al. 2010a, b, c).

Effect of coal mining on frequency, density and abundance has been studied around Moghla coal mines, Kalakote area of Jammu and Kashmir (Charak et al. 2009). Study revealed that pollutants released from open coal mining activities not only effected quantitative distribution but also have effect on the quantitative parameters.

Levels of arsenic (As) and fluoride (F) were determined in an epiphytic lichen *Pyxine cocoes* (Sw.) Nyl., collected from the vicinity of a coal-based thermal power plant of Raebareli, India. Both elements are abundant in lichen thallus, while their substratum contained negligible amount. The As ranged between 8.9 ± 0.7 and 77.3 ± 2.0 µg g^{-1} dry weight in thallus and 1.0 ± 0.0 and 9.7 ± 0.2 µg g^{-1} dry weight in substratum, whereas F ranged between 9.3 ± 0.52 and 105.8 ± 2.3 µg g^{-1} dry weight in thallus; however, it was not detected in the substratum. The quantities of As in thallus increased with decreasing distance from the power plant, but F showed an opposite trend. The distribution of As and F around the power plant showed positive correlation with distance in all directions with better dispersion in western side as indicated by the concentration coefficient ($R2$). The F accumulation patterns in lichens clearly indicate that the coal burning in power plant is the major contributor and has its maximum levels on the downwind side. The analysis of variance and LSD indicated that the As/F concentrations among lichen thallus is significant at $p < 0.01$ % level (Bajpai et al. 2010a, b, c).

The lichen diversity assessment carried out around a coal-based thermal power plant indicated the increase in lichen abundance with the increase in distance from the power plant in general. The photosynthetic pigments, protein and heavy metals were estimated in *Pyxine cocoes* (Sw.) Nyl., a common lichen growing around a thermal power plant for further inference. Distributions of heavy metals from the power plant showed positive correlation with distance for all directions; however, western direction has received better dispersion as indicated by the concentration coefficient $R2$. Least significant difference analysis showed that speed of wind and its direction plays a major role in dispersion of heavy metals. Accumulation of Al, Cr, Fe, Pb and Zn in the thallus suppressed the concentrations of pigments like chlorophyll a, chlorophyll b and total chlorophyll; however, it enhanced the level of protein. Further, the concentrations of chlorophyll contents in *P. cocoes* increased with decreasing the distance from the power plant, while protein, carotenoid and phaeophytisation exhibited significant decrease.

5.3.3 Persistent Organic Pollutants (POPs)

POPs possess toxic properties and resist degradation. POPs bioaccumulate and are transported, through air, water and migratory species, across international boundaries, and deposited far from their place of release, where they accumulate in terrestrial and aquatic ecosystems (Stockholm Convention 2001) (Table 5.12). The Stockholm Convention on Persistent Organic Pollutants (POPs) was adopted in 2001 and revised in 2009 in response to the urgent need for global action to protect human health and the environment from chemicals that are highly toxic and persistent and bioaccumulate and move long distance in the environment. The Convention seeks the elimination or restriction of production and use of all intentionally produced POPs (i.e. industrial chemicals and pesticides). It also seeks the continuing minimisation and, where feasible, ultimate elimination of the releases unintentionally produced POPs such as dioxins and furans (http://www.pops.int/documents/convtext/convtext_en.pdf).

Over the last three decades, organic contaminants have been of increasing importance in environmental monitoring. For the last decades, persistent organic pollutants (POPs) have been found in large concentrations in Arctic areas. These substances accumulate in living organisms and are enriched throughout the food chain (Harmens et al. 2013).

Polychlorinated biphenyl (PCB) is one of the most important environmental toxins of this type. PCB is a group of synthetically produced persistent toxic chlororganic compounds. PCB is stored in the fatty parts of the organism and accumulates in the food chain. Humans, fatty fish and carnivores (such as polar bears) can therefore accumulate concentrations in their bodies that are so high that they are poisoned.

HCB is an interesting chemical in the category of POPs and has long half-lives in air, water and sediment (Mackay et al. 1992) and is extremely persistent in the environment. HCB had several uses in industry and agriculture. HCB was first introduced in 1933 as a fungicide on the seeds of onions, sorghum and crops such as wheat, barley, oats and rye. It is believed that agricultural use of HCB dominated its emissions during the 1950s and 1960s. Its octanol/air and octanol/water partition coefficients are lower than for many other POPs, which indicate it is more likely to undergo environmental re-cycling than for PCBs. Atmospheric degradation of HCB is extremely slow and is not an efficient removal process. In air, HCB is found almost exclusively in the gas phase, with less than 5 % associated with particles in all seasons except winter, where levels are still less than 10 % particle bound (Cortes et al. 1998; Cortes and Hites 2000). Gas phase partitioning results in its transport to great distance in the atmosphere before being removed by deposition or degradation. Van Pul et al. (1998) modelled the atmospheric residence time of HCB. The transport distance (the distance over which 50 % of the chemical is removed) for HCB was calculated to be 10^5 km. Due to this long atmospheric residence time, it is distributed widely on national, regional or global scales. HCB in the troposphere can be removed from the air phase via atmospheric deposition to water and soil (Bidleman et al. 1986; Ballschmitter and Wittlinger 1991; Lane et al. 1992a, b). The hydrophobic nature of HCB results in its preferential partitioning into sediment, soil and plant surfaces (Barber et al. 2005).

Table 5.12 List of Persistent Organic Pollutants (POPs) (according to Stockholm Convention 2001 and 2009 and LRTAP convention 1998 and 2009) and their sources

Persistent Organic Pollutants (POPs)	LRTAP convention	Stockholm convention	Sources
Aldrin	1998	2001	Used as ectoparasiticide, pesticide
Chlordane	1998	2001	Used as ectoparasiticide, insecticide, termiticide and additive in plywood adhesives
DDT (1,1,1-trichloro-2,2-bis (4-chlorophenyl)ethane)	1998	2001	Disease vector control agents for malaria and intermediate in production of dicofol intermediate
Dieldrin	1998	2001	Pesticide
Endrin	1998	2001	None
Heptachlor	1998	2001	Termiticide
Hexachlorobenzene (HCB)	1998	2001	Solvent in pesticide
Mirex	1998	2001	Termiticide
Toxaphene	1998	2001	None
Polychlorinated dibenzo-p-dioxins and dibenzofurans (PCDD/PCDF)	1998	2001	Unintentionally formed and released from thermal processes involving organic matter and chlorine as a result of incomplete combustion or chemical reactions
Polychlorinated biphenyls (PCB)	1998	2001	Unintentionally formed and released from thermal processes involving organic matter and chlorine as a result of incomplete combustion or chemical reactions
Chlordecone	1998	2009	Pesticide
Hexachlorohexane (HCH) including Lindane	1998	2009	Pesticide
Hexabromobiphenyl (HBB)	1998	2009	Industrial
PAHs	1998	2009	Unintentionally formed and released due to incomplete combustion of organic matter
Hexachlorobutadiene	2009	2009	Unintentionally formed and released due to incomplete combustion of organic matter
Pentachlorobenzene	2009	2009	Pesticide
Polybrominated diphenyl ethers (PBDEs)	2009	2009	Unintentionally formed and released due to industrial processes
Perfluorooctane sulfonic acid, its salts	2009	2009	Unintentionally formed and released due to industrial processes
Perfluorooctane sulfonyl fluorides (PFOs)	2009	2009	Unintentionally formed and released due to industrial processes
Polychlorinated naphthalenes	2009	NI	Unintentionally formed and released due to industrial processes
Short-chain paraffins (SCPs)	2009	NI	Unintentionally formed and released due to industrial processes

NI not included

γ-HCH (gamma-hexachlorocyclohexane), also known as the insecticide Lindane, is a chlororganic compound that has been used both as an insecticide in agriculture and as pharmaceutical treatment for head lice and scabies. Lindane is a neurotoxin that primarily affects the nervous system, liver and kidneys in humans. It can also have a carcinogenic effect.

Polychlorinated dibenzo-*p*-dioxins (PCDDs) and polychlorinated dibenzofurans (PCDFs) constitute a family of toxic, persistent and hydrophobic environmental pollutants, which have been shown to accumulate in biota. In Europe, environmental concentrations have increased slowly throughout this century until the late 1980s. These organic compounds have been shown to be carcinogenic in animals and humans. Sources inventories suggest that there are two main sources of PCDD/Fs in the environment: (1) combustion processes and (2) the occurrence as impurities in the manufacture of chlorinated aromatic products. A large number of combustion processes that generate PCDD/Fs occur in urban environments such as burning of coal, wood and petroleum, burning of municipal wastes, domestic burning and metal smelting. Food, especially products that originate from animals, is usually the main source of exposure for PCDD/Fs in humans. Uptake of persistent atmospheric PCDD/Fs in vegetation is the first step in the multi-pathway through the food chain resulting in the contamination of animals and humans.

In order to ensure temporal and spatial representation of pollution measurements, long-term continuous sampling at a large number of sites on broad scale is required. Measurements of PCDD/Fs atmospheric deposition with technical equipment at a regional and national scale are rarely available. Atmospheric analysis of PCDD/Fs is almost restricted to emission sources. Generally, articles concerning PCDD/Fs measurements are made in soil, water or biological organisms.

Lichens have a wide geographical distribution, occurring in rural areas as well as in urban and industrial areas, thus allowing comparison of pollutant concentrations from diverse regions. The morphology of lichens does not vary with seasons and therefore, accumulation can occur throughout the year. Furthermore, they depend mostly on the atmospheric deposition for their nutrition because 'root structures' do not function as in higher plants.

Different lichen species in places with different emission sources were collected at different times of the year in order to test the efficiency of lichens as PCDD/Fs biomonitors and the results showed that total concentration of PCDD/Fs in lichens was more similar to the concentrations reported for animals (top of the food chain) and soils (act as sinks) than those reported for plants. In general, the congeners and homologue profile observed in lichens resemble that of the atmosphere more that of the soil showing that lichens are potential good biomonitors of PCDD/Fs (Augusto et al. 2007).

India has extensive production and usage of organochlorine pesticides (OCPs) for agriculture and vector control. Despite this, few data are available on the levels and distribution of OCPs in the urban atmosphere of India. Passive and active air sampling conducted in seven metropolitan cities, New Delhi, Kolkata, Mumbai, Chennai, Bangalore, Goa and Agra, revealed concentrations (in pg·m^{-3}) ranged between 890 and 17,000 (HCHs), 250 and 6,110 (DDTs), 290 and 5,260 (chlordanes), 240 and 4,650 (endosulfans) and 120 and 2,890 (hexachlorobenzene). HCHs observed in India are highest reported across the globe. Chlordanes and endosulfans are lower than levels reported from southern China. Comparisons with studies conducted in 1989 suggested general decline of HCHs and DDTs in most regions. γ-HCH dominated the HCH signal, reflecting widespread use of Lindane in India. High o,p'-/p,p'-DDT ratios in northern India indicate recent DDT usage. Endosulfan sulphate generally dominated the endosulfan signal, but high values of α/β-endosulfan at Chennai, Mumbai and Goa suggest ongoing usage. Result shows local/regional sources of OCPs within India (Chakraborty et al. 2010).

No data is available on use of lichens for biomonitoring POPs in India except for PAHs (which has been discussed in detail in Sect. 5.1.3).

5.3.4 Peroxyacyl Nitrates (PAN)

Peroxyacyl nitrates (RC(O)OONO$_2$) is generated in air masses polluted by fuel emissions or by biomass burning. They are generally secondary pollutants produced by the photoinduced reactions initiated by the presence of volatile organic pollutants, ozone in atmosphere. They are toxic to the environment and phytotoxic in nature (Teklemariam and Sparks 2004), irritate human eyes and lead to genetic mutation (Kleindienst 1994). PAN act as reservoirs for odd nitrogen compounds, becoming involved in atmospheric circulation and influencing air quality (Singh et al. 1992a, b) on local, regional and global scales. PAN plays an important role in tropospheric chemistry and is a better indicator of photochemical smog than ozone (Zhang et al. 2011). Thermal decomposition rate of PAN is highly temperature dependent, resulting in lifetimes between 1 h at 298 K and about 5 months at 250 K. Thus, PAN can be transported over intercontinental distances in the cold upper troposphere (Singh 1987).

PAN was first detected in the Los Angeles area during smog episodes, exhibiting values of several ppbv (Stephens 1973). PAN amounts in clean air areas are generally much lower, namely, between 50 and 100 pptv only. Widespread southern hemispheric PAN pollution extending from South America to East and South Africa and as well as above the South Pacific has been observed during the annual biomass burning period in South America and South Africa in September and October (Singh et al. 1996, 2000a, b).

Only a few studies have reported PAN in Asia (Lee et al. 2008; Zhang and Tang 1994). More comprehensive studies have been carried out in North America and ambient PAN have been studied in southern California since 1960 (Grosjean 2003).

5.3.5 Ozone (O$_3$)

Ozone (O$_3$) is currently assumed worldwide as the most important air pollutant. Ozone is produced by photochemical reactions of the primary precursors such as hydrocarbons and nitrogen oxides (NOx). Emissions of these precursors are increased by industrialisation and the growing numbers of motor vehicles. Furthermore, O$_3$-producing photochemical reactions are favoured by high temperatures and high light intensities (Lefohn 1991). In many developing countries, urbanisation and industrialisation are increasing (Madkour and Laurence 2002).

Along the past decades, there has been a global increase in the lower tropospheric O$_3$ levels (Emberson et al. 2001) attributed primarily to increases in anthropogenic O$_3$ precursors (Derwent et al. 2002). Concentrations in rural or forested areas are generally as high as or higher than in urban regions (Millán et al. 1992). The identification and characterisation of O$_3$-induced foliar injury symptoms in well-adapted plant species by means of controlled field experiments are of major interest for assessing the risks imposed by air pollutants on local plant species. Such investigations assist in defining areas with phytotoxic concentrations and detecting levels of chronic pollution (Furlan et al. 2008).

In fact, bioindicator organisms must react to both atmospheric concentrations of pollutants and climatic conditions during exposures, the intensity of responses depending on the biological characteristics of each species, enabling the determination of the real amplitude of stress to which the plants and vegetation are exposed (Smith et al. 2003). Field surveys have recorded O$_3$-like injury symptoms on numerous tree, shrub and forbs species in Europe and North America (Furlan et al. 2008). However, little information is available on the effects of ozone on the multitude of native plant species throughout Europe or even North America and much less in tropical regions where O$_3$ concentrations have been increasing.

The effects of O$_3$ on lichens are poorly studied, compared with SO$_2$ and NO$_2$. The studies of toxicity of O$_3$ to lichens seem controversial. O$_3$ causes oxidative damage of cell membranes in lichens as a result of peroxidation of lipid membrane (Conti and Cecchetti 2001). Moreover, a field study with transplanted *Hypogymnia physodes* by Egger et al. (1994a, b) showed a high amount of end products of peroxidation in lichens exposed to a high concentration of O_3. However, Riddell et al. (2010) found no negative response of *Ramalina menziesii* to O_3 fumigations.

Few ozone studies have included bryophytes. Gagnon and Karnosky (1992) have shown that *Sphagnum* species are especially susceptible to ozone, having reduced photosynthesis, reduced growth, loss of colour and symptoms of desiccation, but that there are some remarkable reactive differences among species. Elevated ozone had no effect on germination of *Polytrichum commune* spores at concentrations of 11, 50, 100 and 150 ppb (Bosley et al. 1998), but it stimulated protonematal growth at 50 ppb and gametophore area increased to 189, 173 and 125 % of the controls at 50, 100 and 150 ppb, respectively, compared to that at ambient concentrations (Petersen et al. 1999).

Lichens are adversely affected by peak ozone concentrations as low as 20–60 µg m^{-3} (0.01–0.03 ppm; Egger et al. 1994a, b; Eversman and Sigal 1987). With regard to ozone, most reports of adverse effects on lichens have been in areas where peak ozone concentrations were at least 180–240 µg m^{-3} (0.09–0.12 ppm; Scheidegger and Schroeter 1995; Ross and Nash 1983; Sigal and Nash 1983; Zambrano et al. 2000). Although ozone can, in some cases, damage dry lichens, lichens are generally considered to be less susceptible to ozone damage when dry. Ruoss and Vonarburg (1995), for example, found no adverse effects on lichens in areas of Switzerland with daily summer peaks of 180–200 µg m^{-3} (0.09–0.10 ppm) O_3. They attributed this lack of response to the fact that ozone concentrations never rose above 120 µg m^{-3} (0.06 ppm) when the relative humidity was over 75 %.

Physical and chemical monitoring of air quality is still scarce in the country, even in the more developed south and southeast regions. Alternative methods, such as biomonitoring with sensitive plant species, are expected to offer effective means for identifying ozone-laden areas. Relatively little is known about the ambient levels of O_3 in lichens, especially in India, due to the high cost of using monitoring instruments to assess O_3 levels in urban and rural areas.

5.3.6 Increasing Tourism

Tourism is one of the fastest-growing economic sectors in the world. Similarly, in India, tourism has become one of the major sectors of the economy, contributing to a large proportion of the National Income (up to 6.23 % to the national GDP) and generating huge employment opportunities (8.78 % of the total employment). India witness more than five million annual foreign tourist arrivals and 562 million domestic tourism visits. The 'Incredible India' campaign launched by the Ministry of Tourism highlights natural and culture-rich areas of India, and location of Hindu holy pilgrimages, especially in the Himalayas, also attracts pilgrims (www.incredibleindia.org; www.itopc.org/travel-requisite/tourism-statistics.html).

No doubt tourist activity generates economy but there are pros and cons involved with the development of tourism industry in the country. One of the most important adverse effects of tourism on the environment is increased pressure on the carrying capacity of the ecosystem in each tourist locality. Increased infrastructural developments lead to large scale deforestation and destabilisation of natural landforms. Increased tourist flow leads to increased dumping of solid waste in the vicinity area resulting in disturbances as well as depletion of water and fuel resources. Flow of tourists to ecologically sensitive areas resulted in destruction of rare and endangered species due to trampling, disturbing of natural habitats which directly affect biodiversity, ambient environment and air quality profile of the tourist spot (Lalnunmawia 2010; Shukla 2007).

Lichen biomonitoring studies carried in the Himalayas and Western Ghats (biodiversity-rich areas) indicate that tourist activity has serious impact on the air quality of the area (Nayaka and Upreti 2005a, b). Changing lichen diversity due to infrastructural development is quite evident in urban settlements of Garhwal Himalayas

(Shukla and Upreti 2011a, b). In the holy pilgrimage centre of Badrinath, lichen diversity is dominated by toxitolerant lichens. The lichen family Physciaceae with 10 species is dominant in the area followed by Acarosporaceae and Parmeliaceae with 7 species each. *Lecanora muralis*, *Rhizoplaca chrysoleuca*, species of *Xanthoria* (*X. ulophyllodes, X. elegans* and *X. sorediata*) and *Dimelaena oreina* are abundant in the area.

In Gangotri, Gomukh (origin of holy river Ganga) and Badrinath areas, the occurrence of Physciaceae as the dominant family in these areas indicates nitrophilous conditions prevailing in the holy pilgrimages, which may be due to heavy tourist activity during the holy voyage (Upreti et al. 2004; Shukla and Upreti 2007c).

In the Badrinath area surface coverage of *Xanthoria elegans* and *Lecanora muralis*, known nitrophilous species, is very high. It has been reported by van Herk, (2001) that the presence and dominance of these nitrophilous species indicates the effect of human activity and animal rearing on lichen vegetation. The total absence of fruticose lichens in the area, highly sensitive to environmental alterations, also reflects the deteriorated air quality due to heavy vehicular activity and human activity as a result of pilgrimage.

In a study to observe the PAHs profile in different settlement of Garhwal Himalayas, it was found that out of the four localities the concentration of PAHs and the total carcinogenic PAH percentage (ΣcPAH%) was highest in lichens from Badrinath area, which is located in the inner Himalayas and experienced heavy vehicular and human activity during the 'holy voyage'. PAHs diagnostic ratios utilised to elucidate the probable source of origin of the PAHs indicated pyrolytic origin which is being attributed to cold climate and heavy usage of wood for cooking purpose (Shukla et al. 2010).

Vehicular activity resulting due to tourism adds on to the normal levels of emission due to daily anthropogenic activities (Shukla et al. 2012b). Vehicular activities not only emit significant quantity of metallic pollutant but also are major source of health hazardous compounds, PAHs. In an attempt to characterise, simultaneously, inorganic as well as organic pollutants in a biodiversity-rich area which is heavily influenced by tourist activity and, thus, assess the impact of tourist activity on the ecosystem, the study was carried out with an aim to assess the heavy metal (HM) and polycyclic aromatic hydrocarbons (PAHs) in the air of a biodiversity as well as tourist-rich area of Western Ghats by applying a most frequent-growing lichen *Remototrachyna awasthii* (Hale and Patw.) Divakar and A. Crespo, as biomonitor. Thalli of *R. awasthii* were collected from eight sites of Mahabaleshwar area located in Western Ghats. Total metal concentration (HM) ranged from 644 to 2,277.5 µg g^{-1} while PAHs concentration between 0.193 and 54.78 µg g^{-1}. HM and PAHs concentrations were the highest at Bus stand, while control site (Lingmala Fall) exhibited the lowest concentration of HM as well as PAHs followed by samples from site with little or no vehicular activity (both these sites are having trekking route). It was also evident from this study that vehicular emission played a significant role in the release of HM and PAHs as pollutants in the environment.

B(a)P, the classical chemical carcinogen, is considered to be the useful indicator for cancer risk assessment. The average concentration of B(a)P in Mahabaleshwar ranged from BDL to 1.97. According to the World Health Organization (WHO), B(a)P is considered to be a reliable index for the assessment of total PAHs carcinogenicity. Since B(a)P is easily oxidised and photodegraded, therefore, the PAHs carcinogenic character could be underestimated. For better quantification of carcinogenicity related to the whole PAH factor, BaP equivalent potency (BaPE) index after Yassaa et al. (2001), Mastral et al. (2003) and Cheng et al. (2007) has been calculated. BaPE is quite high at bus stand and Venna Lake, whereas the control sites Lingmala Fall and Wilson Point have 0 values. BaPE index (Fig. 5.3), thus, indicates that the cancer risk is associated with high vehicular activity. Thus, urban population appears to be exposed to significantly higher cancer risk (Bajpai et al. 2013).

5.4 Conclusion

Environmental problems have been aggravated by the rapid expansion of human and industrial activities. In order to solve environmental problems and achieve sustainable development, an integrated effort has to be made to deal with a broad range of environmental issues, from identification of source, managing locally, regionally as well as globally. Biomonitoring is one such cost-effective and reliable method to keep a watch on the environmental problems persisting today.

The present discussion shows that biomonitoring studies not only provide data on the present air quality data but an integrated approach involving physicochemical analysis could establish bioindicators as an integral part of air quality regulatory practices in Asian countries, especially in India, as in the western countries. Biomonitoring data may be effectively utilised in regulatory management practices to reduce emissions of a wide array of toxic pollutants either at local level or by regulatory agencies, making use of natural resources as sentinels of sustainable development.

References

Abdalla K (2006) Health and environmental benefits of clean fuels and vehicles. Keynote presentation. UN DESA, Cairo

Achotegui-Castells A, Sardans J, Ribas A, Peñuelas J (2012) Identifying the origin of atmospheric inputs of trace elements in the Prades Mountains (Catalonia) with bryophytes, lichens, and soil monitoring. Environ Monit Assess. doi:10.1007/s10661-012-2579-z

Adamova LI, Biazrov LG (1991) Heavy natural radionuclides in lichens from different ecosystems of the Western Caucasus (In Russian) – bioindication and biomonitoring. Nauka, Moscow, pp 125–129

Ahmad MN, van den Berg LJL, Shah HU, Masood T, Büker P, Emberson L, Ashmore M (2012) Hydrogen fluoride damage to vegetation from peri-urban brick kilns in Asia: a growing but unrecognised problem? Environ Pollut 162:319–324

Aide M (2005) Elemental composition of soil nodules from two alfisols on an alluvial terrace in Missouri. Soil Sci 170:1022–1033. doi:10.1097/01.ss.0000187351.16740.55

Akimoto H (2003) Global air quality and pollution. Science 302:1716–1719

Al TA, Blowes DW (1999) The hydrogeology of a tailing impoundment formed by central discharge of thickened tailing: implications for tailing management. J Conta Hydrol 38:489–505

Alauddin M (2003) Economic liberalization and environmental concerns: a South Asian perspective. South Asia 26(3):439–453

Alauddin M (2004) Environmentalizing economic development: a South Asian perspective. Ecol Econ 51:251–270

Ammann K, Herzig R, Liebendoerfer L, Urech M (1987) Multivariate correlation of deposition data of 8 different air pollutants to lichen data in a small town in Switzerland. Adv Aerobiol 87:401–406

Aptroot A, van Herk CM (2007) Further evidence of the effects of global warming on lichens, particularly those with *Trentepohlia* phycobionts. Environ Pollut 146(2):293–298

Arb CV, Mueller C, Ammann K, Brunold C (1990) Lichen physiology and air pollution II. Statistical analysis of the correlation between SO_2, NO_2, NO and O_3 and chlorophyll content, net photosynthesis, sulphate uptake and protein synthesis of *Parmelia sulcata* Taylor. New Phytol 115:431–437

Aragón G, Martínez I, Izquierdo P, Belinchón R, Escudero A (2010) Effects of forest management on epiphytic lichen diversity in Mediterranean forests. Appl Veg Sci 13:183–194. doi:10.1111/j.1654-109X.2009.01060.x

Archer DE, Johnson K (2000) A model of the iron cycle in the ocean. Glob Biogeochem Cycle 14:269–279

Asta J, Rolley F (1999) Biodiversitéet bioindication lichénique: qualitéde l'air dans l'agglomération Grenobloise. Bull Int Assoc Fr Lichénol 3:121–126

Asta J, Erhardt W, Ferretti M, Fornasier F, Kirschbaum U, Nimis PL, Purvis O, Pirintsos S, Scheidegger C, Van-Haluwyn C, Wirth V (2002) Mapping lichen diversity as an indicator of environmental quality. In: Nimis PL, Scheidegger C, Wolseley P (eds) Monitoring with lichensmonitoring lichens. Kluwer, Dordrecht, pp 273–279

Augusto S, Catarino F, Branquinho C (2007) Interpreting the dioxin and furan profiles in the lichen *Ramalina canariensis* Steiner for monitoring air pollution. Sci Total Environ 377:114–123

Augusto S, Maguas C, Matos J, Pereira MJ, Soares A, Branquihno C (2009) Spatial modeling of PAHs in lichens for fingerprinting of Multisource atmospheric pollution. Environ Sci Technol 43(20):7762–7769

Augusto S, Maguas C, Matos J, Pereira MJ, Branquihno C (2010) Lichens as an integrating tool for monitoring PAH atmospheric deposition: a comparison with soil, air and pine needles. Environ Pollut 158(2):483–489

Backor M, Loppi S (2009) Interactions of lichens with heavy metals. Biol Plant 53(2):214–222

Bačkor M, Gibalová A, Budová J, Mikeš J, Solár P (2006) Cadmium-induced stimulation of stress protein hsp70 in lichen photobiont *Trebouxia erici*. Plant Growth Regul 50:159–164

Baddeley MA, Ferry BW, Finegan EJ (1972) The effects of sulphur dioxide on lichen respiration. Lichenologist 5:283–291

Baek SO, Field RA, Goldstone ME, Kirk PW, Lester JN, Perry R (1991) A review of atmospheric polycyclic aromatic hydrocarbons: sources, fate and behavior. Water Air Soil Pollut 60:79–300

Baeza A, Del Rio M, Jimenez A, Miro C, Paniagua J (1995) Influence of geology and soil particle size on the surface area/volume activity ratio for natural radionuclides. J Radioanal Nucl Chem 189(2): 289–299

Bajpai R, Upreti DK (2012) Accumulation and toxic effect of arsenic and other heavy metals in a contaminated area of West Bengal, India, in the lichen *Pyxine cocoes* (Sw.) Nyl. Ecotoxicol Environ Saf 83:63–70

Bajpai R, Upreti DK, Mishra SK (2004) Pollution monitoring with the help of lichen transplant technique at some residential sites of Lucknow city, Uttar Pradesh. J Environ Biol 25(5):191–195

Bajpai R, Upreti DK, Dwivedi SK (2009a) Arsenic accumulation in lichens of Mandav monuments, Dhar district, Madhya Pradesh, India. Environ Monit Assess 159:437–442. doi:10.1007/s10661-008-0641-7

Bajpai R, Upreti DK, Dwivedi SK, Nayaka S (2009b) Lichen as quantitative biomonitors of atmospheric heavy metals deposition in Central India. J Atmos Chem 63:235–246

Bajpai R, Upreti DK, Dwivedi SK (2010a) Passive monitoring of atmospheric heavy metals in a historical city of central India by *Lepraria lobificans* Nyl. Environ Monit Assess 166:477–484. doi:10.1007/s10661-009-1016-4

Bajpai R, Upreti DK, Nayaka S (2010b) Accumulation of arsenic and fluoride in lichen *Pyxine cocoes* (Sw.) Nyl., growing in the vicinity of coal-based thermal power plant at Raebareli, India. J Exper Sci 1(4):37–40

Bajpai R, Upreti DK, Nayaka S, Kumari B (2010c) Biodiversity, bioaccumulation and physiological changes in lichens growing in the vicinity of coal based thermal power plant of Raebareli district, north India. J Hazard Mater 174:429–436

Bajpai R, Mishra GK, Mohabe S, Upreti DK, Nayaka S (2011) Determination of atmospheric heavy metals using two lichen species in Katni and Rewa cities. Indian J Environ Biol 32:195–199

Bajpai R, Pandey AK, Deeba F, Upreti DK, Nayaka S, Pandey V (2012) Physiological effects of arsenate on transplant thalli of the lichen *Pyxine cocoes* (Sw.) Nyl. Environ Sci Pollut Res 19:1494–1502

Bajpai R, Shukla V, Upreti DK (2013) Impact assessment of anthropogenic activities on air quality, using lichen *Remototrachyna awasthii* as biomonitors. Int J Environ Sci Technol. doi:10.1007/s13762-012-0156-1

Baker DA (1983) Uptake of cations and their transport within the plants. In: Robb DA, Pierpoint WS (eds) Metals and micronutrients: uptake and utilization by plants. Academic, London, pp 3–19

Balaguer L, Manrique E, Ascaso C (1997) Predictability of the combination effects of sulphur dioxide and nitrate on green algal lichen *Ramalina farinacea*. Can J Bot 75:1836–1842

Ballschmitter K, Wittlinger R (1991) Interhemispheric exchange of hexachlorocyclohexanes, hexachlorobenzene, polychlorobiphenyls, and 1,1,1-trichloro-2,2-bis(p-chlorophenyl) ethane in the lower troposphere. Environ Sci Technol 25:1103–1111

Baptista MS, Teresa M, Vasconcelos SD, Carbral JP, Freitas CM, Pacheo AMG (2008) Copper, nickel, lead in lichens & tree bark transplants over different period of time. Environ Pollut 151:408–413

Barber JL, Sweetman AJ, Wijk D, Jones C (2005) Hexachlorobenzene in the global environment: emissions, levels, distribution, trends and processes. Sci Total Environ 349:1–44

Bargagli R (1998) Trace elements in terrestrial plants: an ecophysiological approach to biomonitoring and biorecovery. Springer, Berlin, p 324

Bargagli R, Nimis PL (2002) Guidelines for the use of epiphytic lichens as biomonitors of atmospheric deposition of trace elements. In: Monitoring with lichens-monitoring lichens, vol 7, pp 295–299

Bari A, Rosso A, Minciardi MR, Troiani F, Piervittori R (2001) Analysis of heavy metals in atmospheric particulates in relation to their bioaccumulation in explanted *Pseudevernia furfuracea* thalli. Environ Monit Assess 69:205–220

Barták M, Solhaug KA, Vráblíková H, Gauslaa Y (2006) Curling during desiccation protects the foliose lichen *Lobaria pulmonaria* against photoinhibition. Oecologia. doi:10.1007/s00442-006-0476-2

Bässler C, Müller J, Hothorn T, Kneib T, Badeck F, Dziock F (2010) Estimation of the extinction risk for high-montane species as a consequence of global warming and assessment of their suitability as cross-taxon indicators. Ecol Indic 10:341–352

Baumard P, Budzinski H, Michon Q, Garrigues P, Burgeot T, Bellocq J (1998) Origin and bioavailability of PAHs in the Mediterranean Sea from mussel and sediment records. Estuarine Coastal Shelf Sci 47:77–90

Beckett RP, Brown DH (1983) Natural and experimentally-induced zinc and copper resistance in the *lichen* genus Peltigera. Ann Bot 52:43–50

Beckett RP, Brown DH (1984) The control of Cd uptake in lichen genus *Peltigera*. J Exp Bot 35:1071–1082

Behling H (1998) Late Quaternary vegetational and climatic changes in Brazil. Rev Palaeobot Palynol 99(143):156

Belandria G, Asta J (1986) Les lichens bioindicateurs: la pollution acide dans la région lyonnaise. Pollut Atmos 109:10–23

Belvermis M, Kılıç Ö, Çotuk Y, Topcuoğlu S (2010) The effects of physicochemical properties on gamma emitting natural radionuclide levels in the soil profile of Istanbul. Environ Monit Assess 163:15–26

Berry WL, Wallace A (1981) Toxicity: the concept and relationship to the dose response curve. J Plant Nutr 3:13–19

Beschel R (1950) Flechten als Alteramasstab Rezenter Moränen. Zeitschrift für Glatscherkunde und Glazialgeiglogie 1:152–161

Biazrov LG (1994) The radionuclides in lichen thalli in Chernobyl and East Urals areas after nuclear accidents. Phyton (Horn, Austria) 34(1):85–94

Biazrov LG, Adamova LI (1990) Heavy metals in lichens of the Caucasusky and Ritzinsky reserves (In Russian) – the reserves of USSR – their real and future. Part 1: Topicals problems of reserve management. Abstracts of the all-union conference, Novgorod, pp 338–339

Bidleman TF, Billings WN, Foreman WY (1986) Vapor-particle partitioning of semivolatile organic compounds: estimates from field collection. Environ Sci Technol 20:1038–1043

Bignal KL, Ashmore MR, Headley AD, Stewart K, Weigert K (2007) Ecological impacts of air pollution from road transport on local vegetation. Appl Geochem 22:1265–1271

Bignal KL, Ashmore MR, Headley AD (2008) Effects of air pollution from road transport on growth and physiology of six transplanted bryophyte species. Environ Pollut 156:332–340

Bird PM (1966) Radionuclides in foods. Can Med Assoc J 94:590–597

Bird PM (1968) Studies of fallout of 137Cs in the Canadian North. Arch Environ Health 17:631–638

Blackman A, Harrington W (2000) The use of economic incentives in developing countries: lessons from international experience with industrial air pollution. J Environ Dev 9:5–44

Blasco M, Domeno C, Nerin C (2008) Lichen biomonitoring as feasible methodology to assess air pollution in natural ecosystems: combined study of quantitative PAHs analysis and lichen biodiversity in the Pyrenees mountain. Anal Bioanal Chem 391:759–771

Boileau LJR, Beckett PJ, Richardsons DHS (1982) Lichens and mosses as monitors of industrial activities associated with Uranium mining in northern Ontario, Canada. Part 1: field procedure, chemical analysis and inter-species comparison. Environ Pollut 4:69–84

Boonpragob K, Nash TH III, Fox CA (1989) Seasonal deposition patterns of acidic ions and ammonium to the lichen *Ramalina menziesii* Tayl. in southern California. Environ Exp Bot 29:187–197

Boudri JC, Hordijk L, Kroeze C, Amann M, Cofala J, Bertok I, Junfeng L, Lin D, Shuang Z, Runquing H, Panwar TS, Gupta S, Singh D et al (2002) The potential contribution of renewable energy in air pollution abatement in China and India. Energy Policy 30:409–424

Branquinho C (2001) Lichens. In: Prasad MNV (ed) Metals in the environment: analysis by biodiversity. Marcel Dekker, New York, pp 117–157

Branquinho C, Brown DH, Magaus C, Catarino CL (1997) Metal uptake and its effects on membrane integrity and chlorophyll fluorescence in different lichen species. Environ Exp Bot 37:95–105

Brewer RF (1960) The effects of hydrogen fluoride gas on seven citrus varieties. Am Soc Hortic Sci 75:236–243

Brifett C (1999) Environmental impact assessment in East Asia. In: Petts J (ed) Handbook of environmental impact assessment, vol 2. Blackwell, Oxford

Brodo IM (1961) Transplant experiments with corticolous lichens using a new technique. Ecology 42:838–841

Brodo IM (1964) Field studies of the effects of ionizing radiation on lichens. Bryologist 67:76–87

Brown DH, Avalos A, Miller JE, Bargagli R (1994) Interactions of lichens with their mineral environment. Crypt Bot 4:135–142

Bryselbout C, Henner P, Carsignol J, Lichtfouse E (2000) Polycyclic aromatic hydrocarbon in highway plants and soils. Evidence for a local distillation effect. Analusis 28(4):290–293

Buccolieri A, Buccolieri G, Dell'atti A, Perrone MR, Turnone A (2006) Natural sources and heavy metal. Annali di Chimica, 96 by Società Chimica Italiana

Budka D, Przybyiowicz WJ, Mesjasz- Przybyiowicz J (2004) Environmental pollution monitoring using lichens as bioindicators: a micro-PIXE study. Radiat Phys Chem 71:783–784

Bunce NJ, Liu L, Zhu J, Lane DA (1997) Reaction of naphthalene and its derivatives with hydroxyl radicals in the gas phase. Environ Sci Technol 31:2252–2259

Butler WL, Kitajima M (1975) Fluorescence quenching in photosystem II of chloroplasts. Biochem Biophys Acta 376:116–125

Button KJ, Rietveld P (1999) Transport and the environment. In: van den Bergh JCJM (ed) Handbook of environmental and resource economics. Edward Elgar, Cheltenham, pp 581–589

C.P.C.B. (2005) Parivesh: proposed limits for Pah in India. Central pollution Control Board, Ministry of Environment and Forest, Delhi – 32. www.cpcb.nic.in

Calderón-Garcidueñas L, Mora-Tiscareno A, Fordham LA, Valencia-Salazar G, Chung CJ, Rodriguez-Alcaraz A et al (2003) Respiratory damage in children exposed to urban pollution. Pediatr Pulmonol 3:148–161

Carignan V, Villard MA (2002) Selecting indicator species to monitor ecological integrity: a review. Environ Monit Assess 78:45–61

Carreras HA, Gudiño GL, Pignata ML (1998) Comparative biomonitoring of atmospheric quality in five zones of Cardóba city (Argentina) employing the transplanted lichen *Usnea* sp. Environ Pollut 103:317–325

Central Pollution Control Board (1999) Parivesh: Newsletter, 6(1), June. CPCB, Ministry of Environment and Forests, Delhi

Chakraborty P, Zhang G, Li J, Xu Y, Liu X, Tanabe S, Jones KC (2010) Selected organochlorine pesticides in the atmosphere of major Indian cities: levels, regional versus local variations, and sources. Environ Sci Technol 44(21):8038–8043

Chaphekar SB (2000) Phytomonitoring in industrial areas. In: Agrawal SB, Agrawal M (eds) Environmental pollution and plant responses. CRC Press, Boca Raton, pp 329–342

Chapin FS, Körner C (1994) Arctic and alpine biodiversity: patterns, causes and ecosystem consequences. Trends Ecol Evol 9:45–47

Charak S, Sheikh MA, Raina AK, Upreti DK (2009) Ecological impact of coal mines on lichens: a case study at Moghla coal mines kalakote (Rajouri), J & K. J Appl Nat Sci 1(1):24–26

Cheng Z et al (2005) Limited temporal variability of arsenic concentrations in 20 wells monitored for 3 years in Araihazar, Bangladesh. Environ Sci Technol 39(13):4759–4766

Cheng J, Yuan T, Wu Q, Zhao W, Xie H, Ma Y, Ma J, Wang J (2007) PM10-bound polycyclic aromatic hydrocarbons (PAHs) and cancer risk estimation in the atmosphere surrounding an industrial area of Shanghai, China. Water Air Soil Pollut 183(1–4):437–446. doi:1007/s11270-007-9392-2

Chetwittayachan T, Shimazaki D, Yamamoto K (2002) A comparison of temporal variation of particle-bound polycyclic aromatic hydrocarbons (pPAHs) concentration in different urban environments: Tokyo, Japan, and Bangkok, Thailand. Atmos Environ 36:2027–2037

Clark BD (1999) Capacity building. In: Petts J (ed) Handbook of environmental impact assessment, vol 2. Blackwell, Oxford

Clark AJ, Landolt W, Bucher JB, Strasser RJ (2000) Beech (*Fagus sylvatica*) response to ozone exposure assessed with a chlorophyll a fluorescence performance index. Environ Pollut 109:501–507

Conti ME, Cecchetti G (2001) Biological monitoring: lichens as bioindicator of air pollution assessment – a review. Environ Pollut 114:471–492

Cortes DR, Hites RA (2000) Detection of statistically significant trends in atmospheric concentrations of semivolatile compounds. Environ Sci Technol 34:2826–2829

Cortes DR, Basu I, Sweet CW, Brice KA, Hoff RM, Hites RA (1998) Temporal trends in gas-phase concentrations of chlorinated pesticides measured at the shores of the Great Lakes. Environ Sci Technol 32:1920–1927

Coskun M, Steinnes E, Coskun M, Cayir A (2009) Comparison of epigeic moss (*Hypnum cupressiforme*) and lichen (*Cladonia rangiformis*) as biomonitor species of atmospheric metal deposition. Bull Environ Contam Toxicol 82:1–5

Crespo A, Divakar PK, Arguello A, Gasca C, Hawksworth DL (2004) Molecular studies on *Punctelia* species of the Iberian Peninsula, with an emphasis on specimens newly colonizing Madrid. Lichenologist 36(5):299–308

Cropper ML, Simon NB, Alberini A, Arora S, Sharma PK (1997) The health benefits of air pollution control in Delhi. Am J Agric Econ 79:1625–1629

Curran TP (2000) Sustainable development: new ideas for a new century. Seminar at the Graduate School of Environmental Studies. Seoul National University, Seoul

Daly GL, Wania F (2005) Organic contaminants in mountains. Environ Sci Technol 39(2):385–398

Das G, Das AK, Das JN, Guo N, Majumdar R, Raj S (1986) Studies on the plant responses to air pollution, occurrence of lichen in relation to Calcutta city. Indian Biol 17(2):26–29

Das P, Joshi S, Rout J, Upreti DK (2012) Shannon diversity index (H) as an ecological indicator of environmental pollution – a GIS approach. J Funct Environ Bot 2(1):22–26

Das P, Joshi S, Rout J, Upreti DK (2013) Impact of anthropogenic factors on abundance variability among Lichen species in southern Assam, north east India. Trop Ecol 54:65–70

Davies L, Bates JW, Bell JNB, James PW, Purvis OW (2007) Diversity and sensitivity of epiphytes to oxides of nitrogen in London. Environ Pollut 146:299–310

Deb MK, Thakur M, Mishra RK, Bodhankar N (2002) Assessment of atmospheric arsenic levels in airborne dust particulates of an urban city of Central India. Water Air Soil Pollut 140:57–71

Delfino RJ, Murphy-Moulton AM, Becklake MR (1998) Emergency room visits for respiratory illnesses among the elderly in Montreal: association with low level ozone exposure. Environ Res Sect A 76:67–77

Dentener FJ, Carmichael GR, Zhang Y, Lelieveld J, Crutzen PJ (1996) Role of mineral aerosol as a reactive surface in the global troposphere. J Geophys Res 101:22869–22889

Derwent R, Collins W, Johnson C, Stevenson D (2002) Viewpoint. Global ozone concentrations and regional air quality. Environ Sci Technol 36:379A–382A

De Sloover J, Le Blanc F (1968) Mapping of atmospheric pollution on the basis of lichen sensitivity. In: Misra R, Gopal B (eds) Proceedings of the symposium on recent advances on tropical ecology. International Society for Tropical Ecology, Varanasi, pp 42–56

Dickerson RR, Kondragunta S, Stenchikov G, Civerolo KL, Doddridge BG, Holben BN (1997) The impact of aerosols on solar ultraviolet radiation and photochemical smog. Science 278:827–830

Dockery DW, Pope CA, Xu X, Spengler JD, Ware JH, Fay ME et al (1993) An association between air pollution and mortality in six US cities. N Engl J Med 329:1753–1759

Domeño C, Blasco M, Sanchez C, Nerin C (2006) A fast extraction technique for extracting polycyclic aromatic hydrocarbons (PAHs) from lichen samples used as biomonitors of air pollution: dynamic sonication versus other methods. Anal Chim Acta 569:103–112

Dubey AK, Pandey V, Upreti DK, Singh J (1999) Accumulation of lead by lichens growing in and around Faizabad, U.P., India. J Environ Biol 20(3):223–225

Dutkiewicz VA, Alvi S, Ghauri BM, Choudhary MI, Husain L (2009) Black carbon aerosols in urban air in South Asia. Atmos Environ 43:1737–1744

Eckl P, Hofmann W, Türk R (1986) Uptake of natural and man-made radionuclides by lichens and mushrooms. Radiat Environ Biophys 25:43–54

Egger R, Schlee D, Turk R (1994a) Changes of physiological and biochemical parameters in the lichen *Hypogymnia physodes* (L.) Nyl due to the action of air-pollutants – a field study. Phyton-Annales Rei Botanicae 34:229–242

Egger R, Schlee D, Türk R (1994b) Changes of physiological and biochemical parameters in the lichen *Hypogymnia physodes* (L.) Nyl. due to the action of air pollutants – a field study. Phyton 34(2):229–242

Ellis KM, Smith JN (1987) Dynamic model for radionuclide uptake in lichen. J Environ Radioact 5:185–208

Ellis CJ, Coppins BJ, Dawson TP (2007) Predicted response of lichen epiphyte *Lecanora populicola* to climate change scenarios in a clean-air region of Northern Britain. Biol Conserv 135:396–404

Emberson LD, Ashmore MR, Murray F, Kuylenstierna JCI, Percy KE, Izuta T, Zheng Y, Shimizu H, Sheu BH, Liu CP, Agrawal M, Wahid A, Abdel-Latif NM, Van Tienhoven M, Bauer LI, Domingos M (2001) Impacts of air pollutants on vegetation in developing countries. Water Air Soil Pollut 130:107–118

Emberson L, Ashmore M, Murray F (eds) (2003) Air pollution effects on crops and forests. Imperial College Press, London

Essington M (2004) Soil and water chemistry – an integrative approach. CRC Press, Boca Raton

Eversman S, Sigal LL (1987) Effects of SO2, O3, and SO2 and O3 in combination on photosynthesis and ultrastructure of two lichen species. Can J Bot 65(9):1806–1818

Faiz A, Weaver CS, Walsh MP (1996) Air pollution from motor vehicles. International Bank for Reconstruction and Development/World Bank, Washington, DC

Farmer AM, Bates JW, Bell JNB (1991) Seasonal variations in acidic pollutant inputs and their effects on the chemistry of stemflow, bark and epiphyte tissues in three oak woodlands in N.W. Britain. New Phytol 118:441–451

Farmer AM, Bates JW, Bell JNB (1992) Ecophysiological effects of acid rain on bryophytes and lichens. In: Bates JW, Farmer AM (eds) Bryophytes and lichens in a changing environment. Clarendon, Oxford

Feige GB, Niemann L, Jahnke S (1990) Lichens and mosses: silent chronists of the Chernobyl accident. Bibl Lichenol 38:63–77

Fendorf SE (1995) Surface reactions of chromium in soils and waters. Geoderma 67:55–71

Fernandez P, Vilanova RM, Grimalt JO (1999) Sediment fluxes of polycyclic aromatic hydrocarbons in European high altitude mountain lakes. Environ Sci Technol 33:3716–3722

Fields RD (1988) Physiological responses of lichens to air pollutant fumigations. In: Nash TH III, Wirth V (eds) Lichens, bryophytes and air quality, Bibliotheca Lichenologica 30. J. Cramer, Berlin/Stuttgart, pp 175–200

Flesher JW, Horn J, Lehner AF (2002) Role of the Bay- and L-regions in the metabolic activation and carcinogenicity of Picene and Dibenz[a,h]anthracene. Polycycl Aromat Compd 22:737–745

Foell W, Green C, Amann M, Bhattacharya S, Carmichael G et al (1995) Energy use, emissions, and air pollution reduction strategies in Asia. Water Air Soil Pollut 85:2277–2282

Foster JB (1993) Let them eat pollution: capitalism and the world environment. Monthly Review, January, pp 10–20

Frati L, Caprasecca E, Santoni S, Gaggi C, Guttova A, Gaudino S, Pati A, Rosamilia S, Pirintsos SA, Loppi S (2006) Effects of NO_2 and NH_3 from road traffic on epiphytic lichens. Environ Pollut 142:58–64

Freitas MC (1994) Heavy metals in *Parmelia sulcata* collected in the neighbourhood of a coal-fired power station. Biol Trace Elem Res 43–45:207–212

Freitas MC (1995) Elemental bioaccumulators in air pollution studies. J Radioanal Nucl Chem 192:171–181

Furlan CM, Moraes RM, Bulbovas P, Sanz MJ, Domingos M, Salatino A (2008) *Tibouchina pulchra* (Cham.) Cogn., a native Atlantic Forest species, as a bio-indicator of ozone: visible injury. Environ Pollut 152:361–365

Gaare E (1990) Lichen content of radiocesium after the Chernobyl accident in mountains in southern Norway. In: Desmet G et al (eds) Transfer of radionuclides in natural and seminatural environments. Elsevier, London/New York, pp 492–501

Gagnon ZE, Karnosky DF (1992) Physiological response of three species of Sphagnum to ozone exposure. J Bryol 17:81–91

Gailey FAY, Smith GH, Rintoul LJ, Lloyd OL (1985) Metal deposition patterns in central Scotland, as determined by lichen transplants. Environ Monit Assess 5:291–309

Galarneau E, Makar PA, Sassi M, Diamond ML (2007) Estimation of atmospheric emissions of six semivolatile polycyclic aromatic hydrocarbons in Southern Canada and the United States by use of an emissions processing system. Environ Sci Technol 41:4205–4213

Garćia AZ, Coyotzin CM, Amaro AR, Veneroni DL, Martínez CL, Iglesias GS (2009) Distribution and sources of bioaccumulative air pollutants at Mezquital Valley, Mexico, as reflected by the atmospheric plant *Tillandsia recurvata* L. Atmos Chem Phys 9:6479–6494

Garty J (1993) Plants as biomonitors. In: Markert B (eds) VCH Verlagsgesellschaft mbh, Germany, pp 193–263

Garty J (2001) Biomonitoring atmospheric heavy metals with lichens: theory and application. Crit Rev Plant Sci 20(4):309–371

Garty J, Galun M, Kessel M (1979) Localization of heavy metal and other elements accumulated in the lichen thallus. New Phytol 82:159–168

Gasparatos D (2012) Fe–Mn concretions and nodules to sequester heavy metals in soils. In: Lichtfouse E et al (eds) Environmental chemistry for a sustainable world, vol 2: remediation of air and water pollution, pp 443–474. doi 10.1007/978-94-007-2439-6_11

Geebelen W, Hoffman M (2001) Evaluation of bio-indication methods using epiphytes by correlating with SO_2-pollution parameters. Lichenologist 33:249–260

Genty B, Briantais JM, Baker NR (1989) The relationship between the quantum yield of photosynthetic electron transport and quenching of chlorophyll fluorescence. Biochem Biophys Acta 990:87–92

George C (2000) Comparative review of environmental assessment procedures and practice. In: Lee N, George C (eds) Environmental assessment in developing and transitional countries. Wiley, Chichester

Gilbert OL (1971) The effect of airborne fluorides on lichens. Lichenologist 5:26–32

Giordani P (2007) Is the diversity of epiphytic lichens a reliable indicator of air pollution? A case study from Italy. Environ Pollut 146:317–323

Giordani P, Brunialti G, Alleteo D (2002) Effects of atmospheric pollution on lichen biodiversity (LB) in a Mediterranean region (Liguria, northwest Italy). Environ Pollut 118:53–64

Giordano S, Sorbo S, Adamo P, Basile A, Spagnuolo V, Cobianchi CR (2004) Biodiversity and trace element content of epiphytic bryophytes in urban and extra-urban sites of southern Italy. Plant Ecol 170:1–14

Gob F, Oetit F, Bravard JP, Ozer A, Gob A (2003) Lichenometric application to historical and subrecent dynamics and sediment transport of a Corsican stream (Figarella River, France). Quat Sci Rev 22:2111–2124

Godinho RM, Wolterbeek HT, Verburg T, Freitas MC (2008) Bioaccumulation behaviour of lichen *Flavoparmelia caperata* in relation to total deposition at a polluted location in Portugal. Environ Pollut 151:318–325

Gombert S, Asta J, Seaward MRD (2002) Correlation between the nitrogen concentration of two epiphytic lichens and the traffic density in an urban area. Environ Pollut 123:281–290

Gorham E (1959) A comparison of lower and higher plants as accumulators of radioactive fall-out. Can J Bot 37:327–329

Goyal R, Seaward MRD (1982) Metal uptake in terricolous lichens. III Translocation in the thallus of *Peltigera canina*. New Phytol 90:85–98

Grabherr G, Gottfried M, Pauli H (1994) Climate effects on mountain plants. Nature 369:448–1448

Gries C, Sanz M-J, Nash TH III (1995) The effect of SO2 fumigation on CO2 gas exchange, chlorophyll fluorescence and chlorophyll degradation in different lichen species from western North America. Cryptogam Bot 5:239–246

Grosjean D (2003) Ambient PAN and PPN in southern California from 1960 to the SCOS97- NARSTO. Atmos Environ 37:S221–S238

Guidotti M, Stella D, Owczarek M, DeMarco A, De Simone C (2003) Lichens as polycyclic aromatic hydrocarbon bioaccumulators used in atmospheric pollution studies. J Chromatogr 985(1–2):185–190

Guidotti M, Stella D, Dominici C, Blasi G, Owazasek M, Vitali M, Protano C (2009) Monitoring of traffic related pollution in a province of central Italy with transplanted lichen Pseudevernia furfuracea. Bull Environ Contam Toxicol 83:852–858

Guttikunda SK, Carmichael GR, Calori G, Eck C, Woo JH (2003) The contribution of megacities to regional sulfur pollution in Asia. Atmos Environ 37:11–22

Haas JR, Bailey EH, Purvis OW (1998) Bioaccumulation of metals by lichens: uptake of aqueous uranium by *Peltigera membranacea* as a function of time and pH. Am Miner 83:1494–1502

Haffner E, Lomsky B, Hynek V, Hallgren JE, Batic F, Pfanz H (2001) Air pollution and lichen physiology. Physiological responses of different lichens in a transplant experiment following an SO_2-gradient. Water Air Soil Pollut 131:185–201

Hafner WD, Carlson DL, Hites RA (2005) Influence of local human population on atmospheric polycyclic aromatic hydrocarbon concentrations. Environ Sci Technol 39:7374–7379

Hale ME (1983) The biology of lichens, 3rd edn. Edward Arnold, London

Halek F, Kianpour-rad M, Kavousi A (2010) Characterization and source apportionment of polycyclic aromatic hydrocarbons in the ambient air (Tehran, Iran). Environ Chem Lett 8:39–44

Han X, Naeher LP (2006) A review of traffic-related air pollution exposure assessment studies in the developing world. Environ Int 32:106–120

Handley R, Overstreet R (1968) Uptake of carrier-free Cs-137 by *Ramalina reticulata*. Plant Physiol 43:1401

Hansen ES (2008) The application of lichenometry in dating glacier deposits. Geografisk Tidsskrift-Danish J Geogr 108(1):143–151

Hanson WC (1967) Cesium-137 in Alaskan lichens, caribou and Eskimos. Health Phys 13:383–389

Hanson WC (1971) Fallout radionuclide distribution in lichen communities near Thule. J Arctic Inst N Am 24(4):269–276

Hanson WC, Eberhardt LL (1971) Cycling and compartimentalizing of radionuclides in northern Alaskan lichen communities. SAEC, COO-2122-5. Memorial Institute of Pacific Northwest Laboratory, Ecos. Department, Battelle, Richland, Washington, DC

Harmens H, Foan L, Simon V, Millis G (2013) Terrestrial mosses as biomonitors of atmospheric POPs pollution: a review. Environ Pollut 173:245–254

Harrison RM, Smith DJT, Luhana L (1996) Source apportionment of atmospheric polycyclic aromatic hydrocarbons collected from an urban location in Birmingham, UK. Environ Sci Technol 30:825–832

Harvey RG, Halonen M (1968) Interaction between carcinogenic hydrocarbons and nucleosides. Cancer Res 28:2183–2186

Hauck M (2008) Epiphytic lichens indicate recent increase in air pollution in the Mongolian capital Ulan Bator. Lichenologist 40(2):165–168

Hauck M (2009) Global warming and alternative causes of decline in arctic-alpine and boreal-montane lichens in North-Western Central Europe. Glob Chang Biol 15:2653–2661.doi:10.1111/j.1365-2486.2009.01968.x

Hawksworth DL (1971) Lichens as litmus for air pollution: a historical review. Int J Environ Stud 1:281–296

Hawksworth DL (1973) Mapping studies. In: Ferry BW, Baddeley MS, Hawksworth DL (eds) Air pollution and lichens. Athlone Press, London, pp 38–76

Hawksworth DL, Rose F (1970) Qualitative scale for estimating sulphur dioxide air pollution in England and Wales using epiphytic lichen. Nature 227:145–148

Heald CL, Jacob DJ, Fiore AM, Emmons LK, Gille JC, Deeter MN, Warner J, Edwards DP, Crawford JH, Hamlin AJ, Sachse GW, Browell EV, Avery MA, Vay SA, Westberg DJ, Blake DR, Singh HB, Sandholm ST, Talbot RW, Fuelberg HE (2003) Asian outflow and transpacific transport of carbon monoxide and ozone pollution: an integrated satellite, aircraft and model perspective. J Geophys Res 108(D24):4804

Heald CL, Jacob DJ, Park RJ, Alexander B, Fairlie TD, Yantosca RM, Chu DA (2006) Transpacific transport of Asian anthropogenic aerosols and its impact on surface air quality in the United States. J Geophys Res 111:14310

Herzig R, Urech M (1991) Flechten als Bioindikatoren. Integriertes biologisches Messsystem der Luftverschmutzung für das Schweizer Mittelland. Bibl Lichenol 43:1–283

Herzig R, Liebendorfer L, Urech M, Ammann K, Cuecheva M, Landolt W (1989) Passive biomonitoring with lichens as a part of an integrated biological measuring system for monitoring air-pollution in Switzerland. Int J Environ Anal Chem 35:43–57

Holopainen T (1984) Types and distribution of ultra structural symptoms in epiphytic lichens in several urban and industrial environments in Finland. Ann Bot Fennici 21:213–229

Holopainen T, Kärenlampi L (1985) Characteristic ultrastructural symptoms caused in lichens by experimental exposure to nitrogen compounds and fluorides. Ann Bot Fenn 22:333–342

Hov Ø (1984) Modelling of the long-range transport of peroxyacetylnitrate to Scandinavia. J Atmos Chem 1:187–202

Hutchinson-Benson E, Svoboda J, Taylor HW (1985) The latitudinal inventory of 137Cs in vegetation and topsoil in northern Canada, 1980. Can J Bot 63:784–791

Hviden T, Lillegraven A (1961) 137Cs and 90Sr in precipitation, soil and animals in Norway. Nature 192:1144–1146

Hyvärinen M, Koopmann R, Hormi O, Tuomi J (2000) Phenols in reproductive and somatic structures of lichens: a case of optimal defence? Oikos 91: 371–375

IEA (International Energy Agency) (1999) World Energy Outlook-1999 insights. Looking at energy subsidies. Getting the Price Right, OCED

Innes JL (1985) Lichenometry. Prog Phys Geogr 9:187–254

Insarov GE (2010) Epiphytic montane lichens exposed to background air pollution and climate change: monitoring and conservation aspects. Int J Ecol Environ Sci 36(1):29–35

Insarov GE, Semenov SM, Insarova I (1999) A system to monitor climate change with epilithic lichens. Environ Monit Assess 55:279–298

Intergovernmental Panel on Climate Change (IPCC) (2001) Climate change 2001: the scientific basis. In: Contribution of working group I to the third IPCC assessment report 944. Cambridge University Press, New York

Iqbal A, Oanh NTK (2011) Assessment of acid deposition over Dhaka division using CAMx-MM5 modelling system. Atmos Pollut Res 2:52–462

Iurian AR, Hofmann W, Lettner H, Türk R, Cosma C (2011) Long term study of Cs-137 concentrations in lichens and mosses. Rom J Phys 56(7–8):983–992

Ivanovich M, Harmon RS (1982) Uranium series disequilibrium – applications to environmental problems. Clarendon, Oxford

Janssen NAH, Schwartz J, Zanobetti A, Suh HH (2002) Air conditioning and source-specific particles as modifiers of the effect of PM_{10} on hospital admissions for heart and lung disease. Environ Health Perspect 110:43–49

Japan Environmental Council (2005) The state of the environment in Asia 2005/2006. Springer, Tokyo, p 3

Jeran Z, Byrne AR, Batic F (1995) Transplanted epiphytic lichens as biomonitors of air-contamination by natural radionuclides around the Zirovski vrh uranium mine, Slovenia. Lichenologist 27(5):375–385

Jeran Z, Jacimovic R, Batic F, Mavsar R (2002) Lichens as integrating air pollution monitors. Environ Pollut 120:107–113

Jerina DM, Thakkar DR, Yagi H, Levin W, Wood AW, Conney AH (1978) Carcinogenicity of benzo(a)pyrene derivatives: the bay region theory. Pure Appl Chem 50:1033–1044

Jorge-Villar SE, Edwards HGM (2009) Lichen colonization of an active volcanic environment: a Raman spectroscopic study of extremophile biomolecular protective strategies. J Raman Spectrosc 41:63–67

Joshi S (2009) Diversity of lichens in Pidari and Milam regions of Kumaon Himalaya. Ph. D. thesis. Kumaon University, Nainital

Joshi S, Upreti DK (2008) Lichenometric studies in vicinity of Pindari Glacier in the Bageshwar district of Uttarakhand, India. Curr Sci 99(2):231–235

Joshi S, Upreti DK, Punetha N (2008) Change in the lichen flora of Pindari Glacier Valley Uttarakhand (India) during the last three decades. Ann For 16(1):168–169

Joshi S, Upreti DK, Das P (2011) Lichen diversity assessment in Pindari glacier valley of Uttarakhand, India. Geophytology 41(1–2):25–41

Joshi S, Upreti DK, Das P, Nayaka S (2012) Lichenometry: a technique to date natural hazards. Sci India Popular Issue V(II):1–16. www.earthscienceindia.info

Jovan S (2008) Lichen bioindication of biodiversity, air quality, and climate: baseline results from monitoring in Washington, Oregon, and California. General technical report PNW-GTR-737. U.S. Department of Agriculture, Forest Service, Pacific Northwest Research Station, Portland, 115 p

Junshum P, Somporn C, Traichaiyaporn S (2008) Biological indices for classification of water quality around Mae Moh power plant. Int J Sci Technol (Thailand, Maejo) 2(01):24–36

Kardish N, Ronen R, Bubrick P, Garty J (1987) The influence of air pollution on the concentration of ATP and on chlorophyll degradation in the lichen *Ramalina duriaei* (De Not.) Bagl. New Phytol 106:697–706

Kathuria V (2002) Vehicular pollution control in Delhi, India. Transp Res Part D 7(5):373–387

Kathuria V (2004) Impact of CNG on vehicular pollution in Delhi – a note. Transp Res Part D 9(5):409–417

Kauppi M (1980) Fluorescence microscopy and microfluorometry for the examination of pollution damage in lichens. Ann Bot Fenn 17:163–173

Khalili NR, Scheff PA, Holsen TM (1995) PAH source fingerprints for coke oven, diesel, and gasoline engines highway tunnels and wood combustion emissions. Atmos Environ 29(4):533–542

Kholia H, Mishra GK, Upreti DK, Tiwari L (2011) Distribution of lichens on fallen twigs of Quercus leucotrichophora and Quercus semecarpifolia in and around Nainital city, Uttarakhand, India. Geophytology 41(1–2):61–73

Kim JW (1990) Environmental aspects of transnational corporation activities in pollution-intensive industries in the Republic of Korea: a case study of the Ulsan/Onsan industrial complexes. In: Environmental aspects of transnational corporation activities in selected Asian and Pacific developing countries. ESCAP/UNCTC Publication Series B, No. 15. United Nations, New York, pp 276–319

Kim TY (1993) A study on the effects of air pollution in China on the Korean peninsula. Masters thesis, Graduate School of Environmental Studies, Seoul National University, Seoul

Kim JW (2006) The environmental impact of industrialization in East Asia and strategies toward sustainable development. Sustain Sci 1:107–114. doi:10.1007/s11625-006-0006-5

Kleindienst TE (1994) Recent developments in the chemistry and biology of peroxyacetyl nitrate. Res Chem Intermed 20:335–384

Kondratyuk SY, Coppins BJ (1998) Lobarion lichens as indicators of the primeval forests of the eastern Carpathians. In: Darwin international workshop, Ukraine Phytosociological Center, Kiev, 25–30 May 1998

Korenaga T, Liu X, Tsukiyama Y (2000) Dynamics analysis for emission sources of polycyclic aromatic hydrocarbons in Tokushima soils. J Health Sci 46(5):380–384

Kreuzer W, Schauer T (1972) The vertical distribution of Cs137 in Cladonia rangiformis and C. silvatica. Svensk Bot. Tidskr 66:226–238

Kricke R, Loppi S (2002) Bioindication: the IAP approach. In: Nimis PL, Scheidegger C, Wolseley PA (eds) Monitoring with lichens – monitoring lichens. Kluwer, Dordrecht, pp 21–37

Kulkarni AV (2007) Effect of global warming on the Himalayan cryosphere. Jalvigyan Sameeksha 22:93–108

Kuusianen M (1996a) Epiphytic flora and diversity on basal trunks of six old-growth forests tree species in southern and middle boreal Finland. Lichenologist 28:443–463

Kuusianen M (1996b) Cyanobacterial macrolichens of Populus tremula as indicators of forest continuity in Finland. Biol Conserv 75:43–49

Kwapulinski J, Seaward MRD, Bylinska EA (1985a) Uptake of 226Radium and 228Radium by the lichen genus Umbilicaria. Sci Tot Environ 41:135–141

Kwapulinski J, Seaward MRD, Bylinska EA (1985b) 137Caesium content of Umbilicaria-species, with particular reference to altitude. Sci Tot Environ 41:125–133

Laden F, Neas LM, Dockery DW, Schwartz J (2000) Association of fine particulate matter from different sources with daily mortality in six US cities. Environ Health Perspect 108:941–947

Lal B, Ambasht RS (1981) Impairment of chlorophyll content in the leaves of Diospyros melanoxylon in relation to fluoride pollution. Water Air Soil Pollut 16:361–365

Lalnunmawia H. (2010) Impact of tourism in India. www.itopc.org/travel-equisite/tourism-statistics.html. Accessed on 28 Nov 2011

Lane DA, Johnson ND, Hanley M, Schroeder WH, Ord DT (1992a) Gas and particle-phase concentrations of alpha-hexachlorocyclohexane, gamma-hexachlorocyclohexane, and hexachlorobenzene in Ontario air. Environ Sci Technol 26:126–133

Lane DA, Schroeder WH, Johnson ND (1992b) On the spatial and temporal variations in the atmospheric concentrations of hexachlorobenzene and hexachlorocyclohexane isomers at several locations in the province of Ontario, Canada. Atmos Environ A 26:31–42

Larsen RS, Bell JNB, James PW, Chimonides PJ, Rumsey FJ, Tremper A, Purvis OW (2007) Lichen and bryophyte distribution on oak in London in relation to air pollution and bark acidity. Environ Pollut 146:332–340

LeBlanc F, Robitaille G, Rao D (1974) Biological response of lichens and bryophytes to environmental pollution in the Murdochville copper Mine area, Quebec. J Hattori Bot Lab 38:405–433

Lee SS (2005) One out of thirty Chinese poisoned by fluoride. http://www.hani.co.kr/kisa/section-004005000/2005/08/p004005000

Lee YH, Shyu TH, Chiang MY (2003) Fluoride accumulation and leaf injury of tea and weeds in the vicinity of a ceramics factory. Taiwanese J Agric Chem Food Sci 41:87–94

Lee G, Jang Y, Lee H, Han JS, Kim KR, Lee M (2008) Characteristic behavior of peroxyacetyl nitrate (PAN) in Seoul megacity, Korea. Chemosphere 73:619–628

Lefohn AS (1991) Surface level ozone exposure and their effects on vegetation. Lewis Publishers, Boca Raton

Li XS, Zhi JL, Gao RO (1995) Effect of fluoride exposure on intelligence in children. Fluoride 28(4):89–192

Li P, Feng XB, Qiu GL, Shang LH, Li ZG (2009) Mercury pollution in Asia: a review of the contaminated sites. J Hazard Mater 168:591–601

Lidén K, Gustavsson M (1967) Relationships and seasonal variation of Cs-137 in lichen, reindeer and man in northern Sweden 1961 to 1965. In: Aberg B, Higate FP (eds) Radioecological concentration processes. Proceedings international symposium, 1966. Pergamon Press, Oxford, pp 193–207

Lisowska M (2011) Lichen recolonisation in an urban-industrial area of southern Poland as a result of air quality improvement. Environ Monit Assess 179(1–4):177–190

Liu K, Colinvaux PA (1988) A 5200-year history of Amazon rain forest. J Biogeogr 15:231–248

Liu J, Diamond J (2005) China's environment in a globalizing world. Nature 435:1179–1186

Liu H, Jacob DJ, Bey I, Yantosca RM, Duncan BN, Sachse GW (2003) Transport pathways for Asian pollution outflow over the Pacific: interannual and seasonal variations. J Geophys Res 108:8786

Lodenius M, Kiiskinen J, Tulisalo E (2010) Metal levels in an epiphytic lichen as indicators of air quality in a suburb of Helsinki, Finland. Boreal Environ Res 15:446–452

Lohani BN, Evans JW, Everitt RR, Ludwig H, Carpenter RA, Tu S-L (1997) Environmental impact assessment for developing countries in Asia. Asian Development Bank, Manila

Loppi S, Frati L (2006) Lichen diversity and lichen transplants as monitors of air pollution in a rural area of central Italy. Environ Monit Assess 114:361–375. doi:10.1007/s10661-006-4937-1

Loppi S, Pirintsos SA (2000) Effect of dust on epiphytic lichen vegetation in the Mediterranean area (Italy and Greece) Isreal. J Plant Sci 48:91–95

Loppi S, Pirintsos SA (2003) Epiphytic lichens as sentinels for heavy metal pollution at forest ecosystem (central Italy). Environ Poll 121:327–332

Loppi S, Pacioni G, Olivieri N, Di Giacomo F (1998) Accumulation of trace metals in the lichen *Evernia prunastri* transplanted at biomonitoring sites in Central Italy. Bryologist 101(3):451–454

Loppi S, Ivanov D, Boccardi R (2002) Biodiversity of epiphytic lichens and air pollution in the town of Siena (Central Italy). Environ Pollut 116:123–128

Lozan JL, Grabl H, Hupfer P (2001) Summary: warning signals from climate in climate of 21st century: changes and risks. Wissenschaftliche Auswertungen, Berlin, pp 400–408

LRTAP Convention (1998) Protocol to the 1979 convention on long-range transboundary air pollution on persistent organic pollutants. http://www.unece.org/env/lrtap/

Mackay D, Wania FA (1995) Global distribution model for persistent organic chemicals. Sci Total Environ 160/161:25–38

Mackay D, Shiu W-Y, Ma K-C (1992) Illustrated handbook of physical-chemical properties and environmental fate for organic chemicals. FL7 Lewis, Boca Raton

Madkour SA, Laurence JA (2002) Egyptian plant species as new ozone indicators. Environ Pollut 120:339–353

Markert BA, Breure AM, Zechmeister HG (2003) Definitions, strategies and principles for bioindication/biomonitoring of the environment. In: Markert BA, Breure AM, Zechmeister HG (eds) Bioindicators and biomonitors. Elsevier, Oxford, pp 3–39

Martin JH (1991) Iron still comes from above. Nature 353:123

Martin JH, Fitzwater SE (1988) Iron deficiency limits phytoplankton growth in the north-east Pacific subarctic. Nature 331:341–343

Martin JR, Koranda JJ (1971) Recent measurements of Cs-137 residence time in Alaskan vegetation. U.S. Atom. Energy Comm. Rep. CONF-71050, pp 1–34

Martin RV, Jacob DJ, Yantosca RM, Chin M, Ginoux P (2003) Global and regional decreases in tropospheric oxidants from photochemical effects of aerosols. J Geophys Res 108:4097

Masclet P, Hoyau V, Jaffrezo JL, Legrand M (1995) Evidence for the presence of polycyclic aromatic hydrocarbons in the polar atmosphere and in the polar ice of Greenland. Analusis 23:250–252

Mason MG, Cameron I, Petterson DS, Home RW (1987) Effect of fluoride toxicity on production and quality of wine grapes. J Aust Inst Agric Sci 53:96–99

Mastral AM, Lopez JM, Callen MS, Garcya T, Murillo R (2003) Spatial and temporal PAH concentrations in Zaragoza, Spain. Sci Total Environ 307:111–124

Mattsson LJS (1974) Cs-137 in the Reindeer Lichen Cladonia alpestris: deposition, retention and internal distribution 1961–1970. Health Phys 28:233–248

Maxwell K, Johnson GN (2000) Chlorophyll fluorescence – a practical guide. J Exp Bot 51:659–668

Mc Cune B (1993) Gradients in epiphytic biomass in three Pseudotsuga-Tsuga forests of different ages in western Oregon and Washington. Bryologist 96:405–411

McGrath SP (1995) Chromium and nickel. In: Alloway BJ (ed) Heavy metals in soils, 2nd edn. Blackie/Academic and Professional, London, pp 152–174

Menard PB, Peterson PJ, Havas M, Steinnes E, Turner D (1987) Lead, cadmium and arsenic in the environment. In: Hutchison TC et al (eds) Environmental contamination. Wiley, New York, pp 43–48

Menon S et al (2002) Climate effects of black carbon aerosols in China and India. Science 297:2250–2253

Meyerhof D, Marshall H (1990) The non-agricultural areas of Canada and radioactivity. In: Desmet G et al (eds) Transfer of radionuclides in natural and semi-natural environments. Elsevier, London/New York, pp 48–55

Miguel AH, Kirchstetter TW, Harley RA, Hering SV (1998) On-road emissions of particulate polycyclic aromatic hydrocarbons and black carbon from gasoline and diesel vehicles. Environ Sci Technol 32:450–455

Millán MM, Artiñnano B, Alonso L, Castro M, Fernádez-Patier R, Goberna J (1992) Meso-meteorological cycles of air pollution in the Iberian Peninsula (MECAPIP) (Air pollution research report 44, EUR N-14834). European Commission, Brussels, DG XII/E-1

Mishra SK, Upreti DK, Pandey V, Bajpai R (2003) Pollution monitoring with the help of lichens transplant technique in some commercial and industrial areas of Lucknow City. Pollut Res 22(2):221–225

Mishra S, Srivastava S, Tripathi RD, Trivedi PK (2008) Thiol metabolism and antioxidant systems complement each other during arsenate detoxification in Ceratophyllum demersum L. Aquat Toxic (Amsterdam, Netherlands) 86:205–215

Moraes RM, Klumpp A, Furlan CM, Klumpp G, Domingos M, Rinaldi MCS, Modesto IF (2002) Tropical fruit trees as bioindicators of industrial air pollution in southeast Brazil. Environ Int 28:367–374

Mukhopadhyaya K, Forssel O (2005) An empirical investigation of air pollution from fossil fuel combustion

and its impact on health in India during 1973–1974 to 1996–1997. Ecol Econ 55:235–250
Naeher LP, Holford TR, Beckett WS, Belanger K, Triche EW, Bracken MB et al (1999) Healthy women's PEF variations with ambient summer concentrations of PM10, PM$_{25}$, SO$_4^{2-}$, H$^+$, and O$_3$. Am J Respir Crit Care Med 60:117–125
Narayan D, Agrawal M, Pandey J, Singh J (1994) Changes in vegetation characteristics downwind of an aluminium factory in India. Ann Bot 73:557–565
Nash TH III (1971) Lichen sensitivity to hydrogen fluoride. Bull Torr Bot Club 98(2):103–106
Nash TH (1976) Sensitivity of lichens to nitrogen dioxide fumigations. Bryologist 79:103–106
Nash TH III (2008) Lichen biology, 2nd edn. Cambridge University Press, Cambridge
Nash TH, Gries C (1991) Lichens as indicators of air pollution. In: Hutzinger O (ed) The handbook of environmental chemistry, vol 4. Part C. Springer, Berlin
Nash TH III, Sommerfield MR (1981) Elemental concentrations in lichens in the area of the four corner power plant, New Mexico. Environ Exp Bot 21:153–162
Nash TH III, Wirth V (1988) Lichens, bryophytes and air quality. Bibl Lichenol 30:1–298
Nayaka S, Upreti DK (2005a) Status of lichen diversity in Western Ghats, India. Sahyadri E-News, p 16. http://wgbis.ces.iisc.ernet.in/biodiversity/sahyadri-news/newsletter/issue16/main_index.htm
Nayaka S, Upreti DK (2005b) Lichen flora of Pune City (India) with reference to air pollution (Abstract). In: IIIrd international conference on plants and environmental pollution, NBRI, Lucknow, 28 Nov–2 Dec 2005
Nayaka S, Upreti DK, Gadgil M, Pandey V (2003) Distribution pattern and heavy metal accumulation in lichens of Bangalore city with special reference to lalbagh Garden. Curr Sci 84(5):674–680
Nayaka S, Singh PK, Upreti DK (2005a) Fungicidal elements accumulated in *Cryptothecia punctata* (Ascomycetes) lichens of an Arecanut Orchard in South India. J Environ Biol 26(2):299–300
Nayaka S, Upreti DK, Pandey V, Pant V (2005b) Manganese (Mn) in lichens growing on magnasite rocks in India. Bull Bri Lic Soc 97:66–68
Nayaka S, Ranjan S, Saxena P, Pathre UV, Dk U, Singh R (2009) Assessing the vitality of Himalayan lichens by measuring their photosynthetic performances using chlorophyll fluorescence technique. Curr Sci 97(4):538–545
Negra C, Ross DS, Lanzirotti A (2005) Oxidizing behavior of soil manganese: interactions among abundance, oxidation state and pH. Soil Sci Soc Am J 69:87–95
Neitlich P, Will-Wolf S (2000) The lichen community indicator in the forest Inventory and Analysis FHM program: using lichen communities to monitor forest health. Poster Forest Health Monitoring Workshop, Orange Beach, Albama, 14–17 Feb 2000
Nevstrueva MA, Ramzaev PV, Ibatullin AA, Teplykh LA (1967) The nature of Cs-137 and Sr-90 transport over the lichen-reindeer-man food chain. In: Radioecology concentration processes. Proceedings of the international symposium, Stockholm, 1966, pp 209–215
Nieboer E, Richardson DHS (1981) Lichens as monitors of atmospheric deposition. In: Eisenreich SJ (ed) Atmospheric pollutants in natural waters. Ann Arbor Science, Ann Arbor, pp 339–388
Nieboer E, Puckett KJ, Richardson DHS, Tomassini FD, Grace B (1977) Ecological and physiochemical aspects of the accumulation of heavy metals and sulphur in lichens. In: Nieboer E, Puckett KJ, Richardson DHS (eds) International conference on heavy metals in the environment. Symposium proceedings, Toranto, vol 2, pp 331–352
Nieboer E, Richardson DHS, Tomassini FD (1978) Mineral uptake and release by lichen: an overview. Bryologist 81:226–246
Niemi GJ, McDonald ME (2004) Application of ecological indicators. Annu Rev Ecol Evol Syst 35:89–111
Nifontova MG (2000) Concentrations of Long-lived Artificial Radionuclides in the Moss-Lichen Cover of Mountain Plant Communities. Russ J Ecol 31(3):182–185
Nikipelov BV, Drozhko EG, Romanov GN, Voronov AS, Spirin DA, Alexakhin RM, Smirnov EG, Suvorova LI, Tikhomirov FA, Buldakov LA, Shvedov VL, Tepyakovi IG, Shilin VP (1990) The Kyshtym accident: close-up (In Russian). Nature (USSR) 5:47–75
Nimis PL (1990) Air quality indicators and indices: the use of plants as bioindicators for monitoring of air pollution. In: Colombo AG, Premazzi G (eds) Proceedings of the workshop on indicators and indices for environmental impact assessment and risk analysis. Joint Research Centre, Ispra, pp 93–126
Nimis PL (1996) Radiocaesium in plants of forest ecosystems. Studia Geobotanica 15:3–49
Nimis PL (1999) Linee guida per la bioindicazione degli effetti dell'inquinamento tramite la biodiversità dei licheni epifiti. In: Piccini C, Salvat S (eds) Atti Workshop Biomonitoraggio Qualita` dell'Aria sul territorio Nazionale, Roma. ANPA, Roma, 1998, pp 267–277
Nimis PL, Castello M, Perotti M (1990) Lichens as biomonitors of Sulphur dioxide pollution in La Spezia (Northern Italy). Lichenologist 22:333–344
Nimis PL, Castello M, Perotti M (1993) Lichens as bioindicators of heavy metal pollution: a case study at La Spezia (N Italy). In: Markert B (ed) Plants as biomonitors, indicators for heavy metals in the terrestrial environment. VCH, Weinheim, pp 265–284
Niriagu JO, Azcue JM (1990) Environmental sources of arsenic in food. Adv Environ Sci Technol 23:103–127
Nriagu JO, Pacyna J (1988) Quantitative assessment of worldwide contamination if air, water and soil by trace metals. Nature 333:134–139
O'Neill MS, Loomis D, Borja-Aburto VH (2004) Ozone, area social conditions, and mortality in Mexico City. Environ Res 94:234–242
Odum EP (1996) Fundamentals of ecology, 1st Indian edn. Natraj Publishers, Dehradun
Ou D, Liu M, Cheng S, Hou L, Xu S, Wang L (2010) Identification of the sources of polycyclic aromatic

hydrocarbon based on molecular and isotopic characterization from the Yangtze estuarine and nearby coastal areas. J Geogr Sci 20(2):283–294

Paakola HE, Miettinen JK (1963) 90Sr and 137Cs in plants and animals in Finnish Lapland during 1960. Ann Acad Sci Fenn Ser A2:125–138

Pacyna JM, Breivik K, Munch J, Fudala J (2003) European atmospheric emissions of selected persistent organic pollutants, 1970–1995. Atmos Environ 37:S119–S131

Panayotou T (2003) Economic growth and the environment. Chapter 2. In: Spring Seminar of the United Nations Economic Commission for Europe. Economic survey of Europe, Geneva, pp 45–72

Pandey GP (1981) A survey of fluoride pollution effects on the forest ecosystem around an aluminium factory in Mirzapur, U.P., India. Environ Conserv 8:131–137

Pandey GP (1985) Effects of gaseous hydrogen fluoride on leaves of Terminalia tomentosa and Buchannania lanzan trees. Environ Conserv 37:323–334

Paoli L, Pisani T, Guttová A, Sardella G, Loppi S (2011) Physiological and chemical response of lichens transplanted in and around an industrial area of south Italy: relationship with the lichen diversity. Ecotoxicol Environ Saf 74(4):650–657

Papastefanou C, Manolopoulou M, Charalambous S (1988) Radiation measurements and radioecological aspects of fallout from the Chernobyl accident. J Environ Radioact 7:49–64

Papastefanou C, Manolopoulou M, Sawidis T (1992) Residence time and uptake rates of 137Cs in lichens and mosses at temperate latitude (40N°). Environ Int 18:397–401

Park SS, Kim YJ, Kang CH (2002) Atmospheric polycyclic aromatic hydrocarbons (PAHs) in Seoul Korea. Atmos Environ 36:2917–2924

Patra AC, Sahoo SK, Tripathi RM, Puranik VD (2013) Distribution of radionuclides in surface soils, Singhbhum Shear Zone, India and associated dose. Environ Monit Assess. doi:10.1007/s10661-013-3138-y

Pauli H, Gottfried M, Reiter K, Klettner C, Grabherr G (2007) Signals of range expansion and contractions of vascular plants in the high Alps: observations (1994–2004) at the GLORIA master site Schrankvogel, Tyrol Austria. Glob Chang Biol 13:147–156

Pawlik-Skowrońska BL, Sanita di Toppi MA, Favali F, Fossati J, Pirszel TS (2002) Lichens respond to heavy metals by phytochelatin synthesis. New Phytol 156:95–102

Perkins DF (1992) Relationship between fluoride contents and loss of lichens near an aluminium works. Water Air Soil Pollut 64:503–510

Pinho P, Augusto S, Branquinho C, Bio A, Pereira MJ, Soares A, Catarino F (2004) Mapping lichen diversity as a first step for air quality assessment. J Atm Chem 49:377–389

Post JE (1999) Manganese oxide minerals: crystal structures and economic and environmental significance. Proc Natl Acad Sci USA 96:3447–3454. doi:10.1073/pnas.96.7.3447

Prasad MNV (1997) Trace metal. In: Prasad MNV (ed) Plant physiology. Wiley, New York, pp 207–249

Preutthipan A, Udomsubpayakul U, Chaisupamongkollarp T, Pentamwa P (2004) Effect of PM_{10} pollution in Bangkok on children with and without asthma. Pediatr Pulmonol 37:187–192

Prospero JM (1999) Long-range transport of mineral dust in the global atmosphere: impact of African dust on the environment of the southeastern United States. Proc Natl Acad Sci USA 96:3396–3403

Puckett KJ (1988) Bryophytes and lichens as monitors of metal deposition. In: Nash TH III (ed) Lichens, bryophytes and air quality, Bibliotheca Lichenologica 30. J. Cramer, Berlin, pp 231–267

Pulak D, Joshi S, Rout J, Upreti DK (2012) Impact of a paper mill on surrounding epiphytic lichen communities using multivariate analysis. Indian J Ecol 39(1):38–43

Purvis OW (2000) Lichens. The Natural History Museum, London

Purvis OW, Dubbin W, Chimonides PDJ, Jones GC, Read H (2008) The multielement content of the lichen Parmelia sulcata, soil, and oak bark in relation to acidification and climate. Sci Total Environ 390:558–568

Rai H, Khare R, Gupta RK, Upreti DK (2011) Terricolous lichens as indicator of anthropogenic disturbances in a high altitude grassland in Garhwal (Western Himalaya), India. Botanica Orientalis. J Plant Sci 8:16–23

Rani M, Shukla V, Upreti DK, Rajwar GS (2011) Periodical monitoring with lichen, Phaeophyscia hispidula (Ach.) Moberg in Dehradun city, Uttarakhand, India. Environmentalist 31:376–381. doi:10.1007/s10669-011-9349-2

Rao DN, LeBlanc F (1967) Influence of an iron sintering plant on corticolous epiphytes in Wawa, Ontario. Bryologist 70:141–157

Ravindra K, Sokhi R, Grieken RV (2008) Atmospheric polycyclic aromatic hydrocarbons: source attribution, emission factors and regulation. Atmos Environ 42:2895–2921. doi:10.1016/j.atmosenv.2007.12.010

Ravindra K, Wauters E, Tyagi SK, Mor S, Van Grieken R (2006) Assessment of air quality after the implementation of compressed natural gas (CNG) as fuel in public transport in Delhi India. Environ Monit Assess 115:405–417

Riddell J, Padgett PE, Nash TH III (2010) Responses of the lichen Ramalina menziesii Tayl. to ozone fumigations. In: Nash TH et al. (eds.) Biology of Lichens-Symbiosis, Ecology, Environmental Monitoring, Systematics and Cyber Applications. Bibliotheca Lichenologica 105:113–123

Rosenfeld D et al (2007) Inverse Relations Between Amounts of Air Pollution and Orographic Precipitation. Science 315:1396–1398

Ross LJ, Nash TH III (1983) Effect of ozone on gross photosynthesis of lichens. Environ Exp Bot 23(1):71–77

Ruoss E, Vonarburg C (1995) Lichen diversity and ozone impact in rural areas of central Switzerland. Cryptogam Bot 5:252–263

Saipunkaew W, Wolseley PA, Chimonides PJ, Boonpragob K (2007) Epiphytic macrolichens as indicators of environmental alteration in northern Thailand. Environ Pollut 146:366–374

Salo A, Miettinen JK (1964) Strontium-90 and Caesium-137 in Arctic vegetation during 1961. Nature 201:1177–1179

Sanità di Toppi L, Musetti R, Vattuone Z, Pawlik-Skowrońska B, Fossati F, Bertoli L, Badiani M, Favali MA (2005) Cadmium distribution and effects on ultrastructure and chlorophyll status in photobionts and mycobionts of *Xanthoria parietina*. Microscop Res Tech 66:229–238

Sanz M-J, Gries C, Nash TH III (1992) Dose-response relationships for SO2 fumigations in the lichens Evernia prunastri (L.) Ach. and Ramalina fraxinea (L.) Ach. New Phytol 122:313–319

Sasaki J, Aschmann SM, Kwok ESC, Atkinson R, Arey J (1997) Product of the gas-phase OH and NO3 Radical-initiated reactions of naphthalene. Environ Sci Technol 31:3173–3179

Satya, Upreti DK (2009) Correlation among carbon, nitrogen, sulphur and physiological parameters of *Rinodina sophodes* found at Kanpur city, India. J Hazard Mater 169:1088–1092. doi:10.1016/j/jhazmat.2009.04.063

Satya, Upreti DK, Patel DK (2012) *Rinodina sophodes*-(Ach.) Massal.: a bioaccumulator of polycyclic aromatic hydrocarbons (PAHs) in Kanpur city, India. Environ Monit Assess 184:229–238

Sawidis T (1988) Uptake of radionuclides by plants after the Chernobyl accident. Environ Pollut 50

Saxena S (2004) Lichen flora of Lucknow district with reference to Air Pollution studies in the area. Ph.D. thesis, Lucknow University, Lucknow

Saxena S, Upreti DK, Sharma N (2007) Heavy metal accumulation in lichens growing in north side of Lucknow city. J Environ Biol 28(1):45–51

Scheidegger C, Schroeter B (1995) Effects of ozone fumigation on epiphytic macrolichens: ultrastructure, CO2 gas exchange and chlorophyll fluorescence. Environ Pollut 88(3):345–354

Seaward MRD (1974) Some observations on heavy metal toxicity and tolerance in lichens. Lichenologist 6:158–164

Seaward MRD (1988) Lichen damage to ancient monuments: a case study. Lichenologist 10(3):291–295

Seaward MRD (1989) Lichens as monitors of recent changes in air pollution. Plants Today 1:64–69

Seaward MRD (1992) Lichens, silent witnesses of the Chernobyl disaster. University of Bradford, Bradford

Seaward MRD (1993) Lichens and sulphur dioxide air pollution field studies. Environ Rev 1:73–91

Seaward MRD (1997) Urban deserts bloom: a lichen renaissance. Bibliotheca Lichenologica 67:297–309

Seaward MRD, Heslop JA, Green D, Bylinska EA (1988) Recent levels of radionuclides in lichens from southwest Poland with particular reference to 134Cs and 137Cs. J Environ Radioact 7:123–129

Sernander R (1926) Stockholms Natur. Almguist and Wiksella, Uppsala

Shirazi AM, Muir PS, McCune B (1996) Environmental factors influencing the distribution of lichen Lobaria oregano and L. pulmonaria. Bryologist 99(1):12–18

Shukla V (2007) Lichens as bioindicator of air pollution. Final technical report. Science and Society Division, Department of Science and Technology, New Delhi. Project No. SSD/SS/063/2003

Shukla V (2012) Physiological response and mechanism of metal tolerance in lichens of Garhwal Himalayas. Final technical report. Scientific and Engineering Research Council, Department of Science and Technology, New Delhi. Project No. SR/FT/LS-028/2008

Shukla V, Upreti DK (2007a) Physiological response of the lichen *Phaeophyscia hispidula* (Ach.) Essl. to the Urban Environment of Pauri and Srinagar (Garhwal), Himalayas. Environ Pollut 150:295–299. doi:10.1016/j.envpol.2007.02.010

Shukla V, Upreti DK (2007b) Heavy metal accumulation in *Phaeophyscia hispidula* en route to Badrinath, Uttaranchal, India. Environ Monit Assess 131:365–369. doi:10.1007/s10661-006-9481-5

Shukla V, Upreti DK (2007c) Lichen diversity in and around Badrinath, Chamoli district (Uttarakhand). Phytotaxonomy 7:78–82

Shukla V, Upreti DK (2008) Effect of metallic pollutants on the physiology of lichen, *Pyxine subcinerea* Stirton in Garhwal Himalayas. Environ Monit Assess 141:237–243. doi:10.1007/s10661-007-9891-z

Shukla V, Upreti DK (2009) Polycyclic Aromatic Hydrocarbon (PAH) accumulation in lichen, *Phaeophyscia hispidula* of DehraDun city, Garhwal Himalayas. Environ Monit Assess 149(1–4):1–7

Shukla V, Upreti DK (2011a) Changing lichen diversity in and around urban settlements of Garhwal Himalayas due to increasing anthropogenic activities. Environ Monit Assess 174(1–4):439–444. doi:10.1007/s10661-010-1468-6

Shukla V, Upreti DK (2011) Statistical correlation of metallic content and polycyclic aromatic hydrocarbon concentration to trace the source of PAH pollution. In: XXXIV All India Botanical Conference, Department of Botany, Lucknow University, Lucknow, Uttar Pradesh, 10–12 October 2011

Shukla V, Upreti DK (2012) Air quality monitoring with lichens in India: heavy metals and polycyclic aromatic hydrocarbon. In: Lichtfouse E, Schwarzbauer J, Robert D (eds) Environmental chemistry for a sustainable world, vol 2, Remediation of air and water pollution. Springer, New York, pp 277–294

Shukla V, Upreti DK, Nayaka S (2006) Heavy metal accumulation in lichens of Dehra Dun city, Uttaranchal, India. Indian J Environ Sci 10(2):165–169

Shukla V, Upreti DK, Patel DK, Tripathi R (2010) Accumulation of polycyclic aromatic hydrocarbons in some lichens of Garhwal Himalayas, India. Int J Environ Waste Manag 5(1/2):104–113

Shukla V, Patel DK, Upreti DK, Yunus M (2012a) Lichens to distinguish urban from industrial PAHs. Environ Chem Lett 10:159–164. doi:10.1007/s10311-011-0336-0

Shukla V, Upreti DK, Patel DK (2012b) Physiological attributes of *Phaeophyscia hispidula* in heavy metal rich sites of Dehra Dun. India J Environ Biol 33:1051–1055

Shukla V, Patel DK, Upreti DK, Yunus M, Prasad S (2013a) A comparison of heavy metals in lichen (*Pyxine subcinerea*), mango bark and soil. Int J Environ Sci Technol 10:37–46. doi:10.1007/s13762-012-0075-1

Shukla V, Upreti DK, Patel DK, Yunus M (2013b) Lichens reveal air PAH fractionation in the Himalaya. Environ Chem Lett. doi:10.1007/s10311-012-0372-4

Sigal LL, Nash TH III (1983) Lichen communities on conifers in southern California: an ecological survey relative to oxidant air pollution. Ecology 64:1343–1354

Sillett SC, Neitlich P (1996) Emerging themes in epiphytic research in Westside forests with special reference to cyanolichens. Northwest Sci 70:54–60

Sillett SC, Mc Cune B, Perk JE, Rambo TR, Ruchty A (2000) Dispersal limitations epiphytic lichen result in species dependent on old growth forests. Ecol Appl 10:789–799

Singh HB (1987) Reactive nitrogen in the troposphere. Environ Sci Technol 21(4):320–327

Singh JS (2011) Methanotrophs: the potential biological sink to mitigate the global methane load. Curr Sci 100(1):29–30

Singh HB, Herlth D, Ohara D, Zahnle K, Bradshaw JD, Sandholm ST, Talbot R, Crutzen PJ, Kanakidou M (1992a) Relationship of peroxyacetyl nitrate to active and total odd nitrogen at Northern High-Latitudes – influence of reservoir species on NOx and O_3. J. Geophys Res Atmos 97:16523–16530

Singh HB, Ohara D, Herlth D, Bradshaw JD, Sandholm ST, Gregory GL, Sachse GW, Blake DR, Crutzen J, Kanakidou MA (1992b) Atmospheric measurements of peroxyacetyl nitrate and other organic Nitrates at high-latitudes – possible sources and sinks. J Geophys Res-Atmos 97:16511–16522

Singh J, Agarwal M, Narayan D (1994) Effect of power plant emissions on plant community structure. Ecotoxicology 3:110

Singh HB, Herlth D, Kolyer R, Chatfield R, Viezee W, Salas LJ, Chen Y, Bradshaw JD, Sandholm ST, Talbot R, Gregory GL, Anderson B, Sachse GW, Browell E, Bachmeier AS, Blake DR, Heikes B, Jacob D, Fuelberg HE (1996) Impact of biomass burning emissions on the composition of the South Atlantic troposphere: reactive nitrogen and ozone. J Geophys Res 101(D19):24203–24219

Singh H, Chen Y, Tabazadeh A, Fukui Y, Bey I, Yantosca R, Jacob D, Arnold F, Wohlfrom K, Atlas E, Flocke F, Blake N, Heikes B, Snow J, Talbot R, Gregory G, Sachse G, Vay S, Kondo Y (2000a) Distribution and fate of selected oxygenated organic species in the troposphere and lower stratosphere over the Atlantic. J Geophys Res 105(D3):3795–3805

Singh HB, Viezee W, Chen Y, Bradshaw J, Sandholm S, Blake D, Blake N, Heikes B, Snow J, Talbot R, Browell E, Gregory G, Sachse G, Vay S (2000b) Biomass burning influences on the composition of the remote South Pacific troposphere: analysis based on observations from PEM-Tropics-A. Atmos Environ 34(4):635–644

Singh N, Ma LQ, Srivastava M, Rathinasabapthi B (2006) Metabolic adaptation to arsenic-induced oxidative stress in *Pteris vittata* L. and *P. ensiformis* L. Plant Sci 170:274–282

Singh J, Dubey AK, Singh RP (2011) Antarctic terrestrial ecosystem and role of pigments in enhanced UV-B radiations. Rev Environ Sci Biotechnol 10(1):63–77. doi:10.1007/s11157-010-9226-3

Sipman HJM (1997) Observations on the foliicolous lichen and bryophyte flora in the canopy of a semi-deciduous tropical forest. Abstracta Botanica 21:153–161

Sloof JE, Wolterbeek BT (1992) Lichens as biomonitors for radiocesium following the Chernobyl accident. J Environ Radioact 16:229–242

Smith DJT, Harrison RM (1998) Polycyclic aromatic hydrocarbons in atmospheric particles. In: Harrison RM, Van Grieken R (eds) Atmospheric particles. Wiley, New York

Smith G, Coulston J, Jepsen E, Prichard T (2003) A national ozone biomonitoring program e results from field surveys of ozone sensitive plants in northeastern forests (1994–2000). Environ Model Assess 87:271–291

Søchting U (2004) *Flavoparmelia caperata* – a probable indicator of increased temperatures in Denmark. Graphis Scripta 15:53–56

Sokolik IN, Toon OB, Bergstrom RW (1998) Modeling of radiative characteristics of airborne mineral aerosols at infrared wavelengths. J Geophys Res 103:8813–8826

Sporn SG, Bos MM, Kessler M, Gradstein SR (2010) Vertical distribution of epiphytic bryophytes in an Indonesian rainforest. Biodivers Conserv 19:745–760. doi:10.1007/s10531-009-9731-2

Srivastava S, Mishra S, Tripathi RD, Dwivedi S, Trivedi PK, Tandon PK (2007) Phytochelatins and antioxidant systems respond differently during arsenite and arsenate stress in *Hydrilla verticillata* (L.f) Royle. Environ Sci Technol 41:2930–2936

Srogi K (2007) Monitoring of environmental exposure to polycyclic aromatic hydrocarbons: a review. Environ Chem Lett 5:169–195. doi:10.1007/s10311-007-0095-0

St. Clair BS, St. Clair LL, Mangelson FN, Weber JD (2002a) Influence of growth form on the accumulation of airborne copper by lichens. Atmos Environ 36:5637–5644

St. Clair BS, St. Clair LL, Weber JD, Mangelson FN, Eggett LD (2002b) Element accumulation patterns in foliose and fruticose lichens from rock and bark substrates in Arizona. Bryologist 105:415–421

State G, Popescu IV, Radulescu C, Macris C, Stihi C, Gheboianu A, Dulama I, Niţescu O (2012) comparative studies of metal air pollution by atomic spectrometry techniques and biomonitoring with moss and lichens. Bull Environ Contam Toxicol 89(3):580–586. doi:10.1007/s00128-012-0713-9

Staudt AC, Jacob DJ, Logan JA, Bachiochi D, Krishnamurti TN, Sachse GW (2001) Continental sources, transoceanic transport, and interhemispheric exchange of carbon monoxide over the Pacific. J Geophys Res 106:32571–32590

Staudt AC, Jacob DJ, Ravetta F, Logan JA, Bachiochi D, Krishnamurti TN, Sandholm S, Ridley B, Singh HB, Talbot B (2003) Sources and chemistry of nitrogen oxides over the tropical Pacific. J Geophys Res 108:8239

Stephens ER (1973) Analysis of an important air pollutant – peroxyacetyl nitrate. J Chem Educ 50:351–354

Subbotina EN, Timofeeff NV (1961) On the accumulation coefficients, characterising the uptake by crust lichens of some dispersed elements from aqueous solutions (Russian, English summary). Bot Z 46:212

Suresh Y, Sailaja Devi MM, Manjari V, Das UN (2000) Oxidant stress, antioxidants, and nitric oxide in traffic police of Hyderabad, India. Environ Pollut 109:321–325

Sutherland WJ, Armstrong-Brown S, Armstrong PR, Brereton T, Brickland J, Campell CD, Chamberlain DE, Cooke AI, Dulvy NK et al (2006) The identification of 100 ecological questions of high policy relevance in the UK. J Appl Ecol 43:617–627

Tarhanen S (1998) Ultrastructural responses of the lichen *Bryoria fuscescens* to simulated acid rain and heavy metal deposition. Ann Bot 82:735–746

Tegen I, Lacis A (1996) Modeling of particle size distribution and its influence on the radiative properties of mineral dust aerosol. J Geophys Res 101:19237–19244

Teklemariam TA, Sparks JP (2004) Gaseous fluxes of peroxyacetyl nitrate (PAN) into plant leaves. Plant Cell Environ 27:1149–1158

Thomas PA, Gate TE (1999) Radionuclides in the lichen-caribou-human food chain near Uranium mining operations in Northern Saskatchewan, Canada. Environ Health Perspect 107(7):527–537

Thormann MN (2006) Lichens as indicators of forest health in Canada. For Chron 82(3):335–343

Thuiller W, Lavorel S, Araújo MB, Sykes MT, Prentice IC (2005) Climate change threats to plant diversity in Europe. Ecology 102:8245–8250

Topcuoğlu S, Dawen AMV, Güngör N (1995) The natural depuration rate of 137Cs radionuclides in a lichen and moss species. J Environ Radioact 29(2):157–162. doi:10.1016/0265-931X(94)00069-9

Trass H (1973) Lichen sensitivity to the air pollution and index of poleotolerance (IP). Fol Crypt Estonia 3:19–22

Tretiach M, Piccotto M, Baruffo L (2007) Effects of ambient NOx on chlorophyll a fluorescence in transplanted *Flavoparmelia caperata* (Lichen). Environ Sci Technol 41:2978–2984

Trivedi RC (1981) Use of diversity Index in evaluation of water quality. In: Zafar AR, Khan MA, Khan KR, Seenayya G (eds) Proceedings of the WHO workshop on biological indicators and indices of environmental pollution. Central Board of the Prevention and Control water Pollution, OSM University

Truscott AM, Palmer SCF, McGowan GM, Cape JN, Smart S (2005) Vegetation composition of roadside verges in Scotland: the effects of nitrogen deposition, disturbance and management. Environ Pollut 136:109–118

Tsibulsky V, Sokolovsky V, Dutchak S (2001) MSC-E contribution to the HM and POP Emission Inventories. Technical note 7/2001, Meteorological synthesizing Centre-East, Moscow. Available from: http://www.msceast.org/publications.html

Tuominen Y, Jaakkola T (1973) Absorption and accumulation of mineral elements and radioactive nuclides. In: Ahamadjan V, Hale M (eds) The lichens. Academic, London, pp 185–223

Tyler G (1989) Uptake, retention and toxicity of heavy metals in lichens. A brief review. Water Air Soil Pollut 47(3–4):321–333

UNSCEAR (1993) Exposure from natural sources of radiation. United Nations, New York

Upreti DK (1994) Lichens: the great benefactors. Appl Bot Abst 14(3):64–75

Upreti DK, Nayaka S (2008) Need for creation of lichen garden and sanctuaries in India. Curr Sci 94(8):976–978

Upreti DK, Pandey V (2000) Determination of heavy metals in lichens growing on different ecological habitats in Schirmacher Oasis, East Antarctica. Spectrosc Lett 33(3):435–444

Upreti DK, Chatterjee S, Divakar PK (2004) Lichen flora of Gangotri and Gomukh areas of Uttaranchal, India. Geophytology 34:15–21

Upreti DK, Nayaka S, Bajpai A (2005) Do lichens still grow in Kolkata city? Curr Sci 88(3):338–339

US EPA, 1998 (1990) Emissions Inventory of Section 112(c)(6) Pollutants: polycyclic organic matter (POM), TCDD, TCDF, PCBs, hexachlorobenzene, mercury, and alkylated lead: Final report. US Environmental Protection Agency Research, Triangle Park. Available from: http://www.epa.gov/ttn/atw/112c6/final2.pdf

US Energy Information Administration (1998) National energy modeling system (NEMS) data base. US Department of Energy, Washington World Bank (1993) Development and the environment. Oxford University Press, Oxford

Usman M, Murata M, Zafar M, Adeel K, Amir NA (2011) A study on correlation between temperature increase and earthquake frequency with emphasis on winter and summer periods, Northern Pakistan. In: 2nd international conference on environmental science and technology IPCBEE, vol 6. IACSIT Press, Singapore

Van der Gon HD, Van het Bolscher M, Visschedijk A, Zandveld P (2007) Emissions of persistent organic pollutants and eight candidate POPs from UNECE–Europe in 2000, 2010 and 2020 and the emission reduction resulting from the implementation of

References

the UNECE POP protocol. Atmos Environ 41: 9245–9261

van Dobben HF, ter Braak CJF (1999) Ranking of epiphytic lichen sensitivity to air pollution using survey data: a comparison of indicator scales. Lichenologist 31(1):27–39

van Dobben HF, Wolterbeek HT, Wamelink GWW, Ter Braak CJF (2001) Relationship between epiphytic lichens, trace elements and gaseous atmospheric pollutants. Environ Pollut 112:163–169

van Geen A et al (2005) Reliability of a commercial kit to test groundwater for arsenic in Bangladesh. Environ Sci Technol 39(1):299–303

Van Haluwyn C, Lerond M (1986) Les lichens et la qualité de l'air. Evolution méthodologique et limites.Ministerè de l'Environnement, Service de la Recherche, des Etudes, et du Traitement de l'Information sur l'Environnement, Paris

van Herk CM (2001) Bark pH and susceptibility to toxic air pollutants as independent causes of changes in epiphytic lichen composition in space and time. Lichenologist 33:419–441

van Herk CM, Aptroot A (1999) *Lecanora compallens* and *L. sinuosa*, two new overlooked corticolous lichen species from western Europe. Lichenologist 31: 543–553

van Herk CM, Aptroot A, van Dobben HF (2002) Long-term monitoring in the Netherlands suggests that lichens respond to global warming. Lichenologist 34:141–154

van Kooten O, Snel JFH (1990) The use of chlorophyll fluorescence nomenclature in plant stress physiology. Photosynth Res 25:147–150

Van Pul WAJ, de Leeuw FAAM, van Jaarsveld JA, van der Gaag MA, Sliggeras CJ (1998) The potential for long-range transboundary atmospheric transport. Chemosphere 37:113–141

Vasconcellos PC, Zacarias D, Pires MAF, Pool CS, Carvalho LRF (2003) Measurements of polycyclic aromatic hydrocarbons in airborne particles from the metropolitan area of Sao Paulo city, Brazil. Atmos Environ 37:3009–3018

Vestergaard N, Stephansen U, Rasmussen L, Pilegaard K (1986) Airborne heavy metal pollution in the environment of a Danish steel plant. Water Air Soil Pollut 27:363–377

Vuille M, Francou B, Wagnon P, Juen I, Kaser G, Mark BG, Bradley RS (2008) Climate change and tropical Andean glaciers: past, present and future. Earth Sci Rev 89:79–96

Wadleigh MA, Blake DM (1999) Tracing sources of atmospheric sulphur using epiphytic lichens. Environ Pollut 106:265–271

Walther GR, Post E, Convey P, Menzel A, Parmesan C, Beebee TJC, Fromentin JM, Hoegh-Guldberg O, Bairlein F (2002) Ecological response to recent climate change. Nature 416:389–395

Watts AW, Ballestero TP, Garder KH (2006) Uptake of polycyclic aromatic hydrocarbon (PAHs) in salt marsh plants *Spartina alterniflora* grown in contaminated sediments. Chemosphere 62:1253–1260

Weinstein LH, Davison AW (2003) Native plant species suitable as bioindicators and biomonitors for airborne fluoride. Environ Pollut 125:3–11

Wenborn MJ, Coleman PJ, Passant NR, Lymberidi E, Sully J, Weir RA (1999) Speciated PAH Inventory for the UK, AEA Technology Environment, Oxfordshire. http://www.airquality.co.uk/archive/reports/cat08/0512011419_REPFIN_all_nov.pdf

Wilbanks TJ, Hunsaker DB, Petrich CH, Wright SB (1993) Potential to transfer the US NEPA experience in developing countries. In: Hildebrand SG, Cannon JB (eds) Environmental analysis: the NEPA experience. Lewis, Boca Raton

Wilf P (1997) When are leaves good thermometers? A new case for leaf margin analysis. Paleobiology 23(3):373–390

Wilhelm M, Ritz B (2003) Residential proximity to traffic and adverse birth outcomes in Los Angeles County, California, 1994–1996. Environ Health Perspect 111:207–216

Wilhm JL (1967) Comparison of some diversity indices applied to populations of benthic macro invertebrates in a stream receiving organic wastes. J Water Pollut Cont Fed 39:1673–1683

Winchester V (2004) Lichenometry. In: Goudie A, Routledge AS (eds) Encyclopedia of geomorphology. Routledge: International Association of Geomorphologists, London/New York, pp 619–620

Wood C (2003) Environmental impact assessment in developing countries: an overview. In: Conference on new directions in impact assessment for development: methods and practice, University of Manchester, pp 24–25

Wolfskeel DW, van Herk CM (2000) *Heterodermia obscurata* nieuw voor Nederland. Buxbaumiella 52:47–50

Wolterbeek HT, Garty J, Reis MA, Freitas MC (2003) Biomonitors in use: lichens and metal air pollution. In: Markert BA, Breure AM, Zechmeister HG (eds) Bioindicators and biomonitors. Elsevier, Oxford, pp 377–419

Wuebbles DJ, Lei H, Lin J (2007) Intercontinental transport of aerosols and photochemical oxidants from Asia and its consequences. Environ Pollut 150:65–84

Xu SS, Liu WX, Tao S (2006) Emission of polycyclic aromatic hydrocarbons in China. Environ Sci Technol 40:702–708

Yassaa N, Meklati BY, Cecinato A, Marino F (2001) Particulate n-alkanes, n-alkanoic acids and polycyclic aromatic hydrocarbons in the atmosphere of Algiers City area. Atmos Environ 35:1843–1851

Zambrano AG, Nash TH III, Herrera-Campos MA (2000) Lichen decline in Desierto de los Leones (Mexico City). Bryologist 103:428–441

Zhang JB, Tang XY (1994) Atmospheric PAN measurements and the formation of PAN in various systems. Environ Chem 1:30–39

Zhang YX, Tao S (2008) Emission of polycyclic aromatic hydrocarbons (PAHs) from indoor straw burning and emission inventory updating in China. Ann N Y Acad Sci 1140:218–227

Zhang YX, Tao S (2009) Global atmospheric emission inventory of polycyclic aromatic hydrocarbons (PAHs) for 2004. Atmos Environ 43:812–819

Zhang M, Song Y, Cai X (2007) A health-based assessment of particulate air pollution in urban areas of Beijing in 200-2004. Sci Total Environ 376:100–108

Zhang JB, Xu Z, Yang G, Wang B (2011) Peroxyacetyl nitrate (PAN) and peroxypropionyl nitrate (PPN) in urban and suburban atmospheres of Beijing, China Atmos. Chem Phys Discuss 11:8173–8206

Zheng M, Fang M (2000) Particle-associated polycyclic aromatic hydrocarbons in the atmosphere of Hong Kong. Water Air Soil Pollut 117:175–189

Zullini A, Peretti E (1986) Lead pollution and moss-inhabiting nematodes of an industrial area. Water Air Soil Pollut 27:403–410

Management and Conservational Approaches

6

Since lichens are widely known for their high sensitivity towards environmental disturbances, both natural and human origin. Therefore, environmental changes result in alteration of habitats and ecosystems at local, regional as well as global scale resulting in loss of lichen biodiversity; extinction of sensitive species invasion of thermophilic species towards higher latitudes. Such changes can be best monitored by lichens as biomonitors. Lichen biomonitoring is not only suitable for monitoring levels of pollutants but also may be utilised as an effective 'early alarms' of climate change and spatio-temporal extent of pollutants along with its health impact. Lichens as indicators possess an undeniable appeal for conservationists and land managers as they provide a cost- and time-efficient means to assess the impact of environmental disturbances on an ecosystem. Information collected with different aims, such as air pollution, climate change, biodiversity and forest continuity studies, may be utilised for conservation purposes. This chapter discusses the need and utility of indicator species especially lichen biomonitoring data in sustainable forest management and conservation.

6.1 Introduction

Environmental problems have been aggravated by the rapid expansion of human and industrial activities. Habitat destruction and fragmentation are major threats for biodiversity (Andrén 1997; Debinski and Holt 2000). For decades, large parts of forest ecosystems have been destroyed or degraded by human activities, resulting in fragmented landscapes (Esseen and Renhorn 1998; Rheault et al. 2003). This has caused radical changes in forests (Valladares et al. 2004), such as an increase of edge effects, microclimatic changes and a loss of the forest environment (Kivistö and Kuusinen 2000; Moen and Jonsson 2003). The loss of the forest environment has subsequently affected forest biodiversity through the decline and disappearance of numerous species (Forman 1995). In order to solve environmental problems and achieve sustainable development, an integrated effort has to be made to deal with a broad range of environmental issues, from identification of source, managing locally, regionally as well as globally. Sustainable development is mainly a concept in which basic human needs are met without destroying or irrevocably degrading the natural systems on which we all depend and managing and conserving the natural environment for future generations (Kates et al. 2005). Biomonitoring is one such cost-effective and reliable method to keep watch on the environmental problems persisting today.

Biological diversity is an important component that governs ecosystem resilience, its dynamic equilibrium and productivity. It is also important for securing and maintaining livelihood of human beings, particularly for the community living in and around forests. Use of certain plant and animal species as bioindicators in monitoring the health of forest ecosystems is relatively new (McGeoch et al. 2002). A critical issue is that practically economically feasible, socially acceptable and environmentally sound feasible methods of management and monitoring of forest resources are difficult to develop.

Bioindicators are defined as an organism or a community that contains information on the quality of the environment (Markert et al. 2003). Bioindicators can integrate pollution over a long period of time. Both plant and animal species have been shown to be useful for bioindication of natural and anthropogenic changes in various environments (Meyer et al. 2012). Biomonitoring programmes may be qualitative, semi-quantitative or quantitative, and are valuable assessment tools that are receiving increased use in monitoring programs of all types. In environmental impact assessment (EIA) studies, there are two types of biomonitoring: (1) surveillance before and after a project is complete and/or before and after a toxic substance enters the ecosystem and (2) surveillance to ensure compliance with regulations and guidelines (Osmond et al. 1995; Plafkin et al. 1989).

Sentinel organisms, or indicator species that accumulate pollutants in their tissues from the surrounding environment or from food, are important biomonitoring devices (Phillips and Rainbow 1993; Kennish 1992). Biochemical, genetic, morphological and physiological changes in certain organisms have been correlated with particular environmental stressors and can be used as indicators. The presence or absence of the indicator or of an indicator species or indicator community reflects environmental conditions. The measurement of atmospheric pollutants using physicochemical techniques alone does not suffice to determine the impact on organisms and the integration of these contaminants on the functioning of ecological systems (Eldridge and Koen 1998). Several components of atmospheric pollutants can have an impact on organisms, and the effect of a mixture of pollutants may be synergistic, antagonistic or additive (Berry and Wallace 1981). Thus, biomonitoring approach based on the sensitivity of organisms is an approach to estimate the effect of complex air pollution on biological communities (Markert et al. 2003).

Bioindicators include biological processes, species or communities which are used to assess quality of the environment and its changes over time. Changes in the environment are mainly attributed to anthropogenic disturbances (e.g. pollution, land-use changes) and/or natural phenomenon (e.g. drought and flood etc), although anthropogenic disturbances are primarily investigated using bioindicators. The widespread development and application of bioindicators has occurred primarily since the 1960s. Over the years bioindicators are utilised to assist in studying all types of environments (i.e. aquatic and terrestrial), using all major taxonomic groups. The use of indicators including lichens has frequently been incorporated into policies and regulations in order to monitor the ecological integrity of different components of the ecosystem (Holt and Miller 2011; Will-Wolf 2010).

Biomonitoring studies confirm the presence of pollutant in the environment and show the relative variation in the amount of pollutants between locations. Thus, indicator species not only contribute towards monitoring degradation but also play an important role in rehabilitation, restoration and ecosystem resilience and thus truly contribute to sustainable management, rather than simply indicating ecosystem changes (Carignan and Villard 2002). They also enable assessment of acceptable degree of habitat modification/changes in the land-use class (Azevedo-Ramos et al. 2006; Kotwal et al. 2008a).

Biomonitoring data also provides spatial distributions of pollutant concentration in indicator species over broad areas which may further be used to identify areas at risk from air pollution or to select reference sites (hot spots) for subsequent instrument monitoring by air samplers (Fig. 6.1). In general, passive biomonitoring studies are of most value as a screening mechanism for establishing a subset of sites where follow-up work (such as instrument monitoring) should be done (Blett et al. 2003).

Management and conservation programmes' primary objectives are to help national forest managers obey federal and state laws and to fulfil agency mandates with regard to the detection and quantification of adverse effects from air pollution on forest ecosystems and resources by:

1. Locating reference sites on national forest lands/an area of interest where inventories

6.1 Introduction

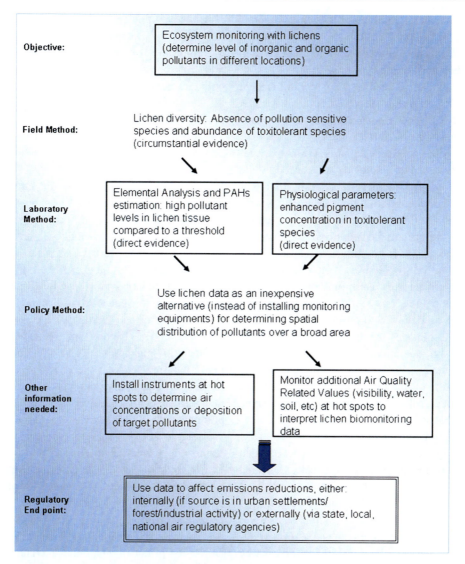

Fig. 6.1 Conceptual diagram showing the effective use of lichen biomonitoring data in ecosystem monitoring (Adapted from Blett et al. 2003)

of lichen communities and chemical analysis of lichen tissue for phytotoxic gases and other pollutants are performed on a regular basis.

2. Monitoring lichen community composition to document and map locations where air quality has improved or deteriorated and to document adverse effects to sensitive lichens as they are highly sensitive to phytotoxic gases and ammonia (NH$_3$) and nitrates (NO$_3$) while less sensitive, but still responsive if levels are sufficiently high, to nitrogen oxides (NOx), ozone (O$_3$) and peroxyacetyl nitrate (PAN).

3. Using chemical analysis to map areas of concern by documenting enhanced levels of sulphur- and nitrogen-containing pollutants and toxic metals in lichen and moss tissue.

4. Building a publicly accessible, unified lichen database, interfaceable with other forest, regional and national databases.

5. Providing analysis and interpretation of biomonitoring data, including thresholds for

enhancement of sulphur, nitrogen and metals in lichen tissue, sensitivities of lichens to air pollutants and site scores for air quality based on lichen community composition.

A secondary objective is to provide current and historical information about the diversity, abundance, distribution and habitat requirements of lichens on national forest lands.

According to Rosso et al. (2000), management studies and implementation need to be focused on populations and habitat needs rather than on individuals; from their calculations, cutting with retention of individual trees surrounded by small buffers could result in avoiding eventual loss of the species from study area (Geiser 2004).

In an effort for conservation practices, the detailed inventories compiled by systematic and standardised protocol from various forest class are prerequisite to observe the associated lichen assemblages, community structure, their age, ecological continuity, past and present forest management as well as observing the spatial contribution of the different epiphytic species to the arboreal lichen flora. The main concern is the sustainable extraction of the needed resources such as timber, which may be achieved by combining design-assisted and model-assisted approaches to interpret systematically collected inventory data on the distribution and ecology of lichens, particularly of rare and vulnerable species (Edwards et al. 2004; Nash 2008).

For present and future management of woodlands and forests, habitat models to forecast the frequency of occurrence of epiphytic lichen species in a forested landscape under different plans have been developed (McCune 2000). Various ecological studies have been undertaken to assess the impact of selective felling, green tree retention and clear-cutting of major areas on diversity and biomass of epiphytic lichens (Nash 2008; Bates and Farmer 1992). In fragmented logging pattern diaspores were found to be less successful in establishing themselves in logged areas as colonisation was species specific. Based on the above study recommendation was made to develop management guidelines based on wide scientific knowledge about the life history characteristics of the species (Hilmo et al. 2005).

Understanding how lichen biodiversity is influenced by changing atmospheric conditions is important for conservation, public policy and environmental health. Climate influences pollutant deposition and fluxes, physiology and the sensitivity of lichens and other organisms to pollution (Bell and Treshow 2002).

Lichens are especially sensitive to forest habitat quality and consequently to management (Kuusinen and Siitonen 1998; Pykälä 2004; Bergamini et al. 2005), because they are poikilohydric and highly sensible to an increase of light intensity (Gauslaa and Solhaug 1996; Nash 1996). Among them, cyanolichens are particularly valuable because they only occur in sites with high humidity (Lange et al. 1988).

Lichens are largely host specific in subtropical to boreal regions, but in tropical rain forest it is not so (Sipman and Harris 1989; Seaward and Aptroot 2003; Dettki and Esseen 1998). Monoculture plantation, selective logging resulting in thinning out forest, leads to loss of phorophytes for lichen colonisation and thus decreases in lichen diversity. Biodiversity is directly proportional to forest productivity (Gjerde et al. 2005; Wolseley et al. 1994). Fragmentation of habitat results in increased irradiance and exposure which could increase, decrease or bring about extinction of a particular species. This phenomenon of decreasing lichen diversity is quite prevalent in tropical rain forests of Assam and Andaman and Nicobar Islands (Rout et al. 2010). Utilisation of lichen biomonitoring data in forest management and conservation has not so far been conducted in India, while use of other indicator species has been carried out (Kotwal et al. 2008a, b). Few studies of lichen vegetation pre- and post-logging have been conducted worldwide (Kantvilas and Jarman 2004).

Lichens differ substantially from higher plants due to their poikilohydrous nature, combined with other physiological processes makes lichen growth particularly susceptible to climatic variations, pollution and other environmental factors which results in changes at genetic, individual, population and community levels.

Lichens have been used as predictive tools for investigating land-forming processes and

6.1 Introduction

rates of environmental change. They have also been used to resolve environmental issues involving management of natural resources such as the effects of fragmentation and habitat alteration, the structure and management of forested stands, the ecological continuity on space and time of the natural or semi-natural forests, effects of development on biodiversity, the effectiveness of conservation practices for rare or endangered species and the protection of genetic resources. Because of their excellence as predictive organisms, lichens have been used in different countries as bioindicators of high-value forests for conservation and to identify important biodiversity sites. Another conservation strategy involves creation of lichen gardens and sanctuaries (Upreti and Nayaka 2008).

While utilising lichens in forest management programmes, standardised protocols are adopted for lichen sampling and data management, and lichen gradient model is designed and standardised to enhance ability to accurately assess patterns and monitor trends region-wide and at larger geographical scales (Will-Wolf 2010).

For effective utilisation of lichens in conservation practices, in Europe and America, Forest Inventory and Analysis (FIA) data is compiled by regular surveys of forests on permanent plots across wide geographical area to inventory status and monitor trends over time. In natural resource assessment inventory refers to one-time assessment of the status of a resource, and monitoring refers to comparison between repeated inventories to evaluate changes in status and identify trends over time. The study of lichen communities in forest ecosystems allows assessment of extent of natural resource contamination, loss of biodiversity, ability to provide ecosystem services and sustainability of timber production. Lichens not only indicate the health of our forests, they also show a clearly established linkage to environmental stressors (Fig. 6.2) (Will-Wolf 2010).

India has 2.1 % of the land mass, about 1 % forest area, 16 % human population and 18 % livestock population of the world. A significant proportion of forest is degraded and lost due to accidental fires leading to loss of forest cover. In India, out of nearly 0.6 million villages, approximately one-third are located near forests for food and fodder. Frequent forest fires during summer (about 55 % forest area), livestock grazing (nearly 270 million livestock graze in 78 % forest area), extensive firewood collection by the communities (more than 300 million m^3), poor productivity (forest biomass 93 tonnes/ha and wood growing stock 47 m^3/ha) and natural regeneration (70 % forest area has poor regeneration) results in loss of forest cover (NFAP 1999).

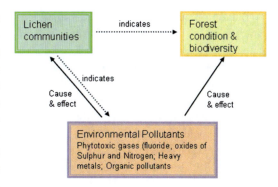

Fig. 6.2 Illustration depicting the cause and effect relation between lichen communities and forest condition with environmental pollutants and use of lichen community changes as an indicator of forest health and level of pollutants (After Will-Wolf 2010)

Despite these factors, the country is bestowed with a rich diversity of plant (49,000) and animal (83,000) species; India is also among 12 megabiodiversity countries of the world and has 25 hot spots of the richest and highly endangered ecoregions of the world (Mayer et al. 2000). India is a low-forest-cover country where the forest cover is 20.6 % (FSI 2005), which is quite less than the country's Forest Policy (1988) target, that is, 1/3 of total geographical area (Kotwal et al. 2008a, b).

Criteria and Indicators (C&I) have been identified as a tool to assess and monitor the sustainable forest management (SFM) globally. Identification of site-specific bioindicators and standardisation of monitoring mechanisms would be very useful in managing forests sustainability. Considering this, the Government of India (Ministry of Environment and Forest) has identified 8 Criteria and 37 Indicators for SFM in the country (Kotwal et al. 2008a).

Data/information collected on identified indicators from the forest management unit is aggregated at province or national level. The values of each indicator can be compared with the norms (standard values) to know the deviations. Periodic collection of such information/data for a period 5–10 years and analysis of such information over a period would indicate the trends towards or away from sustainability. This provides basis for resurrection of the forest system (Kotwal et al. 2008b; Guynn et al. 2004; Hagan and Adrews 2006; Failing and Gregory 2003; Whitman and Hagan 2003).

Diversity of epiphytic lichens responds to air pollutants emitted by pollution source but it requires continuous exposure for a period, and after a long exposure to pollutants changes in species number and composition appears. Initially the changes at physiological level occurs which results in extinction or tolerance of the species in response to the level of pollution, which is ultimately reflected by the change in the species composition (Paoli and Loppi 2008). Therefore, ecophysiological monitoring at individual and/or community level can help to detect early stress symptoms which are important for conservation practices to be implemented.

In India, urban centres mainly city centres are devoid of lichens. In urban settlements some tolerant species dominate the lichen flora mostly dominated by members of *Physciaceae* family (Shukla and Upreti 2007a). It has been observed that the probable adaptation prevalent in the naturally and luxuriantly lichen species growing in these centres may be attributed to increased pigment concentration and protein content. Stress physiology of *Phaeophyscia hispidula* has been investigated, and the result revealed that chlorophyll a, chlorophyll degradation ratio and protein had positive correlation with the increase in pollution (Shukla and Upreti 2007b).

Conservation biology is a recent multidisciplinary science that has developed in response to the late twentieth-century extinction crisis and to address the threats to biodiversity (Geiser et al. 1994). It has two major tasks: (1) investigate human impact on species, populations and biotopes and (2) develop practical approaches to ensure the conservation of species and ecosystems (Geiser and Williams 2002). Conservation biology tries to provide solutions that can be applied in real-world field situations (Maes et al. 2003; Maes and van Swaay 1997).

Thus, biomonitoring studies not only provides data on the present air quality data but an integrated approach involving physicochemical analysis could establish bioindicators as an integral part of air quality regulatory practices in Asian countries as in the Western countries. Biomonitoring data may be effectively utilised in regulatory management practices to reduce emissions of wide array of toxic pollutants either at local level or by regulatory agencies, making use of natural resources as sentinels of sustainable development.

References

Andrén H (1997) Habitat fragmentation and changes in biodiversity. Ecol Bull 46:140–170

Azevedo-Ramos C, Amaral-do BD, Nepstad DC, Filho BS, Nasi R (2006) Integrating ecosystem management, protected areas, and mammal conservation in the Brazilian Amazon. Ecol Soc 11:17

Bates JW, Farmer AM (eds) (1992) Bryophytes and lichens in a changing environment. Clarendon, Oxford

Bell JNB, Treshow M (2002) Air pollution and plant life. Wiley, Chichester

Bergamini A, Scheidegger C, Carvalho P, Davey S, Dietrich M, Dubs F, Farkas E, Groner U, Kärkkäinen K, Keller C, Lökös L, Lommi S, Maguas C, Mitchell R, Rico VJ, Aragon G, Truscott AM, Wolseley PA, Watt A (2005) Performance of macrolichens and lichen genera as indicators of lichen species richness and composition. Conserv Biol 19:1051–1062

Berry WL, Wallace A (1981) Toxicity: the concept and relationship to the dose response review. J Plant Nutr 30:13–19

Blett T, Geiser L, Porter E (2003) Air pollution-related lichen monitoring in national parks, forests, and refuges: guidelines for studies intended for regulatory and management purposes. National Park Service Air Resources Division U.S. Forest Service Air Resource Management Program U.S. Fish and Wildlife Service Air Quality Branch

Carignan V, Villard MA (2002) Selecting indicator species to monitor ecological integrity: a review. Environ Monit Assess 78:45–61

Debinski DM, Holt RD (2000) A survey and overview of habitat fragmentation experiments. Conserv Biol 14:342–355

References

Dettki H, Esseen PA (1998) Epiphytic macrolichens in managed and natural forest landscapes: a comparison at two spatial scales. Ecographysiology 21:613–624

Edwards TC Jr, Cutler DR, Geiser L, Alegria J, McKenzie D (2004) Assessing rarity of species with low detectability: lichens in Pacific Northwest forests. Ecol Appl 14:414–424

Eldridge DJ, Koen TB (1998) Cover and floristics of microphytic soil crusts in relation to indices of landscape health. Plant Ecol 137:101–114

Esseen PA, Renhorn KE (1998) Edge effects on an epiphytic lichen in fragmented forests. Conserv Biol 12:1307–1317

Failing L, Gregory R (2003) Ten common mistakes in designing biodiversity indicators for forest policy. J Environ Manage 68:121–132

NFAP (National Forestry Action Programme) (1999) Government of India, Ministry of Environment and Forests, New Delhi

Forman RF (1995) Land mosaics. The ecology of landscape and regions. Cambridge University Press, New York

FSI (2005) State of forest report 2005, Forest Survey of India, Government of India, Dehradun, India

Gauslaa Y, Solhaug KA (1996) Differences in the susceptibility to light stress between epiphytic lichens of ancient and young boreal forest stands. Funct Ecol 10:344–354

Geiser L (2004) Monitoring air quality using lichens on national forests of the Pacific Northwest: methods and strategy. USDA-forest service pacific northwest region technical paper, R6-NR-AQ-TP-1-04, p 134

Geiser LH, Williams R (2002) Using lichens as indicators of air quality on federal lands. Workshop report. USDA forest service, Pacific Northwest region technical paper R6-NR-AG-TP-01-02. Available on-line at url: http://ocid.nacse.org/research/airlichen/workgroup

Geiser LH, Derr CC, Dillman KL (1994) Air quality monitoring on the Tongass national forest, methods and baselines using lichens. USDA-forest service, Alaska region technical bulletin F10-TB-46. Available on-line at url: http://www.nacse.org/lichenair

Gjerde I, Sætersdal M, Rolstad J et al (2005) Productivity–diversity relationships for plants, bryophytes, lichens and polypore fungi in six northern forest landscapes. Ecography 28:705–720

Guynn DC Jr, Guynn ST, Layton PA, Wigley TB (2004) Biodiversity metrics in sustainable forestry certification programs. J For 102:46–52

Hagan MJ, Adrews W (2006) Biodiversity indicators for sustainable forestry: simplifying complexity. J For 104:203–210

Hilmo O, Holien H, Hytteborn H (2005) Logging strategy influences colonization of common chlorolichens on branches of *Picea abies*. Ecol Appl 15:983–996

Holt EA, Miller SW (2011) Bioindicators: using organisms to measure environmental impacts. Nat Educ Knowl 2(2):8

Kantvilas G, Jarman SJ (2004) Lichens and bryophytes on *Eucalyptus obliqua* in Tasmania: management implications in production forests. Biol Conserv 117:359–373

Kates RW, Parris TM, Leiserowitz AA (2005) What is sustainable development? Goals, indicators, values and practice. Environ Sci Policy Sustain Dev 47(3):8–21

Kennish MJ (1992) Ecology of Estuaries: anthropogenic effects. CRC Press, Boca Raton

Kivistö L, Kuusinen M (2000) Edge effects on the epiphytic lichen flora of *Picea abies* in middle boreal Finland. Lichenologist 32:387–398

Kotwal PC, Kandari LS, Dugaya (2008a) Bioindicators in sustainable management of tropical forests in India. Afr J Plant Sci 2:99–104

Kotwal PC, Omprakash MD, Gairola S, Dugaya D (2008b) Ecological indicators: Imperative to sustainable forest management. Ecol Indic 8:104–107

Kuusinen M, Siitonen J (1998) Epiphytic lichen diversity in old-growth and managed *Picea abies* stands in southern Finland. J Veg Sci 9:283–292

Lange OL, Green TGA, Ziegler H (1988) Water status related photosynthesis and carbon isotope discrimination in species of the lichen genus *Pseudocyphellaria* with green or blue-green photobionts and in photosymbiodemes. Oecologia 75:494–501

Maes D, van Swaay CAM (1997) A new methodology for compiling national Red Lists applied on butterflies [Lepidoptera, Rhopalocera] in Flanders [N.-Belgium] and in The Netherlands. J Insect Conserv 1:113–124

Maes D, Gilbert M, Titeux N, Goffart P, Dennis R (2003) Prediction of butterfly diversity hot spots in Belgium: a comparison of statistically-focused and land use-focused models. J Biogeogr 30:1907–1920

Markert BA, Breure AM, Zechmeister HG (2003) Definitions, strategies and principles for bioindication/biomonitoring of the environment. In: Markert BA, Breure AM, Zechmeister HG (eds) Bioindicators and biomonitors. Elsevier, Oxford, pp 3–39

Mayer N, Millermeier RA, Mittermeirer CG, da Fonseca GAB, Kent J (2000) Biodiversity hotspots for conservation priorities. Nature 403:853–858

McCune B (2000) Lichen communities as indicators of forest health. Bryologist 103:353–356

McGeoch MA, VanRensburg BJ, Botes A (2002) The verification and application of bioindicators: a case study of dung beetles in a savanna ecosystem. J Appl Ecol 39:661–672

Meyer C, Gilbert D, Gillet F, Moskurad M, Franchib M, Bernard N (2012) Using "bryophytes and their associated testate amoeba" microsystems as indicators of atmospheric pollution. Ecol Indic 13:144–151

Moen J, Jonsson BG (2003) Edge effects on Liverworts and lichens in forests patches in a mosaic of Boreal Forest and Wetland. Conserv Biol 17:380–388

Nash TH III (ed) (1996) Lichen biology. Cambridge University Press, Cambridge

Nash TH III (ed) (2008) Lichen biology, 2nd edn. Cambridge University Press, Cambridge

Osmond DL, Line DE, Gale JA, Gannon RW, Knott CB, Bartenhagen KA, Turner MH, Coffey SW et al (1995) WATERSHEDSS: water, soil and hydro – environmental decision support system. http://h2osparc.wq.ncsu.edu

Paoli L, Loppi S (2008) A biological method to monitor early effect of the air pollution. Environ Pollut 155:383–388. doi:10.1016/j.envpol.2007.11.004

Phillips DJH, Rainbow PS (1993) Biomonitoring of trace aquatic contaminants. Elsevier Applied Science, New York

Plafkin JL, Barbour MT, Porter KD, Gross SK, Hughes RM (1989) Rapid assessment protocols for use in streams and rivers: benthic macroinvertebrates and fish. EPA, Washington, DC

Pykälä J (2004) Effects of new forestry practices on rare epiphytic macrolichens. Conserv Biol 18:831–838

Rheault H, Drapeau P, Bergeron Y, Esseen PA (2003) Edge effects on epiphytic lichens in managed black spruce forests of eastern North America. Can J Forest Res 33:23–32

Rosso AL, McCune B, Rambo TR (2000) Ecology and conservation of a rare, old-growth-associated canopy lichen in a silvicultural landscape. Bryologist 103:117–127

Rout J, Das P, Upreti DK (2010) Epiphytic lichen diversity in a reserve forest in south Assam, north India. Trop Ecol 51(2):281–288

Seaward MRD, Aptroot A (2003) Lichens of Silhouette Island (Seychelles). Bibliotheca Lichenol 86:423–439

Shukla V, Upreti DK (2007a) Heavy metal accumulation in *Phaeophyscia hispidula* en route to Badrinath, Uttaranchal, India. Environ Monit Assess 131:365–369. doi:10.1007/s10661-006-9481-5

Shukla V, Upreti DK (2007b) Physiological response of the lichen *Phaeophyscia hispidula* (Ach.) Essl. to the urban environment of Pauri and Srinagar (Garhwal), Himalayas. Environ Pollut 150:295–299. doi:10.1016/j.envpol.2007.02.010

Sipman HJM, Harris RC (1989) Lichens. In: Licth H, Werger MJA (eds) Tropical rain forest ecosystems. Elsevier, Amsterdam, pp 303–309

Upreti DK, Nayaka S (2008) Need for creation of lichen garden and sanctuaries in India. Curr Sci 94(8):976–978

Valladares F, Camarero JJ, Pulido F, Gil-Pelegrín E (2004) El bosque mediterrá neo, un sistema umanizado y diná mico. In: Valladares F (ed) Ecología del bosque mediterráneo en un mundo cambiante. Ministerio de Medio Ambiente, EGRAF, Madrid, pp 13–25

Whitman AA, Hagan JM (2003) Biodiversity indicators for sustainable forestry. Final report National Commission on Science for Sustainable Forestry, Washington, DC

Will-Wolf S (2010) Analyzing lichen indicator data in the forest inventory and analysis program. General technical report, PNW-GTR-818. U.S. Department of Agriculture, Forest Service, Pacific Northwest Research Station, Portland, p 62

Wolseley PA, Moncrieff C, Aguirre-Hudson B (1994) Lichens as indicators of environmental stability and change in the tropical forests of Thailand. Global Ecol Biogeogr Lett 1:116–123

Glossary

Acicular In lichens, said of spores: needlelike, long, very slender and pointed

Acidic rock Quartzite, granite, basalt, sandstone or other rocks that produce no bubbling when a strong acid (usually 10 % HCl) is applied; pH < 7

Adnate Thallus is close to the substrate or the lower surface of the apothecium fused to the substratum

Amphithecium Thalline margin of an apothecium

Ampulliform Flask-like and with a narrow neck

Anthropogenic Pollution generated by human activity

Apical Tip or terminal part of a structure

Apotheciate Bearing apothecia

Appressed The whole underside closely pressed to the substrate or surface; lying flat on and firmly attached to it in appressed apothecia, the base is scarcely constricted

Areolate Having division by cracks into small areas. Descriptive of a thallus marked off into minute usually polygonal areas (areoles) like a jigsaw pattern with pieces together or somewhat separated

Areole A small, flattened part of a lichen thallus separated from the rest of the thallus by deep, narrow to wide cracks or more or less scattered on the substrate

Ascocarp A 'fruiting body' (ascoma) containing fungal ascospores

Ascus (pl. asci) Sac-like vessel or cell (20–100 μm long) of the perfect state of an ascomycetes, containing one or more (most often 8) sexually produced fungal spores

Aseptate Simple, lacking cross walls (septa)

Basic rocks Rocks containing either calcium (calcareous rocks) or magnesium (ultramafic rocks)

Bilocular Divided into two compartments, as a 2-celled spore

Biodeterioration Substances broken down by microorganisms

Blastidium Pseudocorticate budding proliferations of upper or lower phenocortex

Bulbil Multilayered paraplectenchymatous globular outgrowth with only a few algal cells

Calcareous rock Limestone or other rocks containing calcium or lime (calcium carbonate), with pH over 7, vigorously bubbling when treated by a strong acid (usually 10 % HCl)

Capillary Hair-like

Capitate Swollen like a head, knob-like, as in soralia, and tips of paraphyses

Cephalodia (sing. cephalodium) Small, localised colonies of cyanobacteria occurring within or on the surface of lichens in which the primary photobiont is otherwise an alga

Cephalodium (pl. cephalodia) Small (to ca. 0.5–1 mm), delimited, gall-like thallus structure (or tiny thallus)

Chromatography Physico-chemical technique for the identification of metabolic and other chemical products

Cilium (pl. cilia) Short, eyelash-like hair; longish-acute hair-like outgrowth, from the margin or upper surface of lobes or on the margin of the apothecium, consisting of compact strands of hyphae

Conidium (pl. conidia) Asexually formed spore, not originating inside a sporangium

Consoredium Aggregations of incompletely separated soredia

Continuous More or less unbroken, uninterrupted, as in a cortex without pores or cracks

Coralloid Divided up into many short, irregular cylindrical branches, like coral; often brittle

Corticate Having a cortex

Corticolous Growing on bark

Cracked Breaking open in lines or chinks, sometimes exposing the medulla; often irregular and due to age, but sometimes regular and characteristic of a taxon

Crustose Thallus type forming a strongly adherent crust over the substrate (in intimate contact with the substrate), without a lower cortex, rhizines, or umbilicus; often without a distinct or true upper cortex; usually not removable intact

Cyanobacterium An organism related to true bacteria and belonging to the kingdom Monera (prokaryotes, lacking a nucleus and chloroplasts); formerly called blue-green algae

Cyphella (pl. cyphellae) A pore recessed into the lower surface of the thallus (a break in the lower cortex), sharply bounded, concave, cup-like, rounded or ovate

Cyphellae (sing. cyphella) Rimmed, crater-like pores that open into the medulla via the lower surface; characteristic of the genus Sticta

Dactylidium Phenocorticate nonbudding protuberance of upper cortex and photobiont layer

Dense Set close together, compact, often closely interwoven; having the branches or hyphae massed and crowded

Density Density is also an expression of the numerical strength of a species in an area

Desiccation Dry out all the moisture

Dichotomous (= fork-branched) Y-shaped branching

Disc Exposed upper surface of the hymenium in an apothecium, concave to plane or convex, usually pigmented in a characteristic way, often surrounded

Ecorticate Without a cortex

Effigurate Obscurely lobed; radiating at the periphery; used by some authors to include obscurely or even distinctly rosulate or lobate crustose (placodioid) thalli, with elongated marginal lobes

Elliptical Oval or oblong narrowed at each end

Emersed Of perithecia, having only the lower third immersed in the thallus or substrate

Epihymenium Indistinctly delimited uppermost portion of the hymenium, where this differs in appearance from lower part; usually pigmented (often on the swollen tips of the paraphyses)

Epilithic On surface of rock, with little or no penetration between and under the rock particles

Episubstratic Growing upon substrate

Epithecium The layer above the asci, formed by the tips of the paraphyses; in the strict sensor distinct tissue (plectenchyma) of interwoven hyphae on top of hymenium

Epruinose Without pruina

Eucortex A true cortex formed of 'well differentiated tissue'

Exciple (excipulum) Tissue forming the margins or walls of an ascoma

Exo- (Prefix) outside

Farinose Powdery, like flour; used in reference to soredia

Filamentose (filamentous) Hair- or thread-like; a growth form composed of thin hair-like strands of hair- or thread-like; a growth form composed of thin hair-like strands of mycobiont and photobiont

Folicolous Lichens growing on leaves

Foliose Thallus form usually with upper and lower cortices, dorsiventral, flat and somewhat leaf-like; larger than the arbitrarily distinguished squamulose lobes (which are up to 5 mm long and wide)

Foveolate More or less irregularly and delicately pitted; usually used in reference to the upper cortex

Fruticose Thallus form which is usually erect and stalked to rather bushy, shrub-like, tree-like, pendent- and beard-like

Gelatinous Like a jelly, rubbery, slimy, translucent, swelling when wet; in the gelatinous growth form, the thallus is homoiomerous (unstratified), and the distinctions among crustose, foliose and fruticose are often blurred

Globose Smooth and not hairy

Granular Having, composed of or covered by small particles (granules or granule-like particles)

Growth form Habit

Gyrodisc An apothecium (of the genus Umbilicaria) in which the surface of the disc is more or less concentrically fissured

Hapter An aerial organ of attachment formed by the thallus in response to its contact with the substrate

Haustorium A special hyphal branch, especially one within a living cell of the host, for absorption of nutrients

Heteromerous Layered (stratified); thallus form in which more or less distinct tissues (especially a definite algal layer) are present; having the mycobiont and photobiont components in well-marked layers, with photobiont in a more or less distinct one between upper cortex and medulla

Holdfast An algal-free thickened mass of hyphae used for attachment

Hymenium The layer of tissue in which the asci arise; that part of the ascocarp composed of asci and paraphyses (or paraphysoid tissue) in a close arrangement

Hyphae The strands or filaments of the mycelium of fungus; hyphae are the segmented cells that constitute the body of the mycobiont of lichen

Hypo- (Prefix) under

Hypothallus A growth of undifferentiated purely fungal mycelium (the first hyphae of the thallus to grow), sometimes present as a distinct layer below (or on the underside of) the thallus and often projecting beyond it; white to darkly coloured; sometimes thick

Hypothallus (= prothallus) In lichens, a thin, typically dark, tightly appressed weft of fungal threads that in some species develops on the underside of the thallus and may sometimes extend outwards from it, so as to be visible when seen from above

Hypothecium Area of hyaline to pigmented or carbonised tissue in the apothecium immediately below the subhymenium

Imbricate Overlapping, partly covering each other

Immersed Sunken into the thallus or substrate

Inter- (Prefix) between; among

Intra- (Prefix) within, inside

Irregular Uneven, as in lobe margins of foliose lichens

Isidioid soredium Secondarily corticate protuberance produced in soralia-like clusters

Isidium (pl. isidia) A minute (mostly to 0.5–1 mm) outgrowth of the thallus which has a cortex; contains both mycobiont and photobiont

Lacinium (pl. lacinia) A long, slender, linear-elongate thallus lobe

Lamina A thin, flat organ or part, usually the main part or main upper surface of a foliose or squamulose thallus, the blade in contrast to the margin

Lecanorine With a thalline exciple pertaining to an apothecium containing algae at least below the hypothecium and usually having a distinct amphithecium that often also contains algae

Lecideine Without a thalline exciple, pertaining to an apothecium which lacks algae and lacks an amphithecium, and therefore in which the exciples form the apothecial margin

Lenticular Lens-shaped

Leprose Composed more or less entirely of a loosely organised powdery (to finely granular) mass of algal cells and fungal hyphae, without any cortex even in young stages

Lignicolous Growing on wood

Lobe A rounded to linear division of a thallus usually applied to foliose or squamulose forms

Lobule Tiny lobe-like, dorsiventral outgrowths, often occurring along the lobe margins or stress cracks

Loose Lack, lightly attached to more or less free

Macro- (Prefix) long, but commonly used in the sense of mega, i.e. large

Macrolichen Larger lichen of squamulose, foliose or fruticose habit

Macula (pl. maculae) A small pale spot or blotch on the upper or outer surface of a thallus, often due to uneven distribution of photobiont cells below the thalline cortex

Margin Edge or rim, used for the region at the periphery of a zonate to radiately lobed thallus, exciple of an apothecium

Medulla An internal layer of fungal hyphae, below the algal layer, in the thallus or in a lecanorine apothecium; hyphae often more or less loosely interwoven and weakly gelatinized

Mega- (Prefix) large

Micro- (Prefix) small

Microenvironment The environment of a very small specific area

Microlichen Crustose lichen, usually small

Mono- (Prefix) one

Monotypic Of a genus with one species or a family with one genus; in general, applied to any taxon with only one immediately subordinate taxon of one of the principal ranks

Muriform With transverse and longitudinal (or oblique) walls, dividing the spore into more or less numerous (usually 10 or more) chambers, thus appearing like a brick wall

Muriform Pertaining to spores in which both transverse and longitudinal septa are present

Mycelium A mass of hyphae; the thallus of a fungus

Mycobiont Fungal partner in the symbiosis that constitutes lichen

Oval Broadly elliptic, narrowing somewhat from middle to rounded ends

Pachydermatous (of hyphae) Having the outer wall thicker than the lumen

Papillae (sing. papilla) Minute, discrete, typically rounded protruberances of the cortex

Paraphysis (pl. paraphyses) A specialised sterile hypha in the hymenium, thread-like, simple or branched, basally attached, usually more or less vertical (anticlinal); usually relatively thick (1.5 μm or more)

Parathecium (of apothecia) The outside hyphal layer, darker in colour, outside of the hypothecium and inside the amphithecium

Peltate In lichens, referring to a shield-shaped thallus attached to the substrate at a single point

Peri- Around, surrounding

Perithecia (sing. perithecium) In lichens, the minute, flask-shaped ascocarps in which the sexual spores of the fungal partner are produced. Macroscopically, a typical perithecium resembles a tiny dot as seen from above

Perithecium (pl. perithecia) Globose or flask-shaped fungal fruiting body (ascocarp) sessile or more often at least partly immersed in the thallus or in thalline warts, with a single, terminal (central or rarely eccentric) opening (ostiole) and otherwise completely enclosed by a wall

Photobiont The photosynthetic partner in a lichen, consisting of a green alga, a blue-green cyanobacterium or, in some species, both. The lichen fungus derives its carbohydrate requirements from the photobiont

Photophilous Light loving; preferring well-illuminated habitats

Photophobous Light fearing; preferring shaded habitats

Podetia (sing. podetium) The hollow, upright, ascocarp-bearing stalks characteristic of the genus Cladonia

Podetium (pl. podetia) A stalk (more or less elongated, erect, terete portion) of a thallus derived from tissue of apothecial origin (usually the hypothecium and stipe), usually rising from a primary thallus and often bearing apothecia or pycnidia

Pseudocortex A thalline boundary layer in which the hyphae are distinct but not organised into a tissue showing a regular cellular or fibrous structure

Pseudocyphellae (sing. pseudocyphella) Tiny, pale, unrimmed pores in the upper or lower cortex through which the medulla is exposed. In form, pseudocyphellae may be dot-like, angular or irregular

Pustule Blister-like swellings

Pycnidia (sing. pycnidium) In lichens, minute, flask-shaped, asexual spore-producing structures of the fungus, usually imbedded in the thallus and visible from above as a black dot that may occasionally be protuberant

Pycnidium (pl. pycnidia) Neutral term for a minute globose to flask-shaped (pear-shaped) structure, resembling a perithecium and usually immersed in the medulla

Rhizinae In lichens, root-like hairs or bundles of fungal threads that attach the thallus to the substrate

Rhizine(s) (rhizina[ae]) Black to light brown cords of hyphae extending from the lower cortex to the substrate and anchoring the thallus firmly

Rhizohyphae More or less elongated single-row hyphae on the lower surface, for attachment

Rhizoid Hyphal structures on the lower surface anchoring the thallus

Rounded Curved in outline or form; non-technical term for rotund

Saxicolous Growing on (or in) rock (used loosely to include man-made rock-like substrates)

Septate Divided by one or more septa

Septum A wall making a cellular division in a spore or hypha

Septum (pl. septa) A cross wall, especially of a cell or a spore

Sessile Without a stem, stalk or stipe of any kind, sitting closely on the surface, attached directly to the thallus

Siliceous Refers to rock composed mainly of silicon compounds, producing no (or few) bubbles upon application of 10 % HCl

Soralium (pl. soralia) A decorticate area or body of the thallus where soredia are produced

Soredium (pl. soredia) A microscopic group of algal cells and loosely woven hyphae, without a cortex or pseudocortex, which erupt from cracks or pores in the thallus, appear finely powdery to coarsely granular and function as a vegetative reproductive unit

Spore Microscopic reproductive unit (one-celled to many-celled); with lichens, when used without a prefix usually refers to ascospore (or basidiospore), which is haploid and the result of meiosis

Squamule Scale-like thallus or thallus segment (lobe, foliole), usually more or less isodiametric (or at least short), with an entire to flexuous or crenate margin, with or without a lower cortex; intermediate between crustose and foliose

Squamules Small, rounded, often somewhat overlapping lobes, the lower surface of which typically lack a cortex

Squamulose Growth form composed of squamules; frequently forming extensive mats; also used interchangeably with squamulate

Squarrose Branching by many short perpendicular branches from a single main axis; usually in reference to rhizines

Sterile Not producing spores or a sporocarp (at least not by sexual reproduction); pycnidia and pycnospores may be present

Submuriform Pertaining to spores in which both transverse and longitudinal septa are present, though the latter are sparse or poorly developed

Substrate In lichenology, a general term for the surfaces colonised by lichens, whether wood, bark, rock, soil or other

Terricolous Growing on the ground

Thalline exciple Apothecial margin containing algae and derived from the vegetative thallus; usually similar in colour

Thalloconidia (sing. thalloconidium) Minute asexual spores produced on the cortex of some lichens. In Umbilicaria, thalloconidia confer a black, sooty texture to the lower surface and rhizines of several species

Transverse Across the width

True exciple An exciple which lacks algal cells, usually of a different colour than the thallus

Umbilicus A thickened, centrally positioned point of attachment characteristic of some rock-dwelling foliose lichens

Ventral Front or lower surface; the surface facing the axis

Void Unacceptable or empty space between rock

About the Authors

Vertika Shukla is M.Sc. in Organic Chemistry and Ph.D. (2003) in Chemistry of lichens from H.N.B. Garhwal University, Srinagar (Garhwal). Presently, she is working on Indian lichens for monitoring environmental pollution. Dr. Shukla has 10 years of postdoctoral research experience in the field of lichenology and already completed two postdoc fellowships awarded by the Department of Science and Technology (DST, New Delhi), and till now, she has published 41 scientific articles.

Dr. Shukla contributed much on secondary metabolite chemistry, interaction of lichens with the environment, spatio-temporal behaviour of pollutants and role of lichens in bioremediation of atmospheric fallouts.

Rajesh Bajpai is M.Sc. in Environmental Science and Ph.D. in 2009 from Babasaheb Bhimrao Ambedkar (Central) University, Lucknow, and is working as DST Scientist in Lichenology Laboratory of CSIR-NBRI, Lucknow. Dr. Bajpai has published 20 research papers in various national and international journals on accumulation of different metals including arsenic.

Dr. Bajpai has carried out extensive research work on the interaction of lichen with the environment and arsenic pollution and role of lichens in phytoremediation and biodeterioration.

D.K. Upreti, Head of Lichenology Laboratory, National Botanical Research Institute, Lucknow, is Ph.D. (1983) in Botany from Lucknow University under the guidance of Dr.D.D. Awasthi. Dr. Upreti has 30 years of research experience and has published more than 250 research papers in peer-reviewed journals and coauthored three books.

Apart from taxonomy, Dr. Upreti has also carried out extensive research on ecology, lichen chemistry, pollution monitoring, in vitro culture and biodeterioration studies of Indian lichens. Dr. Upreti had been to Antarctica in 1991–1992 and is also the member of the steering committee of CSIR for Antarctic Researches. He is the Indian corresponding member to British Lichen Society and International Association for Lichenology.

Printed by Publishers' Graphics LLC
DBT131006.20.05.57